R. Khosla, N. Ichalkaranje and L. C. Jain

Design of Intelligent Multi-Agent Systems

Studies in Fuzziness and Soft Computing, Volume 162

Editor-in-chief
Prof. Janusz Kacprzyk
Systems Research Institute
Polish Academy of Sciences
ul. Newelska 6
01-447 Warsaw
Poland
E-mail: kacprzyk@ibspan.waw.pl

Further volumes of this series can be found on our homepage: springeronline.com

Vol. 146. J.A. Gámez, S. Moral, A. Salmerón (Eds.)
Advances in Bayesian Networks, 2004
ISBN 3-540-20876-3

Vol. 147. K. Watanabe, M.M.A. Hashem
New Algorithms and their Applications to Evolutionary Robots, 2004
ISBN 3-540-20901-8

Vol. 148. C. Martin-Vide, V. Mitrana, G. Păun (Eds.)
Formal Languages and Applications, 2004
ISBN 3-540-20907-7

Vol. 149. J.J. Buckley
Fuzzy Statistics, 2004
ISBN 3-540-21084-9

Vol. 150. L. Bull (Ed.)
Applications of Learning Classifier Systems, 2004
ISBN 3-540-21109-8

Vol. 151. T. Kowalczyk, E. Pleszczyńska, F. Ruland (Eds.)
Grade Models and Methods for Data Analysis, 2004
ISBN 3-540-21120-9

Vol. 152. J. Rajapakse, L. Wang (Eds.)
Neural Information Processing: Research and Development, 2004
ISBN 3-540-21123-3

Vol. 153. J. Fulcher, L.C. Jain (Eds.)
Applied Intelligent Systems, 2004
ISBN 3-540-21153-5

Vol. 154. B. Liu
Uncertainty Theory, 2004
ISBN 3-540-21333-3

Vol. 155. G. Resconi, J.L. Jain
Intelligent Agents, 2004
ISBN 3-540-22003-8

Vol. 156. R. Tadeusiewicz, M.R. Ogiela
Medical Image Understanding Technology, 2004
ISBN 3-540-21985-4

Vol. 157. R.A. Aliev, F. Fazlollahi, R.R. Aliev
Soft Computing and its Applications in Business and Economics, 2004
ISBN 3-540-22138-7

Vol. 158. K.K. Dompere
Cost-Benefit Analysis and the Theory of Fuzzy Decisions – Identification and Measurement Theory, 2004
ISBN 3-540-22154-9

Vol. 159. E. Damiani, L.C. Jain, M. Madravia
Soft Computing in Software Engineering, 2004
ISBN 3-540-22030-5

Vol. 160. K.K. Dompere
Cost-Benefit Analysis and the Theory of Fuzzy Decisions – Fuzzy Value Theory, 2004
ISBN 3-540-22161-1

Vol. 161. N. Nedjah, L. de Macedo Mourelle (Eds.)
Evolvable Machines, 2005
ISBN 3-540-22905-1

R. Khosla
N. Ichalkaranje
L. C. Jain

Design of Intelligent Multi-Agent Systems

Human-Centredness, Architectures,
Learning and Adaptation

 Springer

Nikhil Ichalkaranje
Professor Lakhmi Jain
Knowledge-Based Intelligent
Engineering Systems Centre
School of Electrical and
Information Engineering
University of South Australia
Adelaide
Mawson Lakes Campus
South Australia SA 5095
Australia
E-mail: Nihil.Ichalkaranje@unisa.edu.au
　　　Lakhmi.Jain@unisa.edu.au

Associate Professor Rajiv Khosla
Business Systems and Knowledge
Modelling Laboratory
School of Business
La Trobe University
Melbourne, Victoria 3086
Australia
E-mail: R.Khosla@latrobe.edu.au

ISSN 1434-9922
ISBN 3-540-22913-2 Springer Berlin Heidelberg New York

Library of Congress Control Number: 2004111008

This work is subject to copyright. All rights are reserved, whether the whole or part of the material is concerned, specifically the rights of translation, reprinting, reuse of illustrations, recitations, broadcasting, reproduction on microfilm or in any other way, and storage in data banks. Duplication of this publication or parts thereof is permitted only under the provisions of the German copyright Law of September 9, 1965, in its current version, and permission for use must always be obtained from Springer-Verlag. Violations are liable to prosecution under the German Copyright Law.

Springer is a part of Springer Science+Business Media
springeronline.com

© Springer-Verlag Berlin Heidelberg 2005
Printed in Germany

The use of general descriptive names, registered names trademarks, etc. in this publication does not imply, even in the absence of a specific statement, that such names are exempt from the relevant protective laws and regulations and therefore free for general use.

Typesetting: data delivered by authors
Cover design: E. Kirchner, Springer-Verlag, Heidelberg
Printed on acid free paper   62/3020/M - 5 4 3 2 1 0

Preface

There is a tremendous interest in the design and applications of agents in virtually every area including avionics, business, internet, engineering, health sciences and management. There is no agreed one definition of an agent but we can define an agent as a computer program that autonomously or semi-autonomously acts on behalf of the user.

In the last five years transition of intelligent systems research in general and agent based research in particular from a laboratory environment into the real world has resulted in the emergence of several phenomenon. These trends can be placed in three categories, namely, humanization, architectures and learning and adaptation. These phenomena are distinct from the traditional logic-centered approach associated with the agent paradigm. Humanization of agents can be understood among other aspects, in terms of the semantics quality of design of agents. The need to humanize agents is to allow practitioners and users to make more effective use of this technology. It relates to the semantic quality of the agent design. Further, context-awareness is another aspect which has assumed importance in the light of ubiquitous computing and ambient intelligence.

The widespread and varied use of agents on the other hand has created a need for agent-based software development frameworks and design patterns as well architectures for situated interaction, negotiation, e-commerce, e-business and informational retrieval. Finally, traditional agent designs did not incorporate human-like abilities of learning and adaptation. Researchers today are using soft computing technologies like neural networks to make agents learn and adapt.

This book attempts to focus on the above trends in the design and evolution of the agent technology. In the rest of this introduction we provide a brief outline of the various chapters in this book.

Soft computing agents today are being applied in a range of areas including image processing, engineering, process control, data mining, internet and others. In the process of applying soft computing agents to complex real world problems three phenomena have emerged. These include their application in distributed environments, humanization of these agents and optimization of their performance. Chapter 1 models these phenomena as part of multi-layered multi-agent architecture. The layered architecture is also motivated by the human-centered approach and criteria outlined in the 1997 NSF workshop on human-centered systems.

Chapter 2 presents the general requirements of the ubiquitous environment and discusses the use of software agent technology to facilitate service provision and composition for ubiquitous client end systems.

Chapter 3 reports a prototype of the developed agent community implementation in the system "KSNet" and a constraint-based protocol designed for the agents' negotiation. An application of the developed agent community to coalition-based operations support and the protocol are illustrated via case studies of a mobile hospital configuration as a task of health service logistics and automotive supply network configuration.

Chapter 4 proposes architectural styles and design patterns for a multi-agent system which adopts concepts from social theories.

The design parameters included are social, intentional, structural, communication and dynamic. An e-business is included in the chapter.

Chapter 5 presents massive multi-agent systems like unsteady software systems whose behaviour develops some emergent characters. These systems are made to learn and to adjust themselves to the situation on environment. Their design is not a usual way of producing all the specifications and a strong validation. They aren't problem solvers. They are systems functioning with a behaviour oriented in the time following their continuous development.

JADE (Java Agent Development Framework) is an "open source" FIPA-compliant software environment to build agent systems. JADE offers an agent middleware to implement efficient FIPA2000 compliant multi-agent systems and supports their development through the availability of a predefined programmable agent model, an ontology development support, and a set of management and testing tools. Chapter 6 describes JADE and its use in three international projects to develop applications in the fields of: corporate memory management, integration of fixed and mobile networks, and integration of Web services.

Chapter 7 introduces ant algorithms and presents their applications. Chapter 8 presents the development of software architecture for agents to communicate and synchronise with the application in a totally transparent way. The design of multi-agent assistant for facilitating e-commerce activities during web browsing is presented.

Chapter 9 presents a set of techniques for selection, exchange and incorporation of information to help a learning agent to achieve its goals. Chapter 10 presents the criteria related to the adaptation and mutation in multi-agent systems.

Chapter 11 explores the possibility of applying reinforcement learning to acquire new high-level actions for animated learning agents. The chosen algorithm is the deterministic version of Q-learning. This allows for easy definition of the task, since only the ultimate goal of the learning agent must be defined. Generated actions can then be used to enrich the animation produced by an animation system. Results achieved when training agents with forward and inverse kinematics control are also demonstrated and compared.

The final chapter presents the basic concepts related to agent technologies. It includes the implementation of an agent-based architecture for information retrieval. It also suggests a new architecture for product search in e-commerce.

We are indebted to the contributors and reviewers for their wonderful efforts. Berend Jan van der Zwaag provided excellent contribution in this book. Thanks are due to the Editorial Staff of Springer-Verlag for their assistance.

Nikhil Ichalkaranje
Rajiv Khosla
Lakhmi Jain

Contents

Chapter 1.   1
Humanization of soft computing agents
Rajiv Khosla, Qiubang Li, and Chris Lai

1 Introduction 1
2 Human-centered system development framework 2
3 Distributed multi-agent architecture 4
   3.1 Problem solving ontology layer 5
   3.2 Optimization layer 9
   3.3 Tool or technology layer 11
4 Human-centered modelling using soft computing
   multi-agent architecture 14
   4.1 Human-centeredness and problem solving agent layer ... 14
      4.1.1 CRM model of Internet-banking 16
      4.1.2 Decomposition phase problem solving agent
           and CRM 17
      4.1.3 Control phase problem solving agent 18
      4.1.4 Decision phase problem solving agent 19
   4.2 Unstained cell image processing 21
   4.3 Human-centeredness and technology agent layer 25
5 Conclusion 26
   References 27

Chapter 2.   31
Software agents for ubiquitous computing
Sasu Tarkoma, Mikko Laukkanen, Kimmo Raatikainen

1 Introduction 31
2 Ubiquitous computing 33
   2.1 Overview 33
   2.2 Wireless networks and roaming 34
   2.3 Client devices 35

	2.4	Location- and context-aware services	36
	2.5	Technology support	37
		2.5.1 Java – the enabling technology for software agents	37
		2.5.2 Other technologies	38
3	Ubiquitous agents	40
	3.1	Overview	40
	3.2	The FIPA architecture	44
	3.3	Agent platforms	46
		3.3.1 JADE-LEAP	47
		3.3.2 FIPA-OS and MicroFIPA-OS	47
	3.4	Proxy-based approaches	48
	3.5	Agent communication	49
	3.6	Events for agents	49
4	Agent-based service provision and deployment	50
	4.1	Agent and service deployment	50
	4.2	Service partitioning based on the environment	52
	4.3	Example scenario: recommendation service	54
	4.4	CRUMPET	57
5	Conclusions	59
	References	60

Chapter 3.   63
Agents-based knowledge logistics
*Alexander Smirnov, Mikhail Pashkin, Nikolai Chilov,
  and Tatiana Levashova*

1	Introduction	63
2	KSNet-approach: major ideas	67
3	Features of agent community in the system "KSNet"	69
4	Communication, interaction and negotiation in the KL system	74
	4.1 Conventional CNP	75
	4.2 Constraint-based negotiation	76
	4.3 Modifications of interaction	77
	4.4 Example of utilizing constraint-based CNP	77

5	Implementation of agent community	84
6	Case study: health service logistics	90
7	Case study: virtual supply network	93
8	Conclusion	94
	Acknowledgements	96
	References	96

Chapter 4. 103
Architectural styles and patterns for multi-agent systems
Manuel Kolp, T. Tung Do, Stéphane Faulkner, and T.T. Hang Hoang

1	Introduction	103
2	Organizational architectural styles	106
	2.1 Applying organizational styles	111
	2.2 Evaluation	114
3	Social patterns	116
	3.1 Modeling social patterns	117
	3.1.1 Social dimension	118
	3.1.2 Intentional dimension	118
	3.1.3 Structural dimension	120
	3.1.4 Communication dimension	125
	3.1.5 Dynamic dimension	126
	3.2 Applying the patterns	128
4	Conclusion	129
	References	130

Chapter 5. 133
Design and behavior of a massive organization of agents
Alain Cardon

1	Introduction	133
2	The systems with particles	136
	2.1 The unsteady systems	138
	2.2 Operators of determination of the behavior for an unsteady system	140

3	The object approach: a very controlled process of construction and run of systems	143
	3.1 The object approach and the software engineering	144
	3.2 Objects and object-oriented design of systems	145
	3.3 Limits of the object approach	147
4	Massive multi-agent systems	148
	4.1 Agents	149
	4.2 Nondeterminism and instability in massive multi-agent systems	152
	4.3 An agentification method for the massive multi-agent systems	155
5	Analysis of the behavior of a massive agent organization: the control problem	162
	5.1 The characterization of an agent organization	163
	5.2 The morphological space, the correspondent of the space of phases for the MMAS	165
	5.3 The organization of morphological agents assuring the representation of the aspectual organization	171
	5.4 Characters of coherent groups	175
	5.5 Evocation agents and self-adaptability of the system	177
6	Entropy and equation of trajectory of MMAS	180
	6.1 Entropy	181
	6.2 Equation of trajectory: a reduction with regard to the morphological analysis	182
	6.3 Validity of the state equations	184
	6.4 Degraded forms of the state equation	185
7	Conclusion	185
	References	187

Chapter 6.   191
Developing agent-based applications with JADE
F. Bergenti, A. Poggi, G. Rimassa, P. Turci and M. Tomaiuolo

1	Introduction	191
2	JADE	193

	2.1 Platform architecture	193
	2.2 Agent architecture	196
3	LEAP	199
	3.1 LEAP architecture	201
4	CoMMA	203
	4.1 CoMMA architecture	204
5	Agentcities	206
	5.1 Network	207
	5.2 Service composition	208
6	Conclusions	210
	Acknowledgments	211
	References	212

Chapter 7.   215
A collective can do better
N.D. Monekosso and P. Remagnino

1	Introduction	215
2	Insect behaviour can be inspiring	217
3	Can nature be mimicked?	219
	3.1 Applying real ant behaviour to computational systems	219
	3.2 Solving classic optimisation problems	222
	3.3 Telecommunications applications	223
	3.4 Robot navigation applications	224
	3.5 Other robotic applications	224
	3.6 Image processing applications	225
4	Combining reinforcement learning and synthetic pheromones	225
	4.1 Reinforcement learning	225
	4.2 Synthetic pheromones and Q-learning	226
5	Cooperative robotic transport	232
6	Conclusions	232
	References	233

Chapter 8. 239
Coordinating multi-agent assistants with an application by means of computational reflection
A. Di Stefano, G. Pappalardo, C. Santoro, and E. Tramontana

1 Introduction 240
2 The motivation for a multi-assistant architecture 243
   2.1 Example: extending a Web browser
       with assistant agents 244
3 The multi-agent reflective architecture 246
   3.1 Computational reflection 247
       3.1.1 Using Javassist 248
   3.2 The architecture 251
   3.3 Coordinator agent 253
   3.4 Coordinator-assistants interactions 257
   3.5 A concrete example: an assistant that highlights
       keywords for a Web browser 259
4 A case study: e-commerce assistants for a Web browser ... 264
   4.1 User profiler assistant 265
   4.2 Data extraction assistant 268
   4.3 Cart manager assistant 270
5 Concluding remarks 272
   References 274

Chapter 9. 279
Learning by exchanging advice
Eugénio Oliveira and Luís Nunes

1 Introduction 279
2 Communicating to improve learning:
   historical notes and review 281
   2.1 Early work on exchange of information
       during learning 281
   2.2 Recent related work 282
3 Advice exchange 283
   3.1 Exchanging information during learning 283

	3.1.1 What type of information?	284
	3.1.2 How to integrate this information with the usual learning process?	285
	3.1.3 When should an agent request/accept information?	285
	3.1.4 Where to get information?	290
4	Experiments	292
	4.1 Predator-prey	294
	4.2 Traffic control	298
	4.3 Learning algorithms	301
5	Results and discussion	304
	5.1 Predator-prey	304
	5.2 Traffic control	307
6	Conclusions and future work	309
	Acknowledgments	311
	References	311

Chapter 10.   315
Adaptation and mutation in multi-agent systems and beyond
Ladislau Bölöni and Dan Cristian Marinescu

1	Introduction	315
2	A taxonomy	317
	2.1 Alternative names	319
	2.2 Classification criteria	319
	2.2.1 The amplitude of the change: weak vs. strong mutability	320
	2.2.2 The granularity of mutation	321
	2.2.3 The continuity of interactions: runtime vs. stoptime	322
	2.2.4 The initiator of the mutation	322
	2.2.5 Mutation technique	323
	2.3 A taxonomy of mutations	324
	2.4 Other classification approaches	325
3	A formal description of mutability	325
	3.1 Agent models and mutability	325
	3.2 A multiplane state machine model of agent behavior	328

    3.3   Modelling agent behavior 329
           3.3.1 Decomposition in the plane. expressing "change" ... 330
           3.3.2 Expressing concurrency 331
    3.4   Mutation operators and invariance properties 332
    3.5   How useful are the invariance properties? 334
4   A software engineering perspective on adaptive
    and mutable agents 336
    4.1   Adding new functionality to the agent 338
    4.2   Removing functionality from an agent 341
    4.3   Adapting to new requirements 343
    4.4   Splitting and merging agents 346
5   Conclusions 347
    References 349

Chapter 11.   355
Intelligent action acquisition for animated learning agents
Adam Szarowicz, Marek Mittmann, Jaroslaw Francik

1   Introduction 355
2   Current state of the art in automatic character animation .. 357
    2.1   General animation architectures 357
    2.2   Physics-based controllers 361
    2.3   Crowd simulation 362
3   Overview of other concepts 363
    3.1   Q-learning 363
    3.2   The agent's senses: collision detection and avoiding 364
    3.3   Agent architectures 366
4   Implementation of the Q-learning 367
5   System implementation and results 370
    5.1   The framework 370
    5.2   Results 372
    5.3   Alternative algorithms 378
6   Conclusions and summary 378
    Acknowledgements 380
    References 380

Chapter 12. 387
Using stationary and mobile agents for information retrieval and e-commerce
Samuel Pierre

1   Basic concepts and background 388
     1.1   Agent and multi-agent systems 388
     1.2   Cooperation and communication mechanisms 390
           1.2.1 Communication among agents 390
           1.2.2 Cooperation among agents 392
     1.3   Mobile agent and mobile code 394
2   Multi-agent architecture for information retrieval 396
     2.1   Mobile agent information retrieval 397
     2.2   Characterization of the architecture 399
     2.3   Experimental number application 408
           2.3.1 Principles of the application 408
           2.3.2 Design choices and modifications
                 to the initial application 409
     2.4   Internet picture retrieval application 410
3   Implementation of the information retrieval architecture .. 411
     3.1   Generic classes and interfaces 411
     3.2   Agents 414
     3.3   Implementation and testing environment 417
4   Evaluation of the information retrieval architecture 419
     4.1   Transportation measures 419
     4.2   Information retrieval scenarios 422
5   Multi-agent architecture for product retrieval 429
     5.1   Description of the problem and general scenario 429
     5.2   Solution and suggested algorithms 431
     5.3   Architecture and agent structure 433
     5.4   Implementation and performance evaluation 436
6   Conclusion 444
     References 446

Chapter 1

Humanization of Soft Computing Agents

Rajiv Khosla, Qiubang Li, and Chris Lai

1 Introduction

Soft computing agents today are being applied in a range of areas including image processing, engineering, process control, data mining, internet and others. In the process of applying soft computing agents to complex real world problems three phenomena have emerged. Firstly, the application of soft computing agents in distributed environments has resulted in merger of techniques from soft computing area with those in distributed artificial intelligence. Secondly, in order to facilitate better understanding and frequent use of the soft computing agent technology, humanization of these agents has become an important design issue (Takagi, 2001). Thirdly, given the approximate and imprecise nature of the solutions provided by soft computing agents, optimization has become an important design issue in an effort to improve the quality of solution provided by soft computing agents. Finally, from a user's perspective, as the soft computing agent technologies have moved out of laboratories into the internet and the real world the need to model and express these technologies in the problem solving context of user. In this chapter we model these four phenomenons as part of multi-layered multi-agent architecture. The layered architecture is also motivated by the human-centered approach and criteria outlined in the 1997 NSF workshop on human-centered systems.

The chapter is organized as follows. Section 2 outlines the human-centered criteria and social, semantic and pragmatic quality dimensions of human-centered systems. Section 3 describes the five layers of a component-based multi-agent soft computing architecture. Section 4 illustrates with the help of real world examples the human-centered aspects of the architecture. Section 5 concludes the chapter.

2 Human-Centered System Development Framework

In Figure 1 four components of the human-centered system development framework are shown. These are the activity-centered analysis, multi-layered multi-agent architecture component, transformation agent component, and multimedia interpretation component. These four components model social, semantic and pragmatic quality dimensions of human-centered systems. These dimensions also model the human-centered criteria outlined in 1997 NSF workshop on human-centered systems and recently by other researchers (2004 IEEE expert systems magazine).

These criteria state that 1) human-centered research and design is problem driven rather than logic theory or any particular technology; 2) human-centered research and design focuses on practitioner's goals and tasks rather than system developer's goals and tasks, and 3) human-centered research and design is context bound. The context criterion relates to social/organizational context and representational context (where people use perceptual as well as internal representations to solve problems).

Human centered analysis must take account of varied social units that structure work and information, organizations and teams, communities and their distinctive social processes and practices.

The underlying system dimension is *social*. The activity-centered analysis component models the social dimension and represents social quality component of the framework (Khosla et. al. 2003). The activity-centered analysis component is also based on human-centered criteria no. 2. The output of the activity-centered analysis component is the human-centered activity model.

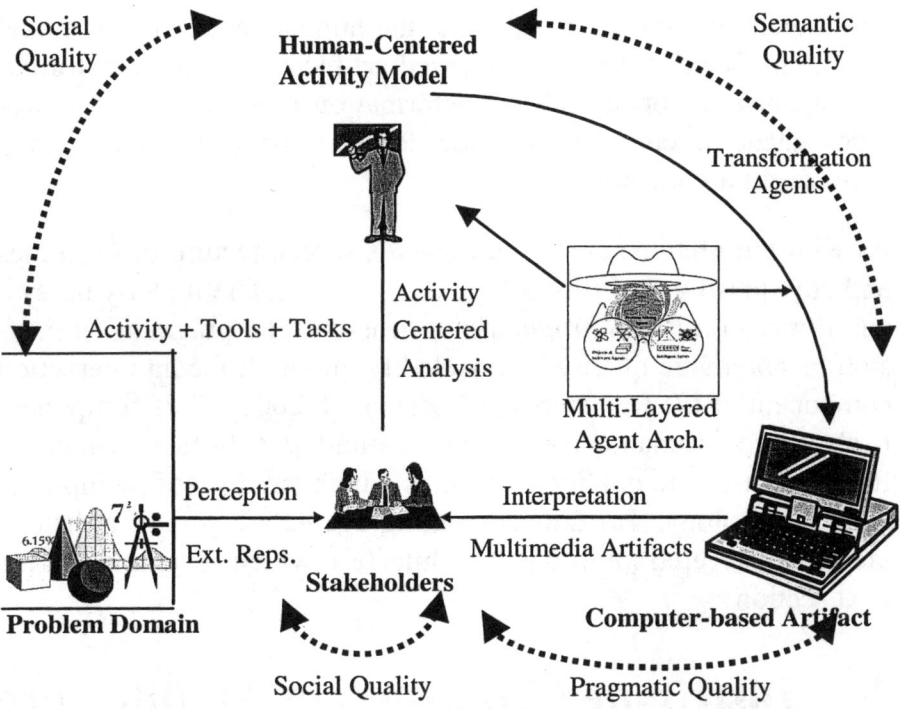

Figure 1. Human-centered system development framework.

The goals for semantic quality are completeness and validity (Linland et al., 1994). Completeness is defined as the degree to which the software system represents each of the concepts in the stakeholder's conceptualizations of the domain of interest. Thus the question whose ideas get put into the design process is an important one for human centered systems. As well, the question of whose problems are being solved is important. systems which seek only to answer a very narrow technical or economic agenda or a set of theoretical technical points do not belong under the "human centered" rubric (Kling and Leigh, 1997). The distributed multi-layered multi-agent architecture component models the semantic quality dimension of the framework and is also based on human-centered criteria number 1. and the is a human-centered domain model. The multi-agent architecture transforms the human-centered task model into human-centered computer based artifact using the transformation agent component. The transformation agent component provides agent based constructs for defining the five layers of the multi-agent architecture.

As with the architecture of buildings, the architecture of machines embody questions of *liveability* and *usability*. The underlying system dimension is the *pragmatic* dimension. The pragmatic dimension or pragmatic quality is modeled by the multimedia interaction component of the framework (Khosla et. al. 2000). This component is also based human-centered criteria number 3. In this chapter we focus on semantic quality of systems which employ soft computing agent technology. The soft computing agents are part of a distributed multi-layered multi-agent architecture which is outlined in the next section.

3 Distributed Multi-Agent Architecture

The multi-layered distributed agent architecture is shown in Figure 2. It is derived from integration of characteristics of soft computing artifacts like fuzzy logic, neural network and genetic algorithm, agents, objects and distributed operating system process model with problem solving ontology model (Khosla et. al. 2000 and 2003). It consists of five layers, namely, the object layer, which defines the data architecture or structural content of an application. The distributed processing agent layer helps to define the distributed processing and communication constructs used for receiving, depositing and processing data in a distributed environment. In the rest of this section we describe the problem solving agent, optimization agent and the technology or tool agent layers respectively. From the perspective of intelligent agent design we illustrate how these three layers facilitate human-centeredness at three distinct levels.

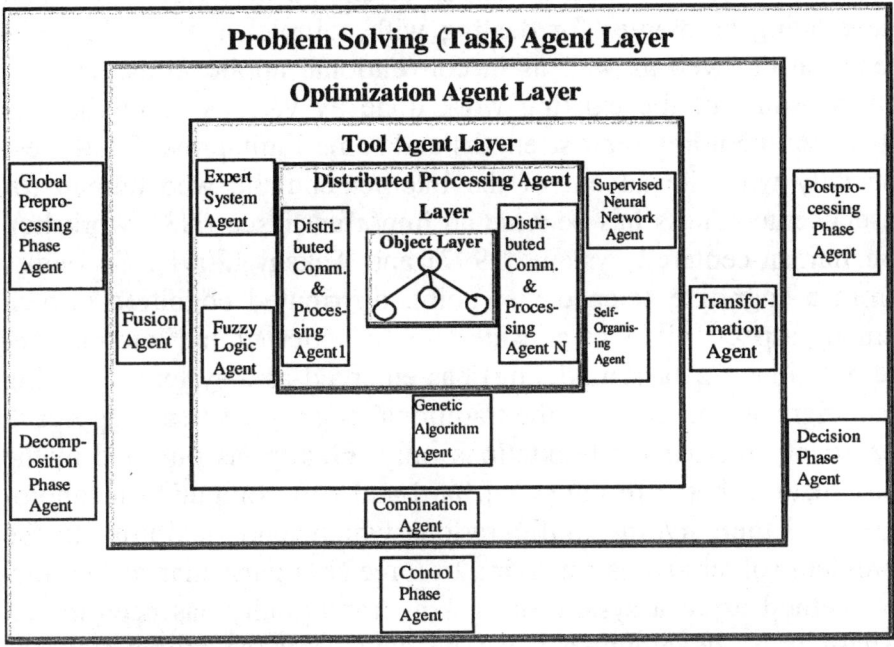

Figure 2. Multi-layered distributed agent ontology for soft computing systems.

3.1 Problem Solving Ontology Layer

The problem solving ontology (task) agent layer defines the constructs related to the problem solving agents namely, preprocessing, decomposition, control, decision and post processing. We discuss the definition of the problem solving agents in this section.

This layer systematizes and organizes practitioners/stakeholder/ user's tasks in the domain under study. It also helps to model practitioner's problem solving approach and tasks in domains which are complex and data intensive. The research on problem solving ontologies or knowledge-use level architectures has largely been done in artificial intelligence. The research at the other end of the spectrum (e.g., radical connectionism, soft computing) is based more on understanding the nature of human or animal behavior rather than developing ontologies for dealing with complex real world problems on the web as well as in conventional applications. Table 1 shows some of the existing work done by various researchers in this area including their strengths and some limitations. Firstly, especially with the advent of the internet and the web human (or user)-centeredness has become an important issue (NSF workshop on human-centered systems 1997) and Takagi (2001). Secondly, from a cognitive science viewpoint, distributed cognition (which among aspects involves consideration of external and internal representations for task modeling) has emerged as a system modeling paradigm as compared to the traditional cognitive science approach based on internal representations only. Finally, as outlined in the last section there are ranges of hard and soft computing technologies and their hybrid configurations which lend flexibility to the problem solver as against trying to force fit a particular technology or method on to a system design (as traditionally has been done). These three developments can be considered as pragmatic constraints. Most existing approaches do not meet one or more of the pragmatic constraints. Besides, from a soft computing perspective most existing ontologies do not facilitate component based modeling at optimization, task and technology levels respectively. Given the approximate and imprecise nature of solutions provided by soft computing agents, these three levels besides facilitating human-centered modeling of soft computing agents (as illustrated in section 4), also facilitate in engineering efficient, effective, and scalable designs of soft computing agents.

Table 1. Strengths and some limitations of some existing approaches.

Approach	Strengths	Some Limitations
Heuristic Classification (inference pattern) Clancey 1985	Good empirical generalization	No distinction between different classification methods. (e.g. weighted evidence combination). Not enough vocabulary. Pragmatic constraints not considered.
Model Based Systems (part-whole, causal, geometric, functional) (Steels 1985; Simmons 1988)	Principled domain models, complete knowledge	Combinatorial explosion, assumes all observations are correct & exact domain theory is known.
Problem Solving Ontologies (between model based and data based) based on Problem Solving Methods - (cover-and-differentiate, propose-test-refine, etc) (McDermott 1988; Fensel 1996, 1997, 2000)	Helps to determine type of domain knowledge required for problem solving, eases the knowledge acquisition bottleneck.	When to stop the knowledge acquisition, when is the system complete. Can lead to deep solution hierarchies. Do not consider the role of external representations or perceptual tasks. Primarily based on knowledge based systems. Do not adequately consider pragmatic constraints like human evaluation, response time, etc.
- Generic Task Based (classification, interpretation, diagnosis & construction/design) (Chandrasekaran 1983, Chandrasekaran & Josephson 1997, Chandrasekaran & Johnson 1993, Steels 1990)	Reuse, basis for interpreting acquired data, can build generic software environments.	Generic task categorization is not generic enough because they (e.g. diagnosis) can be accomplished using many different domain models, and different methods (depending on problem granularity), pragmatic constraints not considered. Tasks only mediated by methods.
Generic Ontology – KADS methodology (domain layer, inference layer, task layer, strategy layer) (Breuker and Weilinga 1989,1991; Weilinga 1993.)	Segregates knowledge modeling into four layers.	System modeling done with low level primitives. Suitable for knowledge based problems only. Does not consider pragmatic constraints.

Some generic goals and tasks associated with each problem solving agents are shown in Table 2. The generic goals and tasks employed by the five problem solving agents are also based on consistent problem solving structures employed by practitioners in numerous complex problem solving situations (e.g., image processing, real-time alarm processing, large scale data and web mining, health informatics, etc.).

Table 2. Some goals and tasks of problem solving agents.

Phase	Goal	Some Tasks
Preprocessing	Improve data quality	Noise Filtering Input Conditioning
Decomposition	Restrict the context of the input from the environment at the global level. By defining a set of orthogonal concepts Reduce the complexity and enhance overall reliability of the computer-based artifact	Define orthogonal concepts
Control	Determine decision selection knowledge constructs within an orthogonal concept for the problem under study.	Define decision selection concepts with in each orthogonal concept as identified by users Determine Conflict Resolution rules between decision selection concepts
Decision	Provide decision instance results in a user defined decision concept.	Define decision instances of interest to the user
Post-processing	Establish outcomes as desired outcomes	Concept validation, Decision instance result validation

The solutions to real world problems are determined by engineers, designers, accountants, sales managers, etc. in a task context (Chandrasekaran 1992; Peerce et al 1997) rather than a technological context. Various soft computing technologies, like fuzzy logic, neural networks and their hybrid configurations (fusion, transformation and combination), propose a technology-based solution to real world problems. The problem solving agent layer of multi-agent soft computing architecture is task oriented in which technological artifacts are considered as primitives for accomplishing various tasks. The use of one or more technological primitives is contingent upon satisfaction of task constraints. The task orientation enables multi-agent soft computing architecture to match a given task to one or more technologies among a suite of technologies rather than match a given technology to tasks in a work activity. Further, the generic tasks of the five problem solving agents have been problem solving agent layer has been grounded in human-centered criteria outlined in the 1997 NSF workshop on human-centered systems.

3.2 Optimization Layer

A generic optimization model is shown in Figure 3. It consists of three components, namely, solution generation, solution quality measurement (dynamic knowledge), and solution re-calibration respectively. The human or environment feedback is used by the solution quality measurement component.

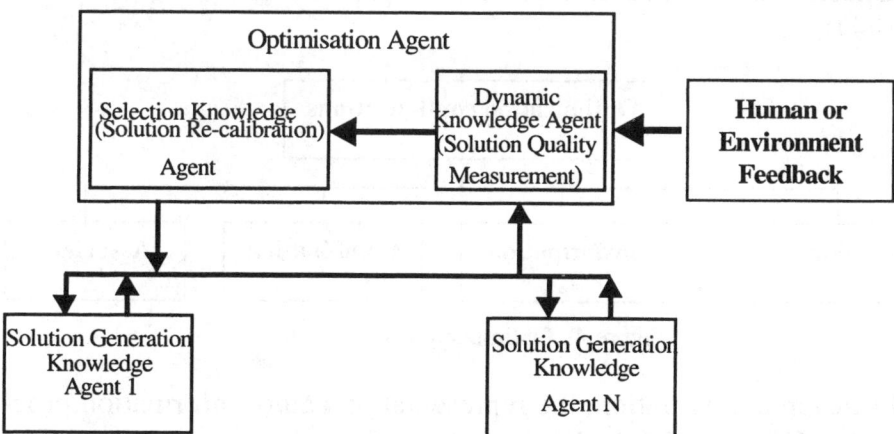

Figure 3. Generic optimization model.

The configuration of the solution quality measurement and solution re-calibration components can be one of the configurations shown in Figure 4. These configurations are also shown in the optimization agent layer of Figure 2. The configuration constructs like fusion, combination, transformation and association technologies which are used for optimizing the quality of solution (e.g., accuracy). In this section we discuss the definition of the optimization agent configurations.

The four most commonly used intelligent technologies are symbolic knowledge based systems (e.g. expert systems artificial neural networks, fuzzy systems and genetic algorithms). The computational and practical issues associated with intelligent technologies have led researchers to start hybridizing various technologies in order to overcome their limitations (Khosla et.al. 1997, 2000, 2003; Chiaberage et. al., 1995). However, the evolution of hybrid configurations is not only an outcome of the practical problems encountered by these intelligent methodologies but is also an outcome of deliberative, fuzzy, reactive, self-organizing and evolutionary aspects of the human information processing system (Bezdek, 1994).

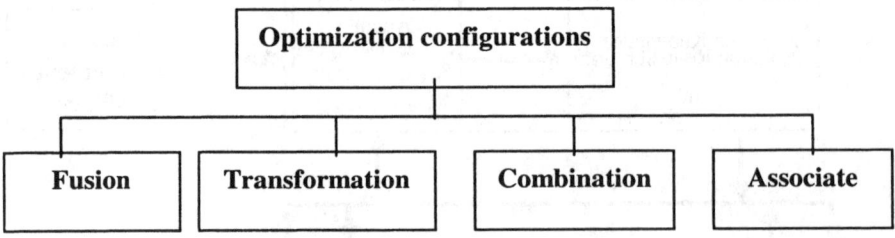

Figure 4. Optimization configurations.

In fusion configuration, the representation and/or information processing features of in technology A are fused into the representation structure of another technology B. From a practical viewpoint, this augmentation can be seen as a way by which a technology addresses its weaknesses and exploits its existing strengths to solve a particular real-world problem.

Transformation configuration is used to transform one form of representation into another. For example, neural nets are used for transforming numerical/continuous data into symbolic rules which can then be used by a symbolic knowledge based system for further processing. Combination configuration involves a modular arrangement of two or more technologies to solve real-world problems.

However, these configurations also suffer from some drawbacks. Fusion and transformation architectures on their own do not capture all aspects of human cognition related to problem solving. For example, fusion architectures result in conversion of explicit knowledge into implicit knowledge, and as a result lose on the declarative aspects of problem solving. Thus; they are restricted in terms of the range of tasks covered by them. The transformation architectures with bottom-up strategy get into problems with increasing task complexity. Therefore the quality of solution suffers when there is heavy overlap between variables, where the rules are very complicated, the quality of data is poor, or data is noisy. The combination architectures cover a range of tasks because of their inherent flexibility in terms of selection of two or more technologies. However, because of lack of (or minimal) knowledge transfer among different modules the quality of solution suffers for the very reasons the fusion and transformation architectures are used. It is useful to associate these architectures in a manner so as to maximize the quality as well as range of tasks that can be covered. The selections of these optimization configurations are contingent upon satisfaction of task constraints and vary from application to application.

3.3 Tool or Technology Layer

The tool or technology agent layer defines the constructs for various intelligent and soft computing tools. Finally, the five layers facilitate a component based approach for agent based software design. The generic agent definition used for defining the agents in various layers is shown in Figure 5. The generic agent definition shown in Figure 5 includes goals which is a desire or desired outcome or state.

```
Name:
Parent Agent:
Goals:
Tasks:
Task Constraints:
Precondition:
Postcondition:
Communicates With:
Communication Constructs:
Linguistic/non-linguistic Features:
Psychological Scale:
Representing(Perceptual) Dimensions:
External Tools:
Internal Tools:
Internal State:
Actions:
```

Figure 5. Generic agent definition.

Tasks are goal directed processes in which people consciously or unconsciously engage. Task constraints are pragmatic constraints imposed by the stakeholders and the environment for successful accomplishment of a task. The task constraints primarily determine the selection knowledge required for selecting a technological artifact. Precondition helps us to define underlying assumptions for task accomplishment. Post-condition defines the level of competence required from the technique or algorithm used for accomplishing the task. The communication constructs employed by the transformation agent. These communication constructs are based on human communicative acts like request, command, inform, broadcast, explain, warn and others. The linguistic and non-linguistic features represent the sensed data from the external environment as well as computed data by the agent. Representing dimension is the physical or abstract dimension used to represent a feature. It can be seen as capturing the perceptual representation or category of a feature. These representing dimensions can be shape, color, distance, location, etc.

Psychological Scale is the abstract measurement property of the physical or abstract dimension of a represented feature. The perceptual dimension (e.g., color, shape, texture, etc.) and psychological scale (e.g., nominal, ordinal, equal interval and ration (Steven, 1957) constructs model the representational context outlined in human-centered criteria no. 3.

Name: Fuzzy Neural Network Agent
Parent Agent: Fuzzy, neural network
Goals: Optimization, adaptation
Tasks (some): Create Fuzzified neural-network model,
 Perform back propagation Test convergence with test data
Tasks Constraints: Normalized Training data available,
 Fuzzified inputs available, optimized Fuzzy-neural model not known
Precondition: Training data normalized, Convergence criteria is known
Post Condition: Converges on global minimum, Optimized fuzzy-neural model
Communicates with: Decision Agent, Distributed process agent
Communication constructs: Inform model parameters to user,
 Receive feedback data from environment
Represented Features: Fuzzified data, training/test set error, convergence criteria
Representing Dimns (Perceptual features): Training / test set error shapes,
 Convergence graphs
Linguistic/nonlinguistic percepts: Fuzzified input, Fuzzy-neural model output
External tools: Simulated data files, Intelligent NN Agent
Internal tools: Sensitivity algorithms
Actions: Feed fuzzified input to network, return training set error,
 return optimized results (e.g., fuzzy rules)

Figure 6. Fuzzy-neural (fusion) agent.

The *parent agent* construct identifies the generic agents in the four agent layers, whose constructs and services have been inherited by a particular application or domain based transformation agent. The *communication with* construct in Figure 6 identifies all the agents and objects that a transformation agent communicates with in the five layers. The *external tools* construct in Figure 6 refers to those computer-based or other tools that are external to the definition of an agent. On the other hand, *internal tools* are those tools that are defined internally by a transformation agent. The *internal state* construct refers to the beliefs of a transformation agent at a particular instant in time. Finally, the actions construct is used to define the sequence of actions for accomplishing various tasks. Figure 6 provides a sample fuzzy-neural network agent definition at optimization level of the architecture.

4 Human-Centered Modelling Using Soft Computing Multi-Agent Architecture

In this section we establish the human-centeredness of the multi-layered agent architecture at the problem solving, optimization and technology levels respectively.

4.1 Human-Centeredness and Problem Solving Agent Layer

The problem solving agent layer is based on consistent problem solving structures employed by practitioners and users in developing solutions to complex problems. These constructs facilitate a problem driven approach as required in human-centered criteria number 1. The five problem solving agents of the problem solving agent layer also represent the five information processing phases (Khosla et. al. 2003), namely, preprocessing, decomposition, control, decision and postprocessing respectively. Further, the component based nature of the five problem solving agents allows them to be used in different problem solving contexts or sequences as shown in Figure 7. Figure 7 shows the decision paths taken by a medical practitioner for diagnosis and prescribing treatments to a patient. In one context the medical practitioner may not need the help of the computerized system for diagnosing a patient's disease and may take decision path from diagnosis to treatment. Whereas, in another context the medical practitioner may not know the diagnosis and thus seek the help of the computerized system from symptoms to diagnosis to treatment.

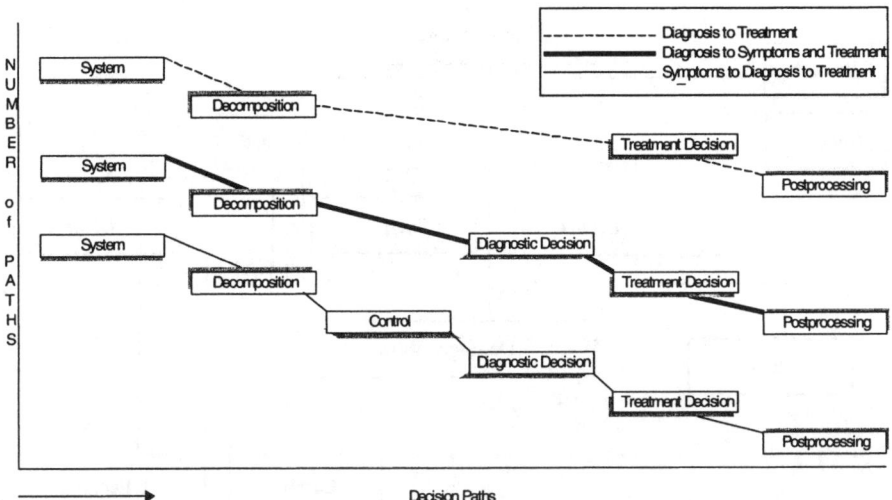

Figure 7. Decision paths taken by a medical practitioner and agent sequence.

In the rest of this subsection we outline how the decision support model of a customer relationship manager in banking and finance is modeled with the help of the problem solving agents of the problem solving agent layer. The problem solving agents allow the CRM the manager to express their data mining needs in terms of the decision making concepts employed by them in CRM domain.

The ability of the financial institutions like banks to collect data far outstrips their ability to explore, analyze and understand it. For that reason, in the past five years banks have moved aggressively towards applying data mining techniques especially in the Customer Relationship Management (CRM) area. Given the cost savings with Internet banking, the banks seem now keen to apply data mining techniques to study online transaction behavior of their clients and improve their on line product offerings. Figure 8 shows a highly simplified data model of a bank with both Internet and branch (face-to-face) banking facilities.

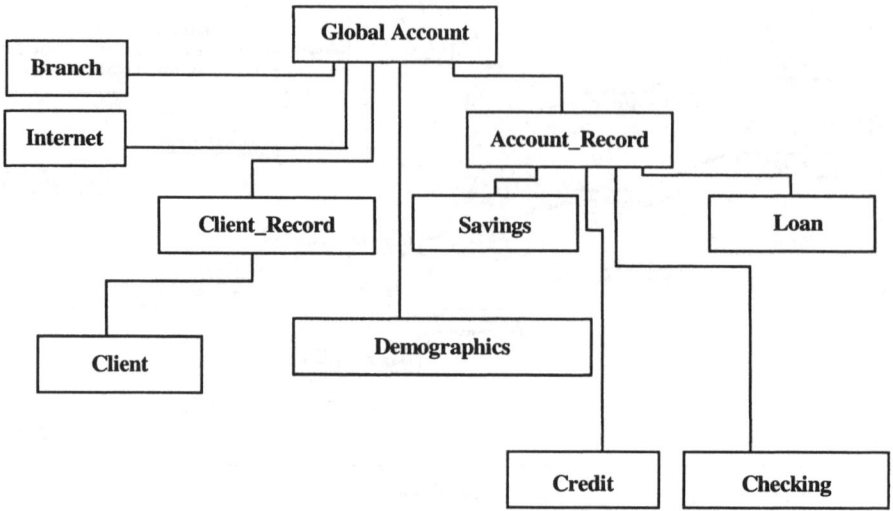

Figure 8. Simplified CRM model of banking domain.

In this application we model the CRM aspects of the internet-banking domain from a Web sponsor's or a CRM manager's viewpoint using the problem solving agent layer.

4.1.1 CRM Model of Internet-Banking

The decision support model of a CRM manager is constructed using the five problem solving agents of the soft computing multi-agent architecture. A brief description of the problem solving agents as applied to CRM in a internet-banking domain is provided next.

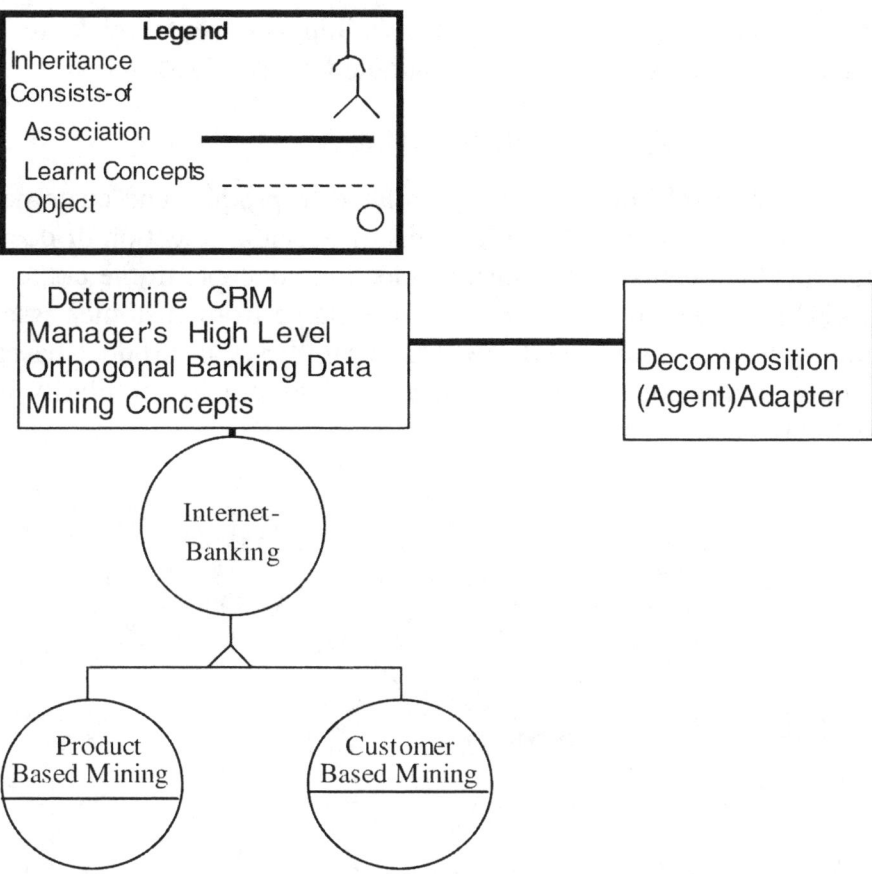

Figure 9. Mapping high level orthogonal CRM concepts employed by CRM manager.

4.1.2 Decomposition Phase Problem Solving Agent and CRM

The goal of the decomposition agent is to restrict input context and reduce domain complexity. In the internet-banking domain it is done by mapping the orthogonal concepts employed by the CRM manager to the decomposition adapter (the term adapter has been adopted for the UML terminology of design patterns and adapters used in software engineering). It is shown in Figure 9. The "product based," and "customer based" data mining concepts relate to a functional decision support model adopted by the CRM manager.

4.1.3 Control Phase Problem Solving Agent

The decision selection knowledge related to product and customer based data mining is defined in terms of decision selection or decision level concepts. The decision selection concepts in the context of CRM are shown in Figure 10. These range from customer association, loan similarity and credit card similarity in product based mining, and account transfer pattern to demographic similarity in customer based mining.

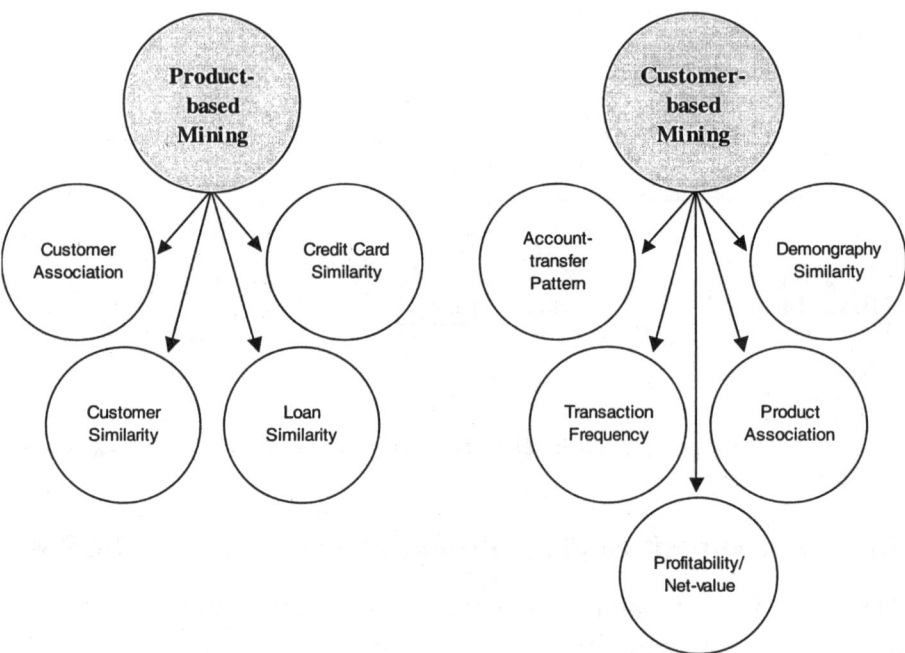

Figure 10. Decision selection CRM concepts.

Preprocessing can occur in any of the three phases (i.e., decomposition, control and decision). In this application the preprocessing adapter is used to filter out the noisy or irrelevant data. For example, removal web-log data of e-commerce server (unrelated to transaction data), erroneous data, data reduction, dimensionality reduction are some of the tasks undertaken by the preprocessing adapter. The tasks employed by the preprocessing agent are both user and software design related.

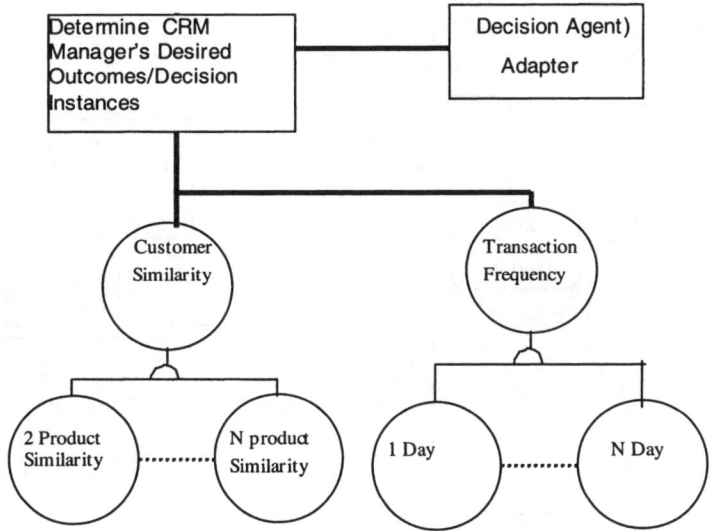

Figure 11. Mapping decision adapter to Internet-banking application.

4.1.4 Decision Phase Problem Solving Agent

The decision agent identifies the desired outcomes or decision instances required by the CRM manager. That is, we must determine the decision instances or events of specific interest to the user. For example, as shown in Figure 11, CRM manager is interested to find out similarity among customers in terms of combination of bank products they purchase from the bank. This will facilitate on-line customization of bank products and higher profitability. On the other hand, by determining the transaction frequency of their customers, the bank can determine transaction costs to different type of customers as well as determine whether existing transaction costs need to be increased or decreased. Figure 11 shows a sample mapping of these concepts with the decision (agent) adapter.

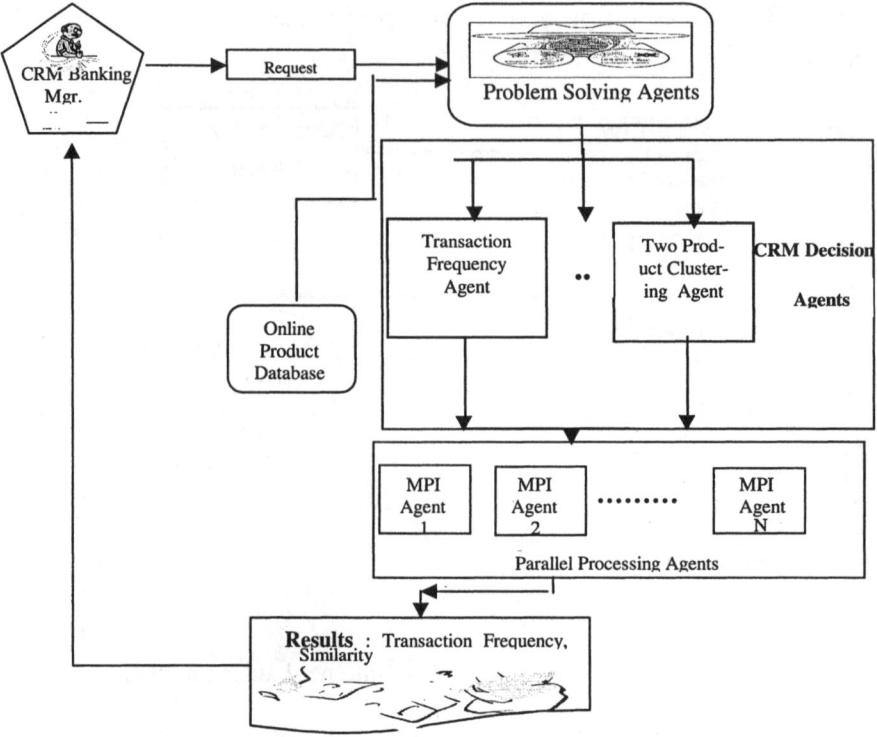

Figure 11. Internet-banking application architecture.

The post-processing agent like the pre-processing agent can be used by any of three problem solving agents (i.e., decomposition, control and decision). The decision employs the post-processing agent to integrate and summarise the product combination and transaction frequency results for the CRM manager.

A partial overview of the agent based design architecture with problem solving agents (which allow to formulate the CRM manager's query), parallel processing agents (related to distributed processing agent layer of the soft computing agent architecture) is shown in Figure 11. Decision agents like "Transaction frequency Agent," and "Two Product Similarity Agent," are shown in expanded form in Figure 11.

4.2 Unstained Cell Image Processing

In general, images of unstained cells (see Figure 12) are characterised by low contrast and edge greyscale gradients on the boundaries between cells and background. Background contains noise and other artefacts. Cells have a non-uniform greyscale level often being textured in a way that is difficult to distinguish from the background. Cells can appear to be "transparent" with respect to the background. There may be clustering of cells that may be touching or overlapping each other. The characteristics of unstained cells present a particularly difficult problem to segmentation. Traditional techniques based on thresholding and edge detection have not achieved acceptable accuracy.

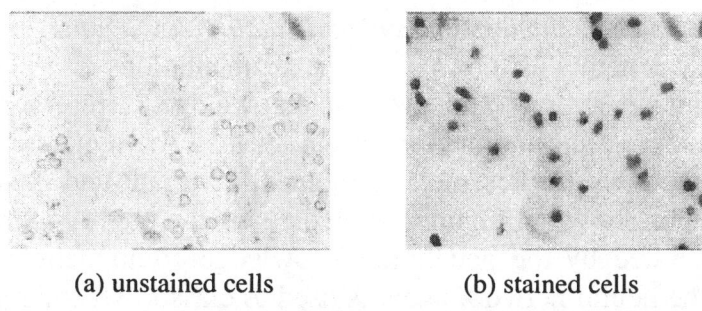

(a) unstained cells   (b) stained cells

Figure 12. Chinese Hamster Ovarian (CHO-K1) samples.

The purpose of this section is to outline the functionality of the various components of the multi-agent soft computing model for segmentation and classification of unstained mammalian cell images. The model shown in Figure 13 is used for predicting (the quality of segmentation) which leads to optimization and improvement in the classification accuracy of unstained cell images.

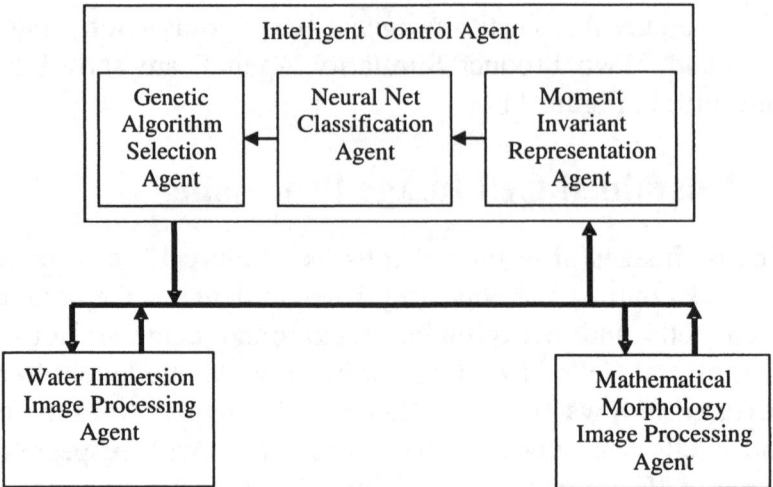

Figure 13. Multi-agent optimization model for unstained cell image processing.

The model shown in Figure 13 consists of three types of knowledge, namely, generation knowledge, dynamic knowledge and selection knowledge respectively. The generation knowledge is represented by *water immersion and morphological agents* in Figure 13). They generate different plausible segmentations of a given image or image region. The dynamic knowledge is represented by neural network and moment invariant agents. The dynamic knowledge component (represented by moment invariant and neural network agents shown in Figure 13) predicts on the quality of the solution generated by the generation knowledge component. In other words, the neural network agent is used to classify whether particular segmentation is acceptable or not acceptable. The moment invariant transformation agent is used to transform the IPA (Image Processing Agent) results to a vector input for the neural network. The neural network is also used for classification of cells in accepted segmented image regions. The genetic algorithm agent is used to determine optimal parameters to be used by water immersion and morphological agents for improving the segmentation of unacceptable segmented regions. The model in Figure 13 represents a combination configuration of soft computing agents in the optimization agent layer shown in Figure 2.

The agent definition of the genetic algorithm agent is shown in Figure 14. Thus our unstained image processing model consists of an Intelligent Control Agent (ICA) – encompassing the genetic algorithm, neural network and moment invariant agents respectively, that interprets the result of its subordinate Image Processing Agents (IPA) – consisting of the water immersion mathematical morphology agents respectively. From the generated segmentations ICA determines the optimum process, and its parameters, to employ next to achieve improved or desired results. The control or feedback loop between the IPA and ICA represents the iterative nature of the model.

Using these two IPA's on various samples the test images like those shown in Figure 14(a), a human could produce good segmentation by selection of technique and parameters based on the results of each technique for a subsequent pass through the selected technique. In our model the ICA provides human operator-like feedback to the image processing agents for improving segmentation and takes over the role of the human operator to automate the process. We have implemented this model for several images of unstained Chinese hamster ovarian samples (Lai, Khosla and Mitsukura 2003). Optimized segmentation of a unstained cells image region is shown in Figure 14.

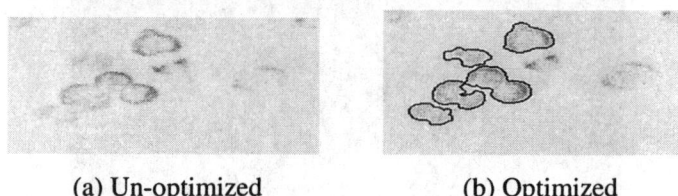

(a) Un-optimized   (b) Optimized

Figure 14. Un-optimized and optimized image segmentation.

A neural network agent (see Figure 15) is trained on examples of acceptable and unacceptable segmentation. The training set is shown in Figure 16. Examples in the class of acceptable segmentation are shown with their boundaries only outlined. The examples in the class of unacceptable segmentation are shown filled.

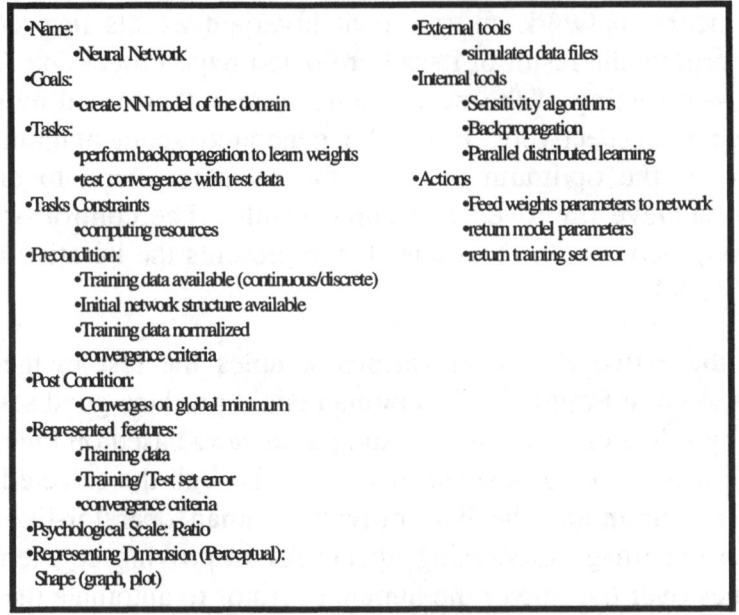

Figure 15. Neural network agent definition.

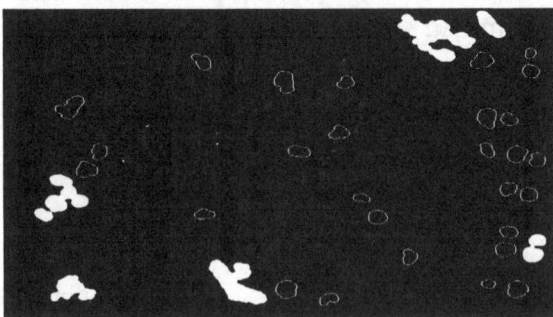

Figure 16. Acceptable and unacceptable segmentations.

4.3 Human-Centeredness and Technology Agent Layer

At the technology level human involvement is encouraged by asking user to define the search parameters or fitness function to be used by a soft computing agent like GA to create the correct combination of shirt design. The shirt design is a part of a real time web based missing person clothing identification system. Figure 17 shows a sample of shirt designs displayed to the user (a person who last saw the missing person). The user selects shirt designs 3 and 5 in the top row of Figure 17 as close to shirt design worn by the missing person before they went missing. This feedback is used as by the GA agent as fitness function. Figure 18 shows the next set of shirt designs (note the revised shirt designs 3 and 5).

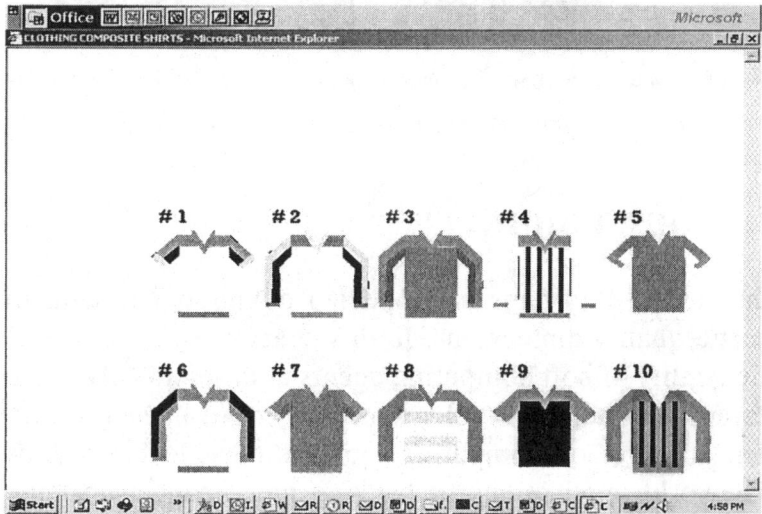

Figure 17. GA-based user-centered shirt design.

Overall, the five layers of the soft computing agent architecture leads to component based distributed (collaboration and competition) soft computing agent design. The component based nature facilitates autonomy in distributed processing, technology and optimization agent layers of the architecture. The problem solving agent layer, among other aspects, facilitates human-centered modeling of user's tasks.

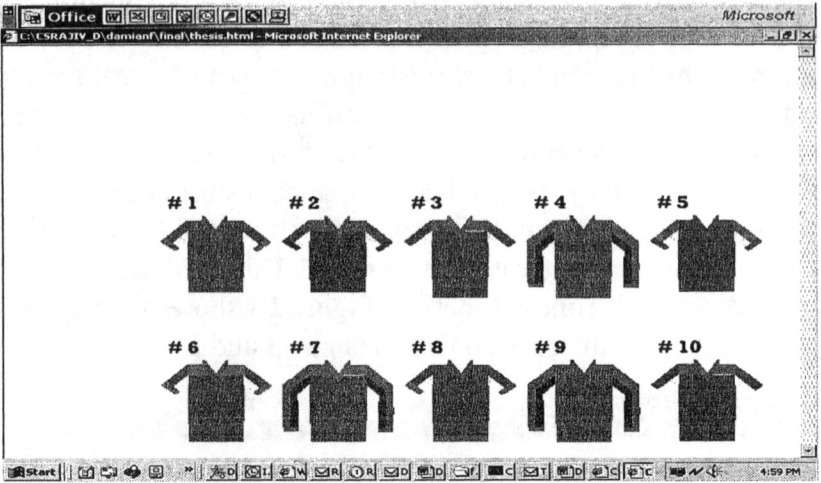

Figure 18. The next set of shirt designs.

5 Conclusion

Human-centered systems are modeled along social, semantic and pragmatic quality dimensions. In this chapter we have modeled semantic quality of soft computing agents at problem solving, optimization and technology level respectively. We have illustrated the humanization of soft computing agents at these levels with the help of real world applications in banking and finance, unstained cell image processing, and real-time web based missing person clothing identification system.

References

Bamford, P. and Lovell, B. (1996), "A water immersion algorithm for cytological image segmentation," *Proceedings of Segment*, Sydney. December

Bezdek, J.C., "What is computational intelligence?' *Computational Intelligence: Imitating Life*, Eds. Robert Marks-II et al., IEEE Press, New York, 1994.

Breuker, J.A. and Weilinga, B.J., (1989) "Model driven knowledge acquisition" in B. Guida & G. Tasso eds. *Topics in the Design of Expert Systems* Springer-Verlag, pp 239 – 280.

Breuker, J.A. and Weilinga, B.J., (1991) *Intelligent Multimedia Interfaces*, Edited by Mark Maybury, AAAI Press, Menlo Park, CA.

Chandrasekaran B. and Josephson, J.R., (1997) "Ontology of tasks and methods", *AAAI 97 Spring Symposium on Ontological Engineering,* March 24-26, Stanford University, CA California, USA.

Chandrasekaran, B. (1983), "Towards taxonomy of problem solving types", *AI Magazine* Vol 4 No. 1, Winter/Spring pp 9-17.

Chandrasekaran, B., Johnson, T.R., and Smith, J.W. (1992) "Task structure analysis for knowledge modeling", *Communication of the ACM* vol.35, no.9, pp.124-137

Chandrasekaran, B., and Johnson, T.R. (1993), "Generic tasks and ask structues: history critique and new directions", *Second Generation Expert Systems,* G.M. Davies, J.P. Krivine and R. Simmons.

Chiaberage, M., Bene. G.D., Pascoli, S.D., Lazzerini, B., and Maggiore, A. (1995). Mixing fuzzy, neural & genetic algorithms in integrated design environment for intelligent controllers, *1995 IEEE Int Conf on SMC,*. Vol. 4, pp. 2988-93.

Clancey, W.J., (1985) "Heuristic classification", *Artificial Intelligence*, 27, 3 (1985), 289-350

Dullmann, D. (1999) "Petabyte databases" *SIGMOD*, ACM

Fensel, D. "Problem-solving methods: understanding, development, description, and reuse", in *Lecture Notes on Artificial Intelligence*, no 1791, Springer-Verlag, Berlin, 2000.

Fensel, D (1997), "The tower-of-adapter method for developing and reusing problem-solving methods," *EKAW*, pp. 97-112

Fensel, D. and Groenboom, R., (1996), "MLPM: defining a semantics and axiomatization for specifying the reasoning process of knowledge-based systems,".*ECAI*, pp 423-427

Khosla, R., and Dillon, T., "Learning knowledge and strategy of a generic neuro-expert system architecture in alarm processing", in *IEEE Transactions on Power Systems*, Vol. 12, No. 12, pp. 1610-18, Dec.1997.

Khosla, R., and Dillon T., *Engineering Intelligent Hybrid Multi-Agent Systems*, Kluwer Academic Publishers, MA, USA August 1997.

Khosla, R., Sethi, I. and Damiani, E., *Intelligent Multimedia Multi-Agent Systems – A Human-Centered Approach*, Kluwer Academic Publishers, MA, USA October 2000.

Khosla, R. Damiani, E., and Grosky, W. (2003), *Human-Centered E-Business*, Kluwer Academic Publishers, MA, USA.

McDermott, J., (1988) Preliminary steps toward a taxonomy of problem solving methods. in *Automated Knowledge Acquisition for Expert Systems*, S. Marcus, Ed., Kluwer Academic, pp. 225-256.

NSF Workshop on *Human-Centered Systems*, February 1997. Final Report

Perrow, C., (1984) *Normal Accidents: Living with High-Risk Technologies* NY: Basic Books

Peerce, J., et al (1997), *Human-Computer Interaction*, Massachusetts: Addison-Wesley Pub

Simmons, R. 1988. "Generate, Test, and Debug: a Paradigm for Solving Interpretation and Planning Problems". Ph.D. diss., AI Lab, Massachusetts Institute of Technology.

Steels, L. (1990) "Components of expertise", *AI Magazine*, 11, 28-49., 11, 28-49

Steels, L., and Van de Velde W. (1985), "Earning in second-generation expert system" in *Knowledge-Based Problem Solving,* ED.J.Kowalik. Englewood Cliffs, NJ.: Prentice-Hall

Steven, S.S. (1957). "On the psychological law", *Psychological Review*, 64(3), 153-181

Takagi, H.K. (2001) "Interactive evolutionary computation: fusion of the capabilities of EC optimization and human evaluation," *Proceedings of the IEEE*, vol. 89, no. 9, September

Weilinga, B.J. Ath. Schreiber and J.A. Breuker (1993) "KADS: a modeling approach to knowledge engineering" *Readings in Knowledge Acquisition and Learning*, eds Buchanan, B.. & Wilkins, D. San Mates California, Morgan Kaufmann, pp 92-116

Chapter 2

Software Agents for Ubiquitous Computing

Sasu Tarkoma, Mikko Laukkanen, Kimmo Raatikainen

Ubiquitous computing draws a picture of a world where people access intelligent and easy-to-use services, irrespective of location and time. Software agent technology is a good candidate for ubiquitous computing, because the agent paradigm is by definition well suited for decentralized, heterogeneous and dynamic environments. This chapter presents the general requirements of the ubiquitous environment, and discusses the use of software agent technology to facilitate service provision and composition for ubiquitous client end systems.

1 Introduction

Ubiquitous computing presents a world where people access intelligent and easy-to-use services, irrespective of location and time. Ideally, these services are adaptive and take into account the desires and requirements of the users. Ubiquitous computing (Weiser 1993), pervasive computing (Satyanarayanan 2001) and the IST Advisory Group's (ISTAG) ambient intelligence (ISTAG 2001) define an invisible computing landscape that consists of myriad computing elements in the physical environment that interact with each other and people in order to provide various services, such as information services, location-based services, multi-mode communication and e-commerce applications. This environment consists of heterogeneous hardware and software, and is dynamic in nature, because people and devices are mobile and communicate using different network technologies, such as Global System for Mobile Communications (GSM), General Packet Radio Service (GPRS), Universal Mobile Telecommunications System (UMTS), wireless LAN (WLAN) and Bluetooth.

A number of core technologies are needed in order to realize the intelligent and adaptive services of tomorrow. Middleware as a generic term refers to a fuzzy set of software services for application developers that are located above the network layer (level 3 in the ISO Open Systems Interconnect (OSI) stack) and below applications. Middleware typically includes services such as messaging and communication, resource discovery, transactions, and storage service. Ubiquitous and mobile computing has created the necessity for mobile middleware, such as Java 2 Micro Edition (J2ME) (Sun 2003), Wireless Application Environment (WAE) (WAP Forum 2002), and the Wireless CORBA specification from OMG (OMG 2003).

Software agent technology is a good candidate for ubiquitous computing, because it is based on asynchronous messaging, and supports the integration and co-operation of a number of autonomous components. Agents may engage in high-level conversation and negotiation over resources and services, and the agent paradigm by definition is well suited for decentralized, heterogeneous and dynamic environments. Agent platforms are an example of middleware that eases the development of agents by providing agent lifecycle management, interoperable communication protocols, and support for ontological knowledge.

This chapter presents the general requirements of the ubiquitous environment, and discusses the use of software agent technology to facilitate service provision for ubiquitous client end systems. We consider the use of agents on the service provider platform as the building blocks for interoperable services, and on small devices, such as Personal Digital Assistants (PDAs) and mobile phones, as a mechanism to access services, support service composition, and allow service logic to be executed on the client in order to cope with network disconnections.

The chapter gives an overview of the environment, and examines current technologies and methodologies to support mobile and ubiquitous computing. We summarize different approaches and give a short overview of the relevant technologies and specifications. We present a case study with MicroFIPA-OS, which is an agent execution environment targeted at PDA devices, and which is derived from the FIPA-OS agent platform (Buckle *et al.* 2002). Both agent systems comply with the FIPA specifications that standardize the agent platform and its external interfaces. The system allows rapid prototyping of FIPA agents on small devices by supporting the FIPA-OS API. We present our experiences with adaptive service composition using an experimental scenario.

2 Ubiquitous Computing

2.1 Overview

The IST Advisory Group's future ambient intelligent scenarios give good examples of future ubiquitous environments. To give an example, we examine the scenario 1: "Maria — The Road Warrior". In this scenario the user, Maria, is on a business trip and arrives at a foreign country. Maria is carrying a personalized communications device, the P-Com, which automates various tasks during her travel from the airport to the hotel: her visa is automatically checked at the immigration, her car rental is arranged beforehand, the P-Com recognizes and personalizes the car for Maria, real-time traffic instructions are provided during the drive, and finally, the hotel room facilities are customized for Maria by interactions between the P-Com and the hotel room computing infrastructure. All of these tasks require the knowledge of the networking infrastructure at a given time and in a particular location. At the airport there may be WLAN hotspots available, and at the immigration there could be local Bluetooth coverage. During the walk from the airport arrival hall and the car garage a GPRS network is available, and once Maria enters the car, there could be a Bluetooth network specific to the car. During the drive to the hotel, traffic information is delivered using a GPRS network, and at the hotel, the hotel room provides a WLAN hotspot to control room facilities and to access the Internet. To make all the different network infrastructures transparent to Maria, the P-Com can make use of the (wireless) network and location information, and based on these, change seamlessly from one network to another.

The scenario points out several characteristics of the ubiquitous environment: multiple (wireless) networks and seamless roaming between them, different kinds of client terminals and network elements, location awareness, and context awareness. We will examine these in more detail in the subsequent sections.

2.2 Wireless Networks and Roaming

The data communications environment – especially wireless data communications – creates many challenges that have not been adequately addressed in today's Internet-based services. Wireless wide-area networks are in a phase of rapid development. High Speed Circuit Switched Data (HSCSD) and GPRS are already on the market and UMTS is about to enter markets. Utilizing WLANs as public "hot spot" access networks to the Internet is currently under implementation.

Each of these networks has its specific data transmission characteristics. Wireless wide-area communications is challenging, because the Quality of service (QoS) (such as line rate, delay, throughput, round-trip time, and error rate) may change dramatically when a user moves from one location to another. For example, when the user roams from a UMTS cell to a GPRS cell, the throughput may drop from 1 Mbits/s down to 24 Kbits/s. In addition, it is foreseen that seamless roaming between different network technologies (e.g., between UMTS and WLAN) will be needed in the near future, leading to an increase in the variability mentioned above. This heterogeneity and high volatility of the environment creates the need for adaptability. Users will demand services that will automatically and transparently adjust to the changes mentioned above.

2.3 Client Devices

The services in the ubiquitous environment may be accessed using a variety of client devices. These devices can be very different in terms of computing functionality, ranging from a mobile phone to a high-end laptop computer. Typically the devices are carried by the user, thus, they can be for instance smart phones, PDA devices, or tablet computers.

The capabilities and limitations of the device dictates the constraints on how the user is able to access the services and what kind of content is provided for the user. The capabilities may be divided into two categories: computational capabilities and interface capabilities. The computational characteristics define what kind of resources the client device has in terms of providing services. Some devices are only capable of acting as an interface to the services running in the network node (e.g. web browsers), whereas more capable devices may run services or parts of the services by themselves. The interface capabilities dictate the characteristics of how the service is displayed to the end-user. For instance, a mobile phone has very limited ways to show high-end images and video, whereas a tablet computer is less limited in this respect.

The capabilities of a device are specified in terms of a device profile, which includes details about its type, display and input devices. An example profile format that may be used to store the terminal characteristics is the W3C Composite Capability / Preference Profile (CC/PP) (World Wide Web Consortium 2003). Using CC/PP a device may register its profile with a service. Content can be adapted to meet the requirements of the device using information stored in the profile, for example, scaling the content to fit the screen and adjusting the color depth of images.

2.4 Location- and Context-Aware Services

Currently, the most commonly used contexts for information services are one-dimensional contexts such as temporal contexts. The user first subscribes information, which can be, for instance, future notifications relating to stock prices, sports, and news. Then, the services send messages and notifications to the user's mobile device according to the a priori subscriptions.

In ubiquitous computing the contexts are multi-dimensional: the nature of a service depends on time (temporal context), location (spatial context), user preferences, and user context. Answers to queries such as "where is the nearest interesting sight?" or "where is this street in relation to my hotel?" can be very different based on the person making the query.

The spatial and temporal aspects become even more dynamic and heterogeneous when taking into account the characteristics of the ubiquitous networking environment, as discussed in Section 2.2. When roaming between different networks and service domains, not only the spatial context, but also network link characteristics change with time and the positional accuracy may also change. For instance, when the bandwidth is low, the mobile user may appreciate less complex but timely content over delayed high-resolution content delivery.

2.5 Technology Support

2.5.1 Java – the Enabling Technology for Software Agents

The benefits of the Java programming language and virtual machine are portability of code and becoming less attached to the operating system and hardware. The choice of the Java platform and its configuration affects performance, memory footprint, portability and reuse of code. The Java 2 Micro Edition (J2ME) is the Java architecture for embedded and consumer devices, such as PDAs, mobile phones, set-top boxes and devices. The architecture consists of a set of standard Java APIs defined through the Java Community Process (JCP) program. J2ME provides a modular framework for building Java applications for various devices with differing characteristics (Sun 2003).

Configurations define the base functionality for a particular range of devices. They consist of a virtual machine and a set of class libraries. Currently J2ME includes two configurations: the Connected Device Configuration (CDC) and the Connected Limited Device Configuration (CLDC). The former is intended for devices such as mobile phones with intermittent network connections, slow processors and limited memory. CLDC is based on the K-virtual machine, which is a scaled down version of the full Java virtual machine. The latter configuration is designed for more powerful devices that have greater network bandwidth, such as set-top boxes and high-level PDAs. CDC includes a full-featured Java virtual machine and a larger subset of the J2SE API.

Configurations are combined with a set of higher level APIs, profiles, that define the application life cycle model, access to device properties, and the user interface. The Mobile Information Device Profile (MIDP) is a profile based on CLDC intended for mobile phones and low-end PDAs. MIDP is currently used in Java-enabled mobile phones.

The Foundation profile is the lowest level profile for CDC that defines the basic API and networking support. The Personal Profile is a CDC profile that supports graphical user interfaces (GUI) using AWT libraries and applets. The Personal Profile replaces the older PersonalJava specification, and provides a migration path for PersonalJava applications to J2ME. Both CLDC and CDC are extensible with optional packages that are specified using the JCP process. For example, optional packages for CDC and CLDC include: Wireless Messaging API (JSR-120), J2ME Web Services (JSR-172), and Bluetooth API (JSR-82).

There are several Java-based messaging solutions for wireless and embedded use. The Java Messaging Service (JMS) is a messaging interface specification that supports both point-to-point and the publish/subscribe model of communication (Sun 2002). JMS is also being used on small devices as a messaging solution, for example the iBus//Mobile[1], which uses a gateway to facilitate the communication between clients and servers.

2.5.2 Other Technologies

CORBA is widely used middleware architecture for facilitating the use of remote objects and components in enterprise applications and services. In addition to the enterprise environment, several small-footprint CORBA ORBs are also available for embedded environments, for example, Orbix/E 2.1 from IONA[2]. The use of

---

[1] http://www.softwired-inc.com/
[2] http://www.iona.com

CORBA on mobile devices supports interoperability and allows to leverage existing CORBA-based services; however, the Remote Produce Call (RPC)-style communication is often synchronous and may not be suitable for all wireless environments, especially communication links with high latency and variability.

Resource discovery frameworks such as Jini (Jini 2003), JXTA (JXTA 2003) and Universal Plug and Play (Chen *et al.* 2001) aim to support resource, device, and service discovery in ad hoc and heterogeneous environments. Profiles, such as CC/PP, and ontologies, such as the Device Ontology specified by FIPA (FIPA 2003), enable reasoning and matching of device capabilities, which is important for adapting services and content according to terminal capabilities.

W3C is currently specifying the Semantic Web and Web Services, which include a number of specifications for defining ontologies (Web Ontology Language), one-way and request-response messaging (SOAP), and specifying web service interfaces (WSDL) (McIlraith *et al.* 2001). For adaptation, it is essential that information about services, specified for example using WSDL and a suitable choreography language, and device capabilities are combined in order to tailor a service for the current environment. The infrastructure needs to provide a mechanism for detecting changes in the environment, which trigger the reconfiguration and re-evaluation of the composed service. In this reconfiguration, new service components may be located that offer better service in terms of QoS or accuracy of content.

3 Ubiquitous Agents

3.1 Overview

Software agent technology provides attractive characteristics for supporting ubiquitous environments: reactivity, pro-activity, autonomy, and co-operation (Jennings *et al.* 1998). Software agents are able to perceive changes in the environment and autonomously react to them. For instance, upon detecting that the throughput of the underlying network connection drops dramatically, an agent specialized in wireless communication may pro-actively make a decision to change from the current network technology to another. Once changed, the agent may inform other agents about the new QoS, who are therefore able to react based on the changing QoS, for instance by applying content adaptation.

Software agents can be specialized for different purposes. Here we use a categorization with the following categories: user agents, middleware agents, and service agents. A user agent serves the human user by automating tasks on user's behalf. The user agent usually runs in the client device, but may also reside on the server side acting as a proxy for the user, when the user's client device is not connected to the network.

Middleware agents hide the complexity of the underlying infrastructure. Examples of middleware agents are network agents, whose tasks could be for instance to enable seamless roaming between multiple wireless networks and to provide information about the QoS of the network connections. Middleware agents may reside on the client device, in the network elements, and on the server-side, depending on their functionality.

Service agents implement the actual service either by themselves or by wrapping some existing legacy service. Service agents usually run on the server-side, but may be distributed partially or totally between the client device and the server-side in order to support efficient service access and disconnected operation. In Section 4 we present a system for service partitioning and composition, and examine how the partitioning decision and service use strategy affect the service access response time with small devices and wireless communication.

Figure 1 illustrates the agent-based ubiquitous environment. Agents are the basic building block for services and applications. Their execution is facilitated by the agent platform, which may be a basic shell intended for a constrained environment, such as a PDA, or a full platform. Ubiquitous environments require scalability from very limited environments, in terms of computational capacity, energy and memory, to very robust and high performance servers. The agent platform uses and hides the underlying middleware, such as CORBA, the Java virtual machine, Jini and JXTA, in order to realize platform services, such as messaging, storage, and discovery of devices and services.

From the mobility viewpoint Figure 1 contains two different kinds of systems:

- Traditional client-server interaction between terminals and fixed-network that provide various services and may support content-adaptation based on the terminal type and other context information.
- Ad hoc or peer-to-peer operation, where the environment is a mixture of devices that have different temporal properties, for example a printing service is available only when a printer is present in the room.

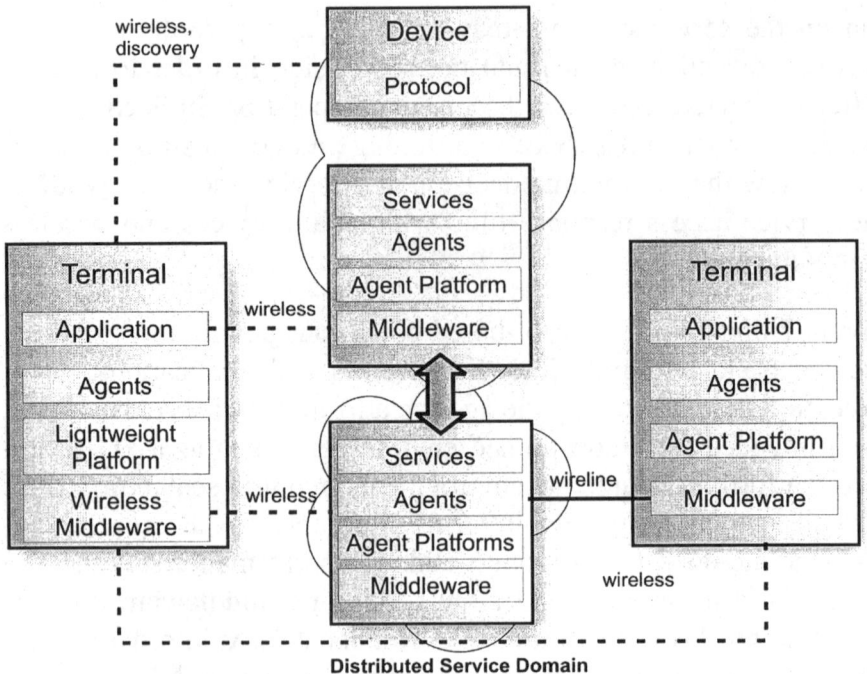

Figure 1. Agents in the ubiquitous environment with two service domains.

In this chapter, we focus on the first item; however, we also discuss the relevant technologies for ad hoc operation and agent systems. The terminal and agents may move between physical areas and logical areas, for example, it may connect to a new access point to access an agent platform and its services. There are several different forms of mobility that may occur in ubiquitous environments: user mobility, terminal mobility, and agent mobility.

In user mobility, the user may become disconnected and change the terminal device. In terminal mobility, the user and the terminal roam to a new location and the access point that is used to access services changes. In order to support mobile devices, mobility support is needed between agent platforms. In addition to mobile terminals, also agents may be mobile and move between execution environments, for example between the platforms or execution environments depicted in Figure 1. Mobile agents are an active research topic, especially in the telecommunications domain. One of the benefits of mobile code is reduced communication cost, because mobile agents may relocate to the location of interest, and perform actions locally. Thus the number of remote interactions may be minimized. On the other hand, mobile agents present various security issues that need to be addressed.

Infrastructure support for mobility is required, because a terminal may move to another location and connect to another access point. Messages should not be lost during disconnections, and the client should be able to continue receiving messages and use services in the new location. Mobile-IP is a network level solution for terminal mobility specified by the IETF (Perkins 2002). Mobile-IP uses a designated home agent, which is a router on mobile node's home link, which keeps track of the location of the mobile host. Packets are tunneled by the home agent to the mobile hosts current address. Wireless CORBA is a middleware level solution for mobile and wireless CORBA objects (OMG 2003). Middleware support for mobility is required in order to provide location transparency for objects, agents, and other components, support efficient and reliable communication in wireless environments, and buffer messages and other data for disconnected operation. In addition, the middleware may support scalability and availability of resources and services.

The ad hoc environment of Figure 1 with terminals and devices connected using wireless links, and terminals connected to servers using wireless or wireline links is challenging, because it changes over time and is highly dynamic. Various computing elements need to be able to discover each other, exchange capabilities and interface descriptions, provide services to other entities and use services available in the network.

3.2 The FIPA Architecture

The FIPA architecture aims to improve agent interoperability by standardizing the external interfaces of agents and agent platforms: message transport protocols, message envelopes, agent communication language (ACL), content languages and interaction protocols (FIPA 2003). Message transport protocol specifies how the agent messages are transferred. The envelope layer allows message transport protocol-independent message handling. The ACL layer defines the language and structure of messages used in agent communication, and the content language describes the content of the messages. On the highest level the interaction protocol defines the interaction pattern, which determines how agents exchange messages. These layers can be structured as a communication stack, which can be seen in Figure 2. In addition to the communication stack, FIPA defines two central components of a FIPA agent platform, the Agent Management System (AMS) and the Directory Facilitator (DF) agents. The AMS provides white pages services and manages the life cycles of the agents residing on the platform. The DF provides agents with yellow page services, i.e. allows agents to advertise their capabilities and to search for agents that match the given constraints. In addition, DFs may be federated to support distributed service discovery between agent platforms.

FIPA does not directly address ubiquitous environments in the specifications, but a few specifications can be seen valuable in this area. The FIPA Nomadic Application Support (NAS) specification deals with agent middleware to support applications in nomadic environment. The FIPA Quality of Service Ontology specification defines an ontology that can be used by agents when communicating about the Quality of Service (QoS). The ontology provides basic vocabulary for QoS.

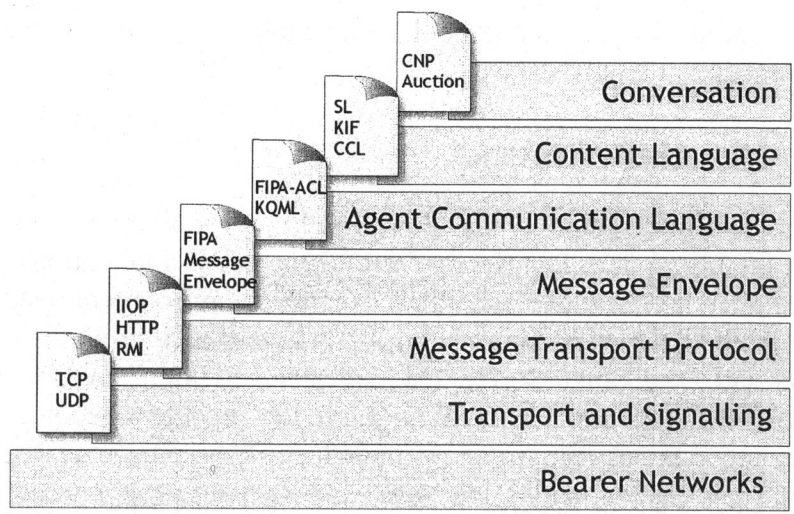

Figure 2. The layered communications model of the FIPA architecture.

The FIPA Ad Hoc Working Group (FIPA Ad Hoc 2003) investigates the use of existing FIPA specifications in ad hoc environments. The environment consists of FIPA compliant agent platforms or fragments of FIPA-compliant platforms running on mobile devices. The working group may also produce new specifications.

The FIPA Agent Message Transport Envelope Representation in the Bit-Efficient Specification and the FIPA ACL Message Representation in the Bit-Efficient Encoding specification deal with message transportation between interoperating agents and also form part of the FIPA Agent Management Specification (FIPA 2003). They contain specifications for the syntactic representation of the message envelope and ACL in a bit-efficient form.

The FIPA Device Ontology specification can be used by agents when communicating about devices. Agents pass profiles that describe device capabilities to each other and validate them against the ontology. Device profiles are needed, for example, in a situation where memory or processing intensive actions take place; agents can negotiate whether some device has enough capabilities to handle a computationally intensive task.

3.3 Agent Platforms

Agent platforms provide high-level APIs and abstract the lower-level interfaces and complexity from agent-based applications. For ubiquitous environments the agent execution environment needs to scale both to high-end systems and also to low-end terminals (Tarkoma and Laukkanen 2002). Ideally, agents written once may be used in different environments without modifying or recompiling code. In practice, resource-constrained environments may require special-purpose code and execution environment programmed in a suitably efficient language.

Java has become a frequently used programming language for agent systems, because Java byte code is portable between various hardware and operating systems, it has many tools and APIs available, and supports features such as code mobility, reflection, and XML. There are many Java-based agent systems available, for example: JADE, FIPA-OS, Grasshopper, and ZEUS[3].

Previous work has been done to investigate the portability and scalability of Java-based agent execution environments. These efforts have focused on the two well-known Java specifications for small devices: Connected Device Configuration (CDC) and Connected Limited Device Configuration (CLDC). JADE-LEAP targets at CLDC, and MicroFIPA-OS targets at the PersonalJava specification, which is an older specification. PersonalJava is equivalent to the CDC with the Personal profile. The former Java environment is available in many mobile phones, and the latter is available for many PDAs and is very similar to the JDK 1.1 API.

---

[3] Publicly available agent platform implementations:
http://www.fipa.org/resources/livesystems.html

3.3.1 JADE-LEAP

JADE-LEAP is a distributed FIPA-compliant agent platform that allows agents to be executed on desktop computers and small devices, such as PDAs and mobile phones that support CLDC and MIDP (Bergenti and Poggi 2002). The platform supports mobile agents, and has been extended to support different transport protocols and message formats. The agent platform may consist of several hosts distributed over the network. Each host runs an agent container that connects to a main container that hosts the AMS and DF. JADE-LEAP supports the use of a split container for low-end devices. A container is split into two parts: the front-end and the back-end, and only the front-end part is executed on the small device. The split container approach requires a permanent connection between the container parts and agent mobility is not supported.

3.3.2 FIPA-OS and MicroFIPA-OS

MicroFIPA-OS is an agent development toolkit and platform based on the FIPA-OS agent toolkit. The system targets medium to high-end PDA devices that have sufficient resources to execute a PersonalJava compatible virtual machine. MicroFIPA-OS supports the use of FIPA-OS components such as AMS and DF, and facilitates the rapid prototyping of agents in PDA environments without modifying agent code with tradeoffs between portability, and performance and resource consumption.

Previous performance measurements with MicroFIPA-OS indicate that high-end PDAs are capable of supporting multiple agents on a single device with a cost in memory and system latency. The wireless connections are the critical part of the system, because they are considerably slower than message passing within devices. Since the environment is constrained by the capabilities of the device and the limitations of the wireless link, the agent execution environment and the applications need to decide what data is transmitted, where and when (Tarkoma and Laukkanen 2002).

The architecture of the system is extensible by plugging in components that either replace or extend the architecture. An example of this kind of contribution is Nomadic Application Support (NAS), which provides support for wireless environments. The main contributions in NAS are the agent naming management, the wireless message transport protocol and bit-efficient encodings for the envelopes and ACL messages (Laukkanen, Helin, and Laamanen 2002).

MicroFIPA-OS does not implement mechanisms for resolving agent naming issues. It is the task of the FIPA-specified AMS to enforce naming policies within a platform. NAS solves this by introducing the concept of a terminal ID. The terminal ID is a unique identifier, which is added to an agent's name. The terminal ID is required in order to prevent addressing conflicts in the case when agents with the same name are running on different devices that are part of the same domain. Moreover, by binding communication end-points to a persistent terminal ID, the possible network address changes can be handled seamlessly.

3.4 Proxy-Based Approaches

The general approach for supporting disconnected and mobile environments is to use a proxy component on the fixed-network side that represents the client, may adapt content according to client capabilities, and supports message buffering for disconnected operation (Rao *et al.* 2001). The client-proxy-server pattern is also commonly used to facilitate content adaptation. A content-adaptation proxy may adapt content, for example HTML-pages, according to device capabilities.

3.5 Agent Communication

Agents rely on asynchronous message delivery, which should be efficient and reliable. Since agents generally communicate by sending and receiving messages, it is reasonable to expect that the agent execution environment should minimize communication latency both within a terminal and between network hosts. Message-based communication is well suited to ubiquitous environments, because it is asynchronous in nature and message queuing makes it easier to support disconnected operation.

In many wireless and firewall scenarios it is not possible to push data to the terminal. For example, server-side push is not possible with MIDP devices that support only HTTP[4]. If the mobile device is behind a NAT (Network Address Translation), it is typically not publicly addressable. This prevents the pushing of messages to the device, and the only way to allow bi-directional messaging is to let the mobile device open a persistent connection or poll the server for new messages. An implementation of the FIPA Nomadic Application Support specification was developed in the EU IST CRUMPET project (Poslad *et al.* 2001) based on the persistent connection approach for both MicroFIPA-OS and JADE (Laukkanen, Helin, and Laamanen 2002).

3.6 Events for Agents

Agent communication is usually directed and the recipients of a message are determined before sending the message. Distributed events are a generic enabler for de-coupled communication, and facilitate the notification of changes in the computing environment to interested components.

---

[4] Datagram and sockets support is optional in MIDP 2.0. They are not available in MIDP 1.0.

Agents should be able to detect changes in their environment, in the local system, and also in the distributed environment at run-time. These changes need to be notified to other agents and interested components in a scalable fashion. Distributed events and content-based information delivery is a good candidate for supporting and enhancing agent systems in ubiquitous environments, because it supports anonymous one-to-many communication and the filtering of information. Distributed events may be provided by the agent platform as a generic service, or an agent may provide this service and route events. If an event-system supports expressive filter semantics, the monitoring and notification functionality of an agent may be delegated to the event service and distributed over the event system, which makes the implementation and execution of agents simpler (Tarkoma 2003).

4 Agent-Based Service Provision and Deployment

The ubiquitous computing environment is heterogeneous and dynamic in nature. Adaptation is needed in order to meet the requirements of the changing environment. Users require continuous and high-quality service irrespective of time and location. In this section we examine the local and external communication in MicroFIPA-OS and present a service composition scenario that is adaptable to the operating context.

4.1 Agent and Service Deployment

Currently, most of the agent platforms are long-lived, stationary and support the execution of a number of agents. The requirements of the ubiquitous environment are different from the traditional desktop environment. The ubiquitous environment created by the current and forthcoming small devices requires that agents can be deployed on different terminals. This creates a number of different configurations typically using either the client-server or peer-to-peer model.

The client-server configurations can be categorized into browsers and partial platforms. Browser-based approaches execute only a browser or a similar user interface on the terminal that interfaces resources on the fixed network side provided by service agents. The clients are connected with the agent domain using various protocols, usually HTTP, and access services provided by the server-side agent system using a proxy or a gateway. If most of the service functionality is implemented at the server-side, the client-side system may become inoperable when disconnected and may suffer from the long round-trip times of wireless links. Partial platform approaches support the execution of one or many agents on the terminal; however, part of the functionality is provided by the fixed network agent platform. This may require persistent connections between the two systems.

Partial platforms allow the execution of part of the service logic on the client-side, and the client may use the server-side system more freely without relying on the communication patterns programmed in a proxy. This kind of operation bridges the agents on the small device with the fixed network agent world. Agent code may be potentially reused and deployed on both clients and servers, and agent communication language is used throughout the domain. The possibility of executing client-side logic improves response times when network access is not necessary or feasible, and may support partial functionality when the client system is disconnected.

In peer-to-peer operation, the whole platform is potentially mobile and interoperates with other platforms using standardized protocols. This approach requires that the terminal host a full-featured platform with the necessary support services. Full platforms host the necessary white pages and yellow pages services for managing and locating agents.

The agent communication language (ACL) is required for supporting interoperable communication between heterogeneous agents. If the agents on a terminal do not support ACL they need to use proprietary intermediaries in order to facilitate the communication. In addition to ACL support, which the basic requirement for any agent platform, agent development may be improved by incorporating the notion of tasks or behaviors, concurrent operation, and conversations, which ease the management of high-level interaction protocols between an agent and its peers.

MicroFIPA-OS and JADE-LEAP support these two models for creating and running agents. The general usage scenario for MicroFIPA-OS is the partial platform scenario, where agents use the AMS and DF of a long-lived and stabile fixed network platform. However, the system also supports the possibility of running an independent system on the small device, including AMS and DF, and interfacing other FIPA systems with the HTTP transport. Moreover, MicroFIPA-OS supports two agent-programming modes: rapid prototyping and the minimal mode. In the first mode application programmers that use the FIPA-OS API may reuse their existing code developed for the FIPA-OS platform. In the second mode, developers use ACL, envelopes, and content language parsers provided by the toolkit, but write their own message and task handling code.

4.2 Service Partitioning Based on the Environment

Agent-based services are usually implemented using a number of agents that co-operate in the realization of the service functionality. We have investigated service composition or partitioning that for each component of the service the system, a middleware agent or a user agent, evaluates whether to use it locally or remotely. It is assumed that a service consists of a number of stateless agents, and some may be local and some external. External agents may be required, because many service elements cannot be provided locally because of their nature. On the other hand, it is not reasonable to use components externally if they are locally available and the execution environment is fast enough.

In this work, the goal of adaptive service composition is to minimize the service latency for end-users. Other metrics that need to be taken into account in the composition decision are local system resources, possible server load, and the QoS and characteristics of the communication link. The service composition decision can be made at start-time or at run-time. Start-time decisions may become unoptimal when the environment changes. Figure 3 illustrates the continuous and dynamic service composition and adaptation process. When the execution environment changes, the system evaluates the current service access strategy and possibly re-configures the system using a control mechanism.

In order to make a composition decision, we need to have a model of the application behaviour: the type, volume and frequency of interactions. The modeling of application behaviour is challenging and may require feedback and input from the authors of the application and run-time performance monitoring and modeling. The execution environment may support service composition transparently using information stored in application profiles. The application profiles need to include information about the availability, location and requirements of various components. In addition, a middleware profile is needed that contains information on the expected behaviour of the middleware platform, the agent platform in this case. Internal latencies may be substantial, which motivates combining the middleware profile and the application profile for the service usage decision. We have used a decision matrix and the Weighted

Sum Model (WSM) in order to find the optimal usage strategy that minimizes service latency (Tarkoma and Laukkanen 2003).

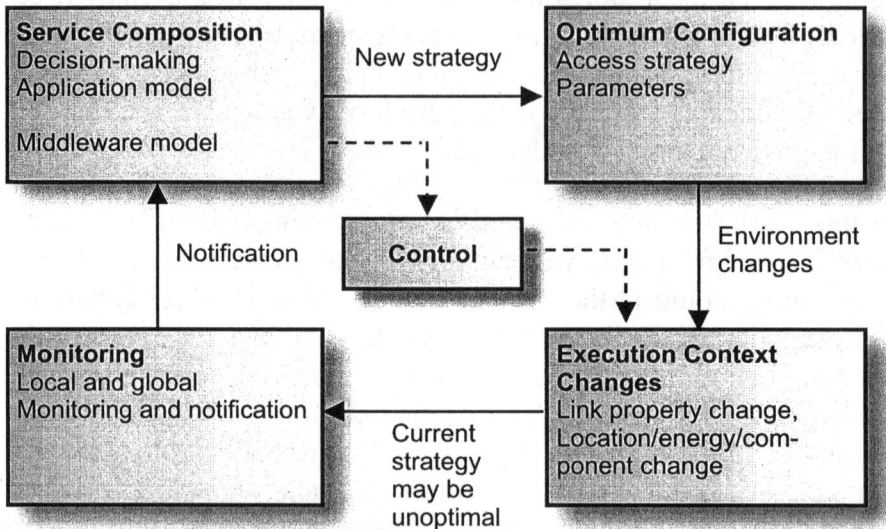

Figure 3. Service composition in a dynamic environment.

4.3 Example Scenario: Recommendation Service

The experimental scenario used to examine service composition in wireless environments and small devices consists of a location-based recommendation service. The service takes the location of the client, accesses an external database that contains location specific information, such as nearby restaurants and their menus, and creates a map image of the neighborhood with the nearby location highlighted. The results presented in this section are based on a prototype Java implementation of the components of the map service, and the MicroFIPA-OS platform. The service configuration is presented in Figure 4 and the system consists of the following components (Tarkoma and Laukkanen 2003):

- Client agent that controls the service usage and has knowledge of its spatial location,
- Proxy DB-agent that caches map information and retrieves information from spatial-information database located on the fixed network,
- Proximity agent that calculates distances between objects and returns the identifiers of nearby objects,
- Database-agent that manages spatial information (coordinates and descriptions of objects), and
- Map-agent that takes a vector of locations and their descriptions as input and an image (GIF-format) of the neighbourhood.

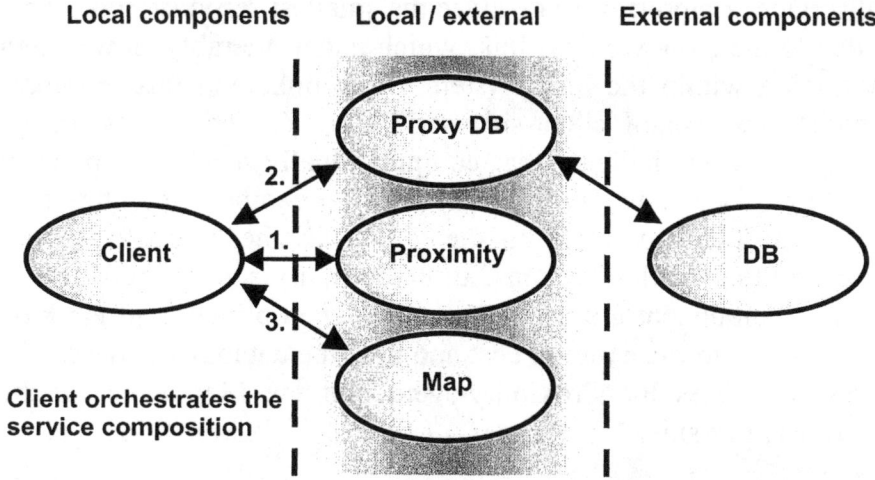

Figure 4. Overview of the service configuration space.

In Figure 4, the client agent contacts the other agents in order to provide the recommendation service for the user. The client-agent composes the recommendation service by contacting the agents that implement different parts of the service. First, the client sends the coordinates to the Proximity agent that returns a vector containing identifiers of the nearby locations (1). After this, the client sends this vector of identifiers of the locations to the Proxy DB/Database-agent (2), which returns the names and descriptions of the objects, and other relevant information. In the final phase, the client sends the data to the Map-agent (3), which creates an image of the nearby locations and their names with the coordinates highlighted on the map.

We created and examined decision matrices for 50x50 and 200x200 pixels map generation for WLAN and GPRS on two mobile systems: a laptop and a Compaq iPAQ H3630 PDA with Linux and JDK 1.1 on both systems. We have assumed that the agents are stateless and that they are locally and externally available. Figure 5 presents a summary of the results for the iPAQ. For the laptop, local service component usage gives the smallest response time. This is due to the slow wireless link, which is considerably slower than messaging within the local system. Slow links and fast terminals motivate the use of client-side service logic. The results on the iPAQ, however, indicate that computationally heavy components should be used remotely, such as the Map component in this scenario. All-local operation on the iPAQ has the highest response time, and the smallest response time (optimal) is achieved when all but one component are used externally. The optimal configuration is very close to the external cost and uses the database proxy on the client, and uses the Proximity-agent and the Map-agent on the fixed-network side.

These results indicate that neither the all-external nor the all-local configurations are optimal for a small device given that the service is composed from the client-side and the communication link has high latency. Moreover, the results motivate using local components and service logic with slow links and reasonably powerful client systems, such as GPRS and the laptop used in this experiment. This approach may also be used to implement partial fault tolerance, in which part of the service functionality may still be provided even if network connectivity is lost.

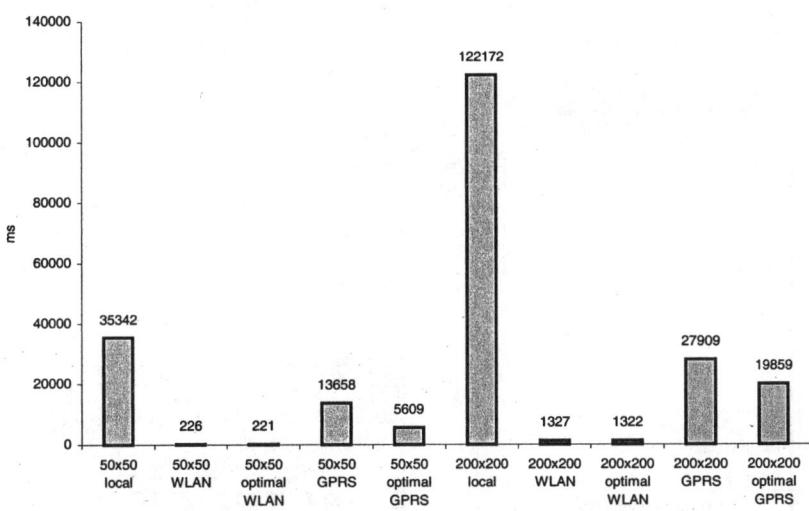

Figure 5. Local, remote, and optimal latencies for service access on the iPAQ.

4.4 CRUMPET

The CRUMPET project investigated the development and deployment of agent-based services for the tourism sector. In particular, the project investigated the use of agent technology as a suitable approach for developing scalable, seamlessly accessible nomadic services (Poslad *et al.* 2001). The CRUMPET architecture uses terminal agents running on MicroFIPA-OS to access the tourism-related services, and monitor and control the Quality of Service (QoS) of the wireless connection. The architecture uses server-side agents to manage spatial information and adapt content for user terminals. Figure 6 presents the recommendation service using the CRUMPET system. The client-side user interface is a small-footprint web browser. An agent formats and manages the content and queries. In addition, the system supports proactive tips and notifications, and location tracking on a map. The service is constructed using similar generic service elements that we presented in the previous section and used in the experimental adaptive service composition scenario.

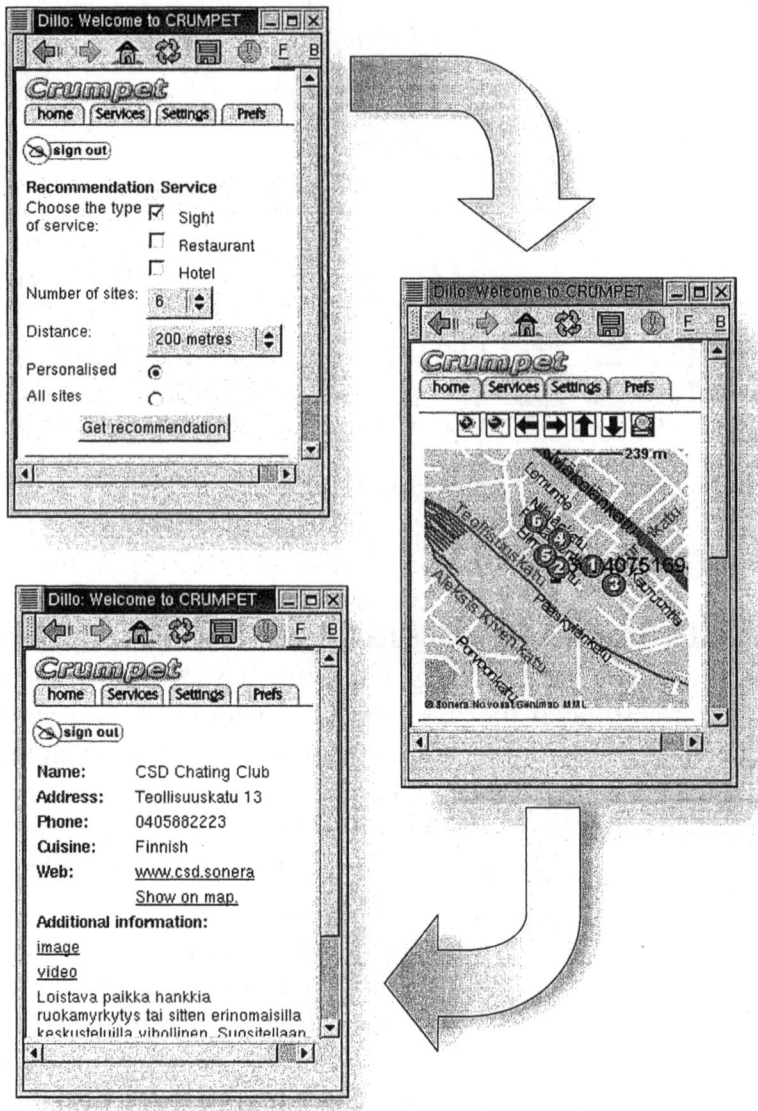

Figure 6. The CRUMPET Recommendation Service.

5 Conclusions

The agent paradigm is a good candidate for ubiquitous computing, because it is based on asynchronous messaging, and supports the integration and co-operation of a number of autonomous components. In recent years, agent standards and interoperable agent platform implementations have emerged. Currently several platforms support wireline and wireless communication, and the support for mobile and ad hoc computing is an active research topic. Current standard wireless middleware, peer-to-peer systems such as JXTA, and network-level solutions for supporting mobility are the building blocks for ubiquitous agent systems.

In this chapter we have presented an overview of the ubiquitous environment, examined the FIPA architecture and two agent platforms that support the execution of agents on small devices. The environment is challenging and the tradeoffs are between software complexity, development time and cost, and device limitations (energy, memory, computing power). This calls for adaptive and context-aware solutions for meeting the requirements of ubiquitous computing. We discussed agent deployment on small devices, and presented adaptive service partitioning and experimental results from the recommendation service scenario.

Currently, it is possible to develop agents for server-side deployment, and also execute them on terminals; however, there is a price on application performance and latency. Wireless service access and intermittent connections motivate the use of client-side service logic, especially with devices that have reasonable computing power, such as laptops. The benefits of adaptive agent-based service composition are reduced latency, portability of code, and continuous service use even in the case that network connectivity is not available. Some applications may require constant network access, but many services may be provided by using local resources and caching of server-side resources.

References

Bergenti F. and Poggi A. (2002), "LEAP: A FIPA Platform for Handheld and Mobile Devices," *Intelligent Agents VIII*, Springer, pp. 436-446.

Buckle, P., Moore, T., Robertshaw, S., Treadway, A., Tarkoma, S., and Poslad, S. (2002), "Scalability in Multi-agent Systems: The FIPA-OS Perspective," *Foundations and Applications of Multi-Agent Systems,* pp. 110-130, Lecture Notes in Artificial Intelligence (LNAI) 2403, Springer.

Chen, H.,Joshi, A., and Finin, T. (2001), "Dynamic Service Discovery for Mobile Computing: Intelligent Agents Meet Jini in the Aether," *Cluster Computing,* Volume 4, Number 4, pp. 343-354.

FIPA Ad Hoc (2003), *FIPA Ad Hoc Working Group homepage.* Available at: http://www.fipa.org/activities/ad_hoc.html

Foundation for Intelligent Physical Agents (FIPA). (2003), *The FIPA Architecture*, Geneva, Switzerland. Available at http://www.fipa.org

ISTAG (2001), "Scenarios for Ambient Intelligence in 2010," *European Commission, User-friendly Information Society, the IST Advisory Group (ISTAG).* At: http://www.cordis.lu/ist/istag-reports.htm

Jennings, N., Sycara, K., and Wooldridge, M. (1998), "A Roadmap of Agent Research and Development," *Autonomous Agents and Multi-Agent Systems, pp. 275-306,* Kluwer Academic Publishers.

Jini, Sun Microsystems (2003), *Jini technology homepage.* Available at: http://wwws.sun.com/software/jini/

JXTA (2003), *Project JXTA Home Page.* Available at: http.//www.jxta.org

Laukkanen M., Helin H., and Laamanen H. (2002), "Supporting Nomadic Agent-Based Applications in the FIPA Agent Architecture," In C. Castelfranchi and W. L. Johnson, editors, *Proceedings of the First International Joint Conference on Autonomous*

Agents & Multi-Agent Systems (AAMAS 2002), Bologna, Italy, pp. 1348-1355. ACM Press, New York, NY, USA.

McIlraith, S., Son T., and Zeng, H. (2001), "Semantic Web Services. IEEE Intelligent Systems," *Special Issue on the Semantic Web*, Volume 16, No. 2, pp. 46-53, March/April.

Object Management Group (OMG) (2003), "Wireless Access & Terminal Mobility in CORBA," version 1.0.

Perkins, C. (ed.). (2002), "IP Mobility Support for IPv4," *IETF RFC 3344*.

Poslad S., Laamanen H., Malaka R., Nick A., Buckle P., and Zipf, A. (2001), "CRUMPET: Creation of User-friendly Mobile Services Personalised for Tourism," *Proceedings of the Second International Conference on 3G Mobile Communication Technologies (3G-2001)*, London, UK.

Rao H., Chen Y., Chang D., and Chen M. (2001), "iMobile: A Proxy-based Platform for Mobile Services," *The First ACM Workshop on Wireless Mobile Internet (WMI)*, Rome, Italy.

Satyanarayanan, M. (2001), "Pervasive computing: vision and challenges," *IEEE Personal Communications*, Volume 8, Issue 4, pp. 10-17.

Sun Microsystems (2002), "Java Message Service (JMS) specification," version 1.1.

Sun Microsystems (2003), *J2ME homepage*. Available at: http://java.sun.com/j2me/

Tarkoma S. and Laukkanen, M. (2003), "Adaptive Agent-based Service Composition for Wireless Terminals," *Cooperative Information Agents (CIA)*, Helsinki, Finland. Lecture Notes in Artificial Intelligence, Vol. 2782, pp. 16-29.

Tarkoma, S. (2003), "Distributed Event Dissemination for Ubiquitous Agents," *AMAS Agent Track. Proceedings of the 10th ISPE International Conference on Concurrent Engineering: Research and Applications*, pp. 105-110.

Tarkoma, S. and Laukkanen, M. (2002), "Facilitating Agent Messaging on PDAs," *Fourth International Workshop on Mobile*

Agents for Telecommunication Applications (MATA-2002), Barcelona, Spain. Lecture Notes in Computer Science (LNCS) 2521.

WAP Forum (2002), "Wireless Application Environment," *WAP Specification*. Available at: www.wapforum.org

Weiser, M. (1993), "Some computer science issues in ubiquitous computing," *Communications of the ACM*, Volume 36, Issue 7, pp. 75-84.

World Wide Web Consortium (2003), "Composite Capability/Preference Profiles (CC/PP). Structure and Vocabularies 1.0," *W3C Proposed Recommendation*.

Chapter 3

Agents-Based Knowledge Logistics

Alexander Smirnov, Mikhail Pashkin,
Nikolai Chilov, and Tatiana Levashova

A research carried out in the framework of the knowledge logistics lies in the base of the chapter. As a result of the research an approach addressing the knowledge logistics problem was developed. The approach considers the problem as a problem of configuring a knowledge source network that is assumed to consist of distributed heterogeneous knowledge sources. An implementation of the approach was put into practice through its realization in the system "KSNet". Distribution and heterogeneity of the knowledge sources determine a distributed and scalable character to the problem of the network configuring. Such nature of the problem causes for the system to have a multi-agent architecture. This chapter presents a prototype of the developed agent community implementation in the system "KSNet" and a constraint-based protocol designed for the agents' negotiation. An application of the developed agent community to coalition-based operations support and the protocol are illustrated via case studies of a mobile hospital configuration as a task of health service logistics and automotive supply network configuration.

1 Introduction

Decision making in a wide range of e-applications require using the open information environment. This leads to the constantly increasing importance of knowledge for decision making (Figure 1) and expansion of e-applications dealing with knowledge storage & sharing, and based on intensive use of Internet-technologies and such standards as XML, RDF (S), DAML+OIL, etc. As a result it is possible to speak about an evolution of the information environment from "regular" (with fixed interactions between knowledge sources) to "intelligent" (with flexible configuration of knowledge source network in which humans are involved). All these have led to an appearance of the new scientific direction in the knowledge management called *Knowledge Logistics* (KL). KL stands for acquisition of the right knowledge from distributed sources, its integration and transfer to the right person within the right context, at the right time, for the right purpose.

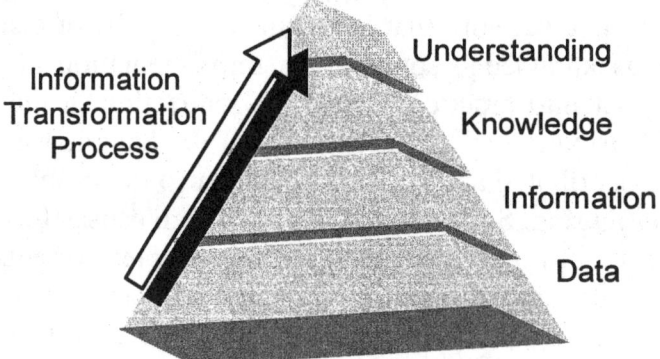

Figure 1. Conceptual framework of information support.

Described here approach is based on representation of KL problem as a problem of network configuration whose nodes represent elements of a global information environment. These elements include end-users/customers, loosely coupled knowledge sources (experts, knowledge bases, repositories, documents, etc.), and information/knowledge management tools. Such the network was called "Knowledge source network" or "KSNet" and the approach was called "KSNet-approach". Based on this approach, the architecture of KL system called "KSNet" was developed (Smirnov et al., 2003a, Smirnov et al., 2003b).

The approach is based on the propositions that knowledge as a resource is characterized by cost, location, access time and lifetime, and a knowledge worker is an owner of knowledge. It utilizes principles of ambient intelligence implying synergistic use of knowledge from different sources in order to complement insufficient knowledge and to obtain new knowledge. The approach is based on such advanced technologies as ontology management, intelligent agents, soft computing, user profiling, knowledge mapping, virtual reality, groupware, constraint satisfaction, etc. Multi agent architecture is widely used in the knowledge management systems since knowledge is located in the distributed sources and presented in different formats. Utilizing intelligent agents in the system "KSNet" is motivated by a distributed and scalable nature of the problem. In the chapter the developed constraint-based contract net protocol for agent negotiation and a prototype of agent community implemented in the system "KSNet" is presented.

The KL system scenario is based on individual user requirements, available knowledge sources, and context analysis in the open information environment. An applicability of the KSNet-approach is presented via a case study based on the Binni scenario in the area of e-health (portable hospital configuration for a given location taking into account a current situation in the location's region) and an e-business case study for automotive supply network.

A research topic dealing with knowledge is Knowledge Management (KM). It is defined as a complex set of relations between people, processes and technologies bound together with the cultural norms, like mentoring and knowledge sharing, which constitute the organization's social capital (Rasmus, 2000). KM consists of the following tasks: knowledge discovery (knowledge entry, capture of tacit knowledge, knowledge fusion (KF), etc.), knowledge representation (knowledge base development, knowledge sharing and reuse, knowledge exchange, etc.), knowledge mapping (identifying knowledge sources (KSs), indexing knowledge, making knowledge accessible) (Jarrar et al., 2002, Dzbor et al., 2000, Park et al., 1997, Vail, 1999). There are a number of different approaches proposed and tools developed for these tasks solving based on the algorithms of data searching and retrieving in large databases, technologies of data storing and representation, etc. Among them the following ones can be pointed out: SearchServer/KnowledgeServer (Hummingbird, 2002), Text-To-Onto (Maedche and Staab, 2000), etc. for knowledge search and retrieval from different types of documents; Disciple-RKF (Tecuci et al., 2002), EXPECT (Blythe et al., 2001), Trellis (Trellis, 2002), COGITO (Cogito, 2002), TKAI (Cheah and Abidi, 2001), etc. for knowledge acquisition from experts and tacit knowledge revealing; OntoEdit (Ontoprise, 2003), Protégé (Protégé-2000, 2000), OntoLingua (Ontolingua, 2001), etc. for ontology engineering; HPKB (Pease et al., 2000), AKT (AKT, 2002), etc. for knowledge base organization and development; KRAFT (Visser et al., 1999), InfoSleuth (Nodine and Unruh, 1997), etc. for knowledge and information integration. The above approaches are targeted at pertinent, clear, correct information and knowledge processing and timely delivering to locations of need for global situational awareness and ability to predict development of going on processes at the level of understanding. Since KL addresses these issues as a whole it can significantly facilitate the processes of KM (Smirnov et al., 2003c).

The possible application domains of KL belong to the following areas: (i) large-scale dynamic systems with distributed operations in uncertain and rapidly changing environment (Adams et al., 2000, Howells et al., 1999), (ii) focused and Web-enhanced logistics operations (DARPA, 2002), (iii) markets via partnerships with different organizations, (Cunningham, 2001, Kim et al., 1997), etc.

For all of the above areas it is possible to describe management systems as an organizational combination of people, technologies, procedures and information/knowledge.

2   KSNet-Approach: Major Ideas

Ontology management is one of the major technologies the KSNet-approach is based on. In the approach FIPA (FIPA, 2002) ontology definition is used which is based on the following postulates: (1) an ontology is an *explicit specification* of a structure of a certain *domain;* (2) an ontology includes *a vocabulary* for referring to a subject area, and *a set of* logical statements expressing the *constraints* existing *in the domain* and restricting the interpretation of the vocabulary; and (3) an ontology provides a vocabulary for representing and communicating *knowledge* about some topic, as well as *a set of relationships and properties* that hold for the *entities* denoted by that vocabulary.

The following ontology types for the system were defined: (i) top-level ontology providing notation for ontology representation in the system; (ii) domain ontology representing static knowledge about a particular domain in terms of the domain; (iii) tasks & methods ontology describing problem-solving knowledge in terms of a domain or high-level terms that are general for several domains; (iv) application ontology describing an application domain in terms of domain and tasks & methods ontologies; (v) preliminary KS ontology describing KS in KS's terms and the top-level ontology notation; (vi) KS ontology containing correspondence between terms of KS and application ontology; (vii) preliminary request ontology describing user request in user's terms (which are used by the user for requests input) and the top-level ontology notation; and (viii) request ontology containing correspondence between terms of preliminary request ontology and application ontology. The ontologies are stored in a common ontology library that allows sharing and reusing them.

The conceptual scheme of the user-oriented ontology-driven methodology developed for the KSNet-approach is presented in (Figure 2). It takes into account such modern requirements to knowledge management systems as (i) flexibility, (ii) learning from user, (iii) integrity, (iv) velocity, (v) open connectivity, (vi) reasoning and (vii) customizability.

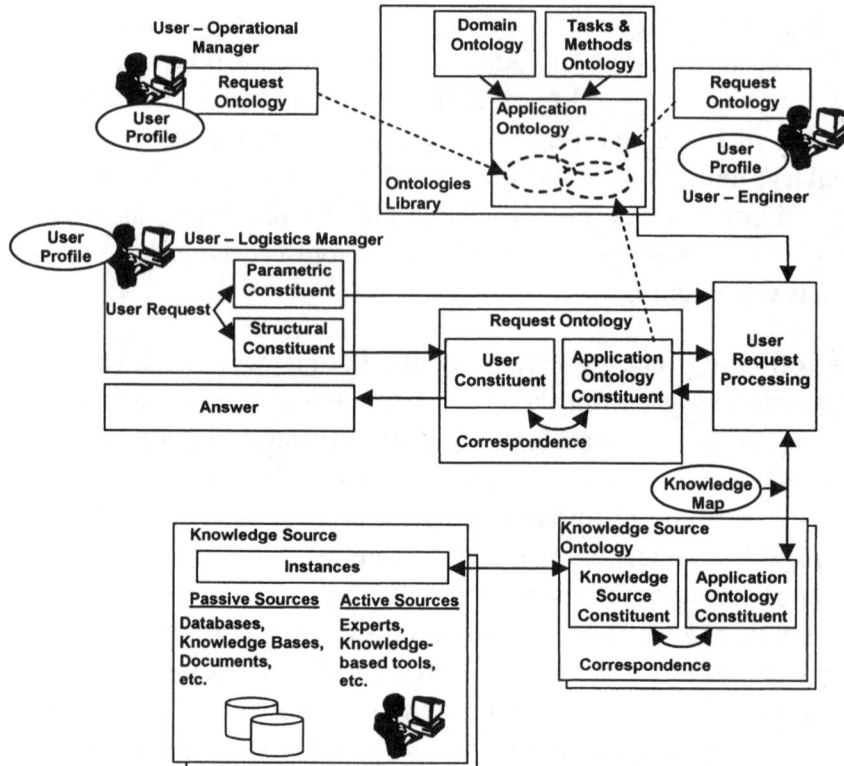

Figure 2. Conceptual scheme of the developed user-oriented ontology-driven KL methodology.

The system works in terms of a common vocabulary. Application ontology is based on domain, tasks and methods ontologies stored in the ontology library. Each user/user group works in terms of an associated expandable request ontology and thereby with a part of application ontology pertinent to the user/ user group. User profiles are used during interactions to provide for an efficient personalized service. To find knowledge a user inputs a request into the system via interface forms. Every user request consists of two parts: (i) structural constituent (containing the request semantic - terms and relations between them), and (ii) parametric constituent (containing additional user-defined constraints). For the request processing, an auxiliary KS network configuration is built defining when and what KSs are to be used for the request processing in the most efficient way. For this purpose a knowledge map including information about locations of KSs is used. Translation between the system's and KS's notations & terms is performed using KS ontologies. As the system's knowledge notation the formalism of object-oriented constraint networks was chosen (Smirnov et al., 2003d).

3 Features of Agent Community in the System "KSNet"

Because of the distributed nature of the problem the KSNet-approach assumes an agent-based architecture. A multi-agent system architecture, based on FIPA Reference Model (FIPA, 2002) as an infrastructure for definition of agent properties and functions, was chosen as a technological basis for the system "KSNet" since it provides standards for heterogeneous interacting agents and agent-based systems, and specifies ontologies and negotiation protocols to support interoperability in specific application areas. FIPA-based technological kernel agents used in the system are: wrapper (interaction with KSs), facilitator ("yellow pages" directory service for the agents), mediator (task execution control), and user agent (interaction with users). The following problem-oriented agents specific for KL, and scenarios for their collaboration were developed: translation agent (terms translation between different vocabularies), knowledge fusion (KF) agent (knowledge fusion/integration operation performance), configuration agent (efficient configuring of KS network), ontology management agent (ontology operations performance), expert assistant agent (interaction with experts), and monitoring agent (KSs verifications). The community of agents is represented in Figure 3 according to the above described conceptual scheme of KSNet-approach. Table 1 describes some special features of the agents, used in the system "KSNet".

The agent structure containing the following modules was developed for the system "KSNet": (i) identifying, (ii) functional, and (iii) knowledge repository. Identifying module contains such parameters as unique identifier, creation date and time, type, etc. Functional module contains a set of procedures to be executed by an agent this module belongs to. Knowledge repository contains special information, such as history of the agent's contacts, temporary results, new knowledge, etc.

The main system tasks and techniques assigned to the problem-oriented agents are presented in Figure 4. These techniques were tested via developed research software prototypes of the corresponding problem oriented agents.

Agents of the system "KSNet" have the following characteristics: (i) benevolence – willingness to help each other, (ii) veracity – agents do not process knowingly false information, and (iii) rationality – willingness to achieve the goal and not to avoid it.

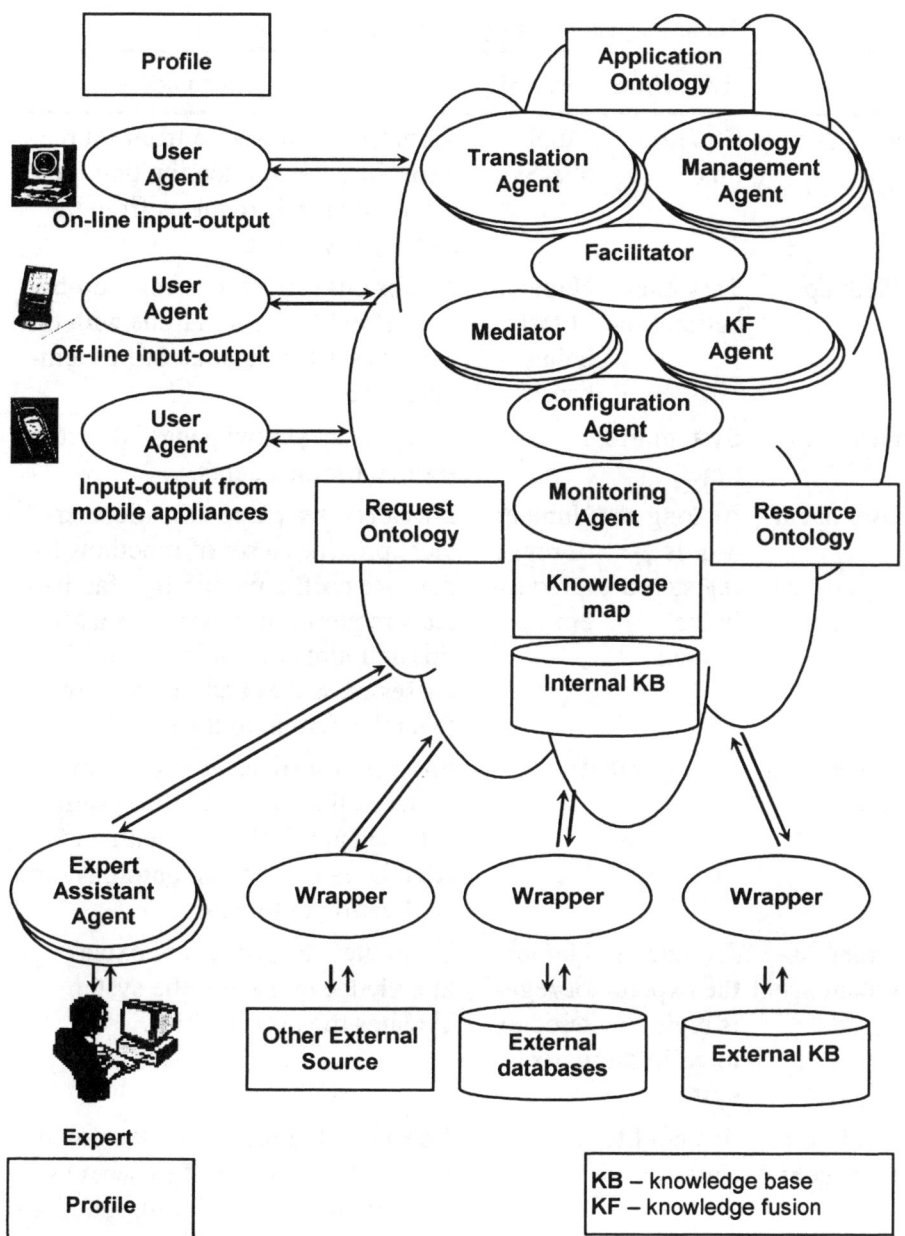

Figure 3. Developed agent-based architecture for the KSNet-approach.

Table 1. Features of agents of the system "KSNet".

Agent	Life time	Quantity	General tasks
Wrapper	KSs life time	Number of KSs	Translates knowledge from source terms into the application ontology terms and sends requests from the system to sources.
Mediator	Task execution time	Number of tasks being processed	Tracks out task processing step-by-step. Provides negotiations with the expert assistant agents. Stores temporary results.
Facilitator	System life time	1	Provides a "yellow pages" directory service for the agents.
User agent	As long as user is registered in the system	Number of registered users	Provides user personalization service: provides a set of functions for the user profile processing, facilitates request input, provides a set of tips and hints for the user, and passes messages and information from the system to the user.
Translation agent	System life time	1	Provides for translation of terms between the users and the system, between application domain and KSs. Uses the request ontology, and application ontology.
Expert assistant agent	As long as the expert is registered in the system	Number of registered experts	Facilitates the process of expert knowledge entry into the system. Updates the user profiles.
Configuration agent	System life time	1	Confiures the KS network using the knowledge map and the user profiles. Performs scheduling functions. Negotiates with the KF agents and the wrappers.

Agents-Based Knowledge Logistics

Agent	Life time	Quantity	General tasks
KF agent	System life time	Varying	Obtains knowledge from the mediator and processes it. Generates new knowledge. Validates it. Interacts with the monitoring agent.
Monitoring agent	System life time	1	Provides a set of functions for diagnostics of the system knowledge base and external KSs.
Ontology management agent	System life time	1	Provides a set of functions for ontology engineering and management. Checks correspondence between KS and request ontologies and application ontology.

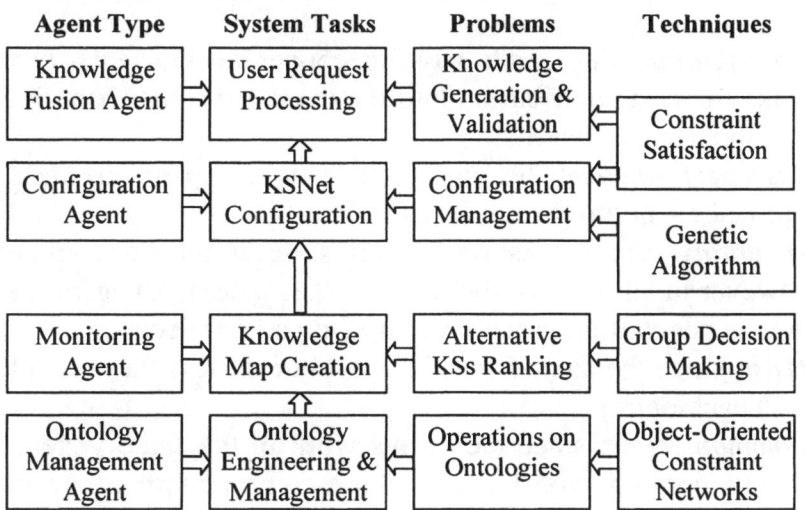

Figure 4. Main problem-oriented agents' tasks and techniques.

4 Communication, Interaction and Negotiation in the KL System

Most multi-agent systems require using agent negotiation models to operate. The negotiation models are based on the negotiation protocols defining basic rules so that when agents follow them, the system behaves as it is supposed to. In the system "KSNet" the negotiation is required in two scenarios: (i) negotiation between *configuration agent* and *wrappers* during KS network configuration and (ii) negotiation for expert knowledge conformation during distributed direct knowledge entry (Smirnov et al., 2002).

The main specifics of the KSNet-approach related to making a choice of the agent negotiation model were formulated as follows:

1. *Contribution*: the agents have to cooperate with each other to make the best contribution into the overall system's benefit – not into the agents' benefits;
2. *Task performance*: the main goal is to complete the task performance – not to get profit out of it;
3. *Mediating*: the agents operate in a decentralized community, however in all the negotiation processes there is an agent managing the negotiation process and making a final decision;
4. *Trust*: since the agents work in the same system they completely trust each other;
5. *Common terms*: since the agents work in the same system they share a common vocabulary and use common terms for communication.

Among the main negotiation protocols the following ones can be selected: voting, bargaining, auctions, general equilibrium market mechanisms, coalition games, and contract net protocol (CNP) (Sandholm, 1999). Based on the analysis of these protocols and the above requirements to them, CNP (Smith, 1980) was chosen as a basis for the negotiation model in the KSNet-approach. As it can be seen from Table 2 this protocol meets most of the requirements.

Table 2. Comparison of negotiation protocols for KSNet approach.

Criteria \ Protocols	Voting	Bargaining	Auctions	General Equilibrium Market Mechanisms	Coalition Games	Contract Nets
Contribution	☑	☑	☐	☐	☐/☑	☑
Task performance	☐/☑	☑	☐	☐	☐	☐/☑
Mediating	☐	☐	☑	☐	☐	☑
Trust	☑	☑	☑	☑	☐	☑
Common terms	☑	☑	☑	☑	☑	☑

Legend: ☑ - supported; ☐ - not supported; ☐/☑ - weakly supported.

4.1 Conventional CNP

CNP is one of basic coordination strategies between agents in multi-agent systems. It was originally introduced by Randall Davis and Reid G. Smith (Davis and Smith, 1983).The main features of this protocol are (i) *managers* (*initiators* in FIPA) divide tasks, (ii) *contractors* (*participants* in FIPA) bid, (iii) *manager* makes contract for the lowest bid, (iv) there is no negotiation of bids. The UML sequence diagram of FIPA-based CNP is presented in Figure 5. Since CNP is a basic protocol any particular multi-agent system requires some modifications for CNP to be implemented (Payne et al., 2002, Sandholm and Lesser, 1997). In the following two sections the modifications made for original CNP required for a system "KSNet" (based on the KSNet-approach) are described. Improvements to the conventional CNP concern a formalism for agents' knowledge representation and for communications between agents, and a scenario of the agents' interaction.

4.2 Constraint-Based Negotiation

Since the system "KSNet" uses object-oriented constraint networks for knowledge representation this formalism was also chosen for representation of agents' knowledge and for communications between agents. Two types of constraints were defined for agents: "local" and "global". Each contractor agent deals with the local constraints describing its limitations, e.g. time of the task execution cannot be less than some value ($time \geq time_{lower\,bound}$). Manager also deals with the local constraints and the main purpose of these constraints is a definition of requirements for the task execution, e.g. the task execution cannot last longer than some time ($time \leq time_{upper\,bound}$). Manager can also have objectives such as minimization of the task execution costs ($costs \rightarrow min$). Global constraints describe constraints defined by the agents' community, e.g. resource limitations. The constraints can concern such parameters as time, costs, reliability, agents/resources availability/unavailability, network traffic, etc. For constraints processing the technologies of constraint satisfaction and propagation are used.

Thus, a generic call for proposals from a manager to contractors has the following form:

```
Objective (optional)
Constraints (optional)
Content (required)
```

Objective is optional and used for meeting manager's constraints such as minimization of costs for a task processing. Constraints are also optional and used for a similar purpose, namely to meet requirements for a task execution. Content is a message itself including functions to be performed by contractors and some other parameters.

Contractors' proposals besides a content part contain constraints corresponding to manager's objective and constraints. This will be illustrated by the example given at the end of this section. If contractors cannot meet the requirements of the manager they still can propose the closest possible parameters and it is up to the manager to decide whether to accept the proposal or not.

The negotiation process terminates when an acceptable solution is found or no improvement is achieved at the current iteration.

4.3 Modifications of Interaction

Modifications of the CNP-based negotiation model include introduction of additional messages and features required for KL related tasks, most important of which are presented in (Table 3) and illustrated in Figure 7 to Figure 9 via UML diagrams.

Table 4 represents agents' roles in the two major scenarios of the system "KSNet". The first scenario of a KS network configuration is executed as a part of user request processing. The second scenario "Direct knowledge entry" consists of two subscenarios: execution and conformation. In these subscenarios expert assistant agents change their roles from "service provider" to "contractor".

4.4 Example of Utilizing Constraint-Based CNP

To compare the results of the conventional CNP and constraint-based CNP the following example is considered. The experiments were made using the implemented research prototype of the system "KSNet" described in the section "Implementation of the Agent Community".

Configuration agent (CA) is supposed to obtain knowledge from three wrappers (W1, W2, and W3) with the time of knowledge acquisition being minimal. It is also preferable for the configuration agent to choose a cheaper deal among the deals with the same time.

Table 3. Changes in features of the conventional CNP required for the system "KSNet".

Feature \ Protocols	Conventional CNP	Modified CNP for the system "KSNet"
Iterative negotiation	-	the negotiation process can be repeated several times until acceptable solution is achieved
Conformation	-	concurrent conformation between manager and contractors
Available messages	fixed set of 8 messages (Figure 5)	flexible set: new messages specific for KF process and corresponding to FIPA *Request* and *Confirm* communicative acts, and message *Clone* not corresponding to any FIPA communicative act are included (this message requests facilitator to create new instances of ("clone") mediators, user agents or expert assistant agents)
Participants roles	manager and contractors	manager and two types of contractors: (i) "classic" contractors negotiating proposals, and (ii) auxiliary service providers not negotiating but performing operations required for KL (e.g., ontology modification, user interfacing, etc.)
Role changing	-	agents can change their roles during a scenario

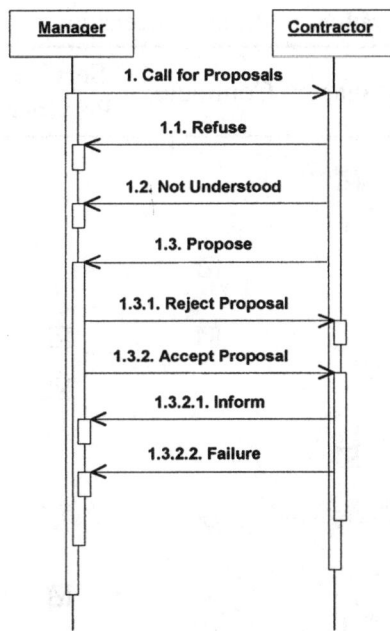

Figure 5. UML sequence diagram of FIPA-based CNP.

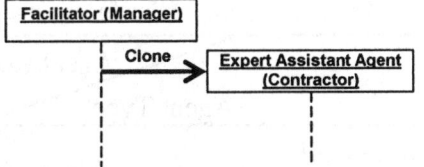

Figure 6. Example of new available messages.

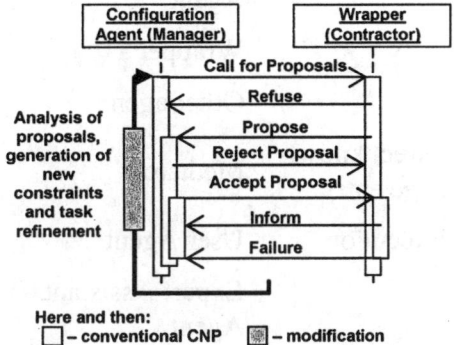

Figure 7. Example of the constraint-based iterative negotiation.

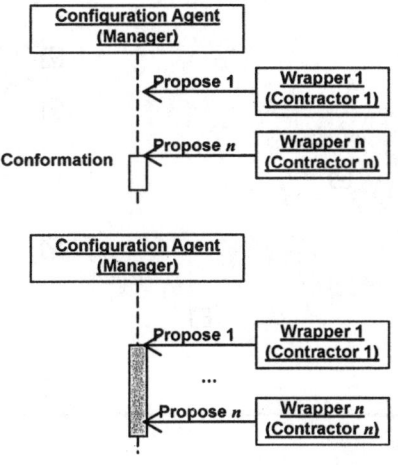

Figure 8. Example of conformation.

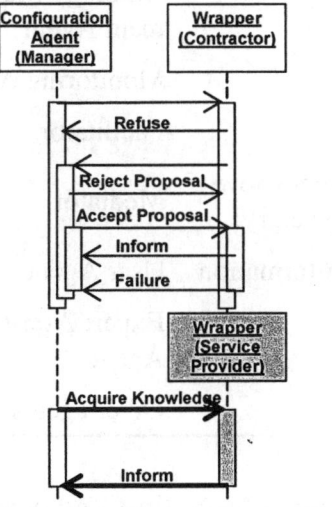

Figure 9. Example of new agents' roles and role changing.

Table 4. Agents' roles for the main operations of the major scenarios.

Operation	Agent Type	Manager	Contractor	Service Provider
KS network configuration	Configuration Agent	☑		
	Expert Assistant Agent		☑	
	Wrapper		☑	☑
	Other agents			☑
Direct knowledge entry: Execution	Mediator	☑		
	User Agent		☑	
	Expert Assistant Agent			☑
	Translation Agent			☑
	Ontology Management Agent			☑
	Monitoring Agent			☑
	Facilitator			☑
Direct knowledge entry: Conformation	Mediator	☑		
	User Agent			☑
	Expert Assistant Agent		☑	
	Other agents			☑

These criteria (time and costs) were chosen because they are widely used. This goal can be described as follows:

Agents-Based Knowledge Logistics

```
CA: time → min, costs → min
```

The wrappers can make different offers such that the costs inversely depend on the time of knowledge delivery. This dependency is described by a table function given below:

```
W1: 30min/$15
W2: 15min/$20; 25min/$10; 45min/$5; …
W3: 50min/$25; 60min/$15; 70min/$10; …
```

The resulting time and costs are calculated as follows:

```
time  = max(timeW1, timeW2, timeW3)
costs = sum(costsW1, costsW2, costsW3)
```

The first scenario is performed in accordance with the conventional CNP (Figure 10). The configuration agent sends calls for proposals to all the wrappers concurrently. Besides description of the task to be performed, each call contains additional constraints. In this case these constraints will contain the following:

```
time → min
```

The offers from wrappers will contain the following:

```
W1: 30min/$15
W2: 15min/$20
W3: 50min/$25
```

The result will be 50 min and $60.

The second scenario is similar to the first one but here the configuration agent does not send calls for proposals concurrently but consequently (Figure 12).

The first wrapper receives the following: `time → min` and replies with `30min/$15`.

The configuration agent analyses this offer and sends to the second wrapper the following request: `costs → min AND time ≤ 30`. Here the objective has become a constraint and a new objective is added. The wrapper replies with `25min/$10`.

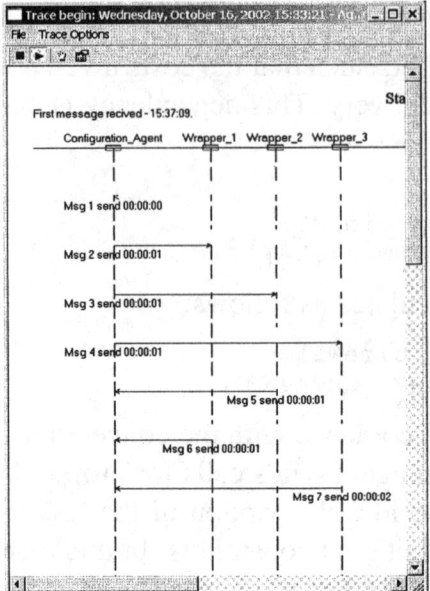

Figure 10. Experimentation: scenario 1.

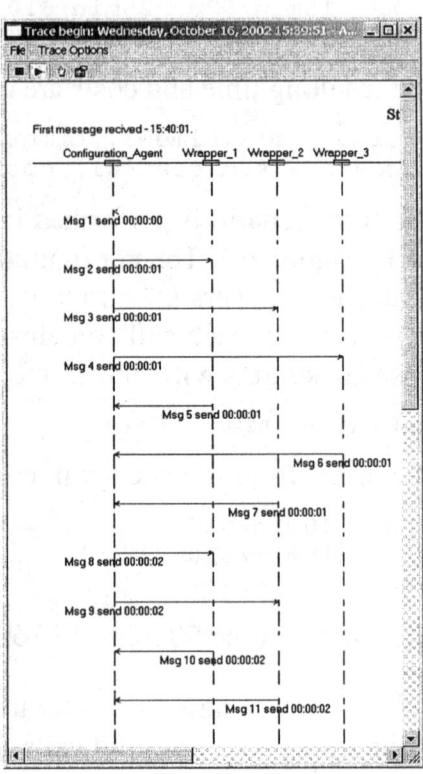

Figure 11. Experimentation: scenario 3.

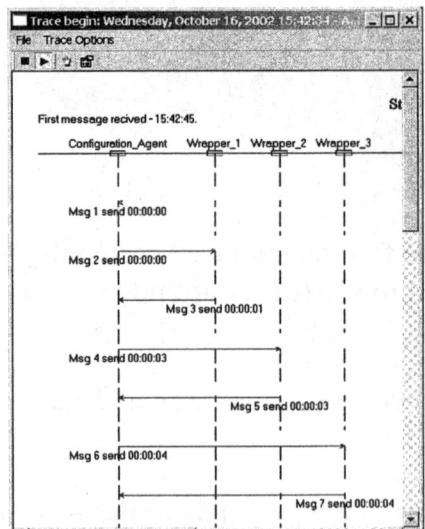

Figure 12. Experimentation: scenario 2.

The configuration agent analyses the offers and sends to the third wrapper the same request: `time ≤ 30 AND costs → min`. The wrapper replies with `50min/$25`. Here the wrapper cannot meet the requirements and it returns the best possible proposal.

The result is 50 min and $50:

```
W1: 30min/$15
W2: 25min/$10
W3: 50min/$25
```

The third scenario is performed in accordance with the presented in this section modified CNP. It contains concurrent conformation and iterative negotiation (Figure 11). At the first iteration the configuration agent sends calls for proposals to all the wrappers concurrently as in the first scenario, and the offers from the wrappers are the same:

```
time → min

W1: 30min/$15
W2: 15min/$20
W3: 50min/$25
```

After this the configuration agent analyses the results and sends new calls to the wrappers 1 and 2:

```
time ≤ 50 AND costs → min
```

The wrappers reply as follows:

```
W1: 30min/$15
W2: 45min/$5
```

The result is 50 min and $45:

```
W1: 30min/$15
W2: 45min/$5
W3: 50min/$25
```

As it can be seen, the proposed here constraint-based CNP on the one hand allows achieving better solutions than the conventional

CNP. In other words, given U_C, U_M – solution's utility for conventional and modified constraint-based CNP respectively, $U_M \geq U_C$ holds. On the other hand, it is obviously, that negotiation time for the proposed protocol increases. It can be seen that, given $T_{i=1...n}$ – response time of contractors (n – is the number of participating contractors), T_{man} – manager's response time, and T_C, T_M – negotiation time for conventional and modified constraint-based CNP respectively, $T_C \leq T_M \leq T_C + T_{max} + T_{man}$ holds, where $T_{max} = \max_{i=1}^{n} T_i$. This difference does not directly depend on the number of participating agents but it depends on T_{man} that in turn may depend on the task complexity and thereby on the number of participating agents. This dependency should be estimated for any particular case of the protocol usage.

5 Implementation of Agent Community

The agent community was implemented as a component of the integrated research prototype (Figure 13). The research prototype of the system "KSNet" has a client-server architecture. This architecture was chosen in accordance with the following reasons: (i) minimization of requirements to user computers (Web-based application allows user to have only HTML-compatible Web-browser and an access to the Internet because the procedures are executed on the server – not on the user's computer), (ii) a requirement of processing large amounts of information received from distributed KSs on the central (server) computer, and (iii) specifics of the agent community implementation.

In accordance with up-to-date technologies and standards the information kernel for the system "KSNet" was built as shown in Figure 14. In accordance with notation of object-oriented constraints network, knowledge in the system is represented by an aggregate of interrelated classes, their attributes, attributes' domains, and relations between them. An object scheme for working with the knowledge and a database structure for its internal storage are designed based on this notation. An access to the database is performed via ODBC as a standard data access mechanism under MS Windows. Remote access to the stored knowledge is performed via common HTTP Internet protocol. Knowledge is represented by either interactive HTML+VRML JavaScript enabled pages for users or a format based on DAML+OIL for knowledge-based tools.

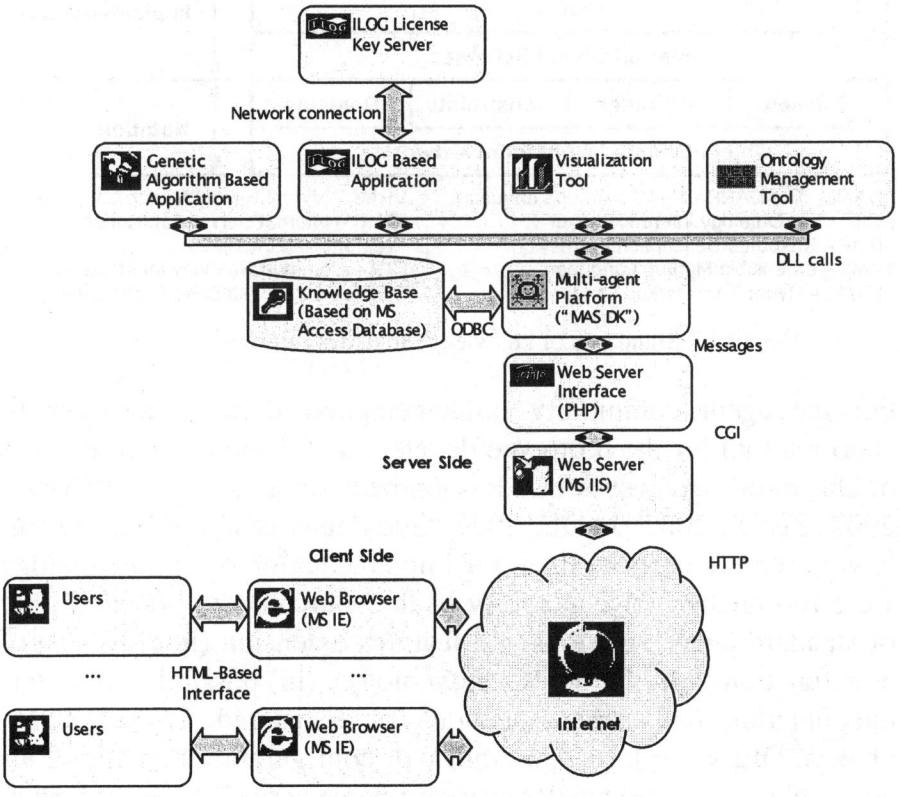

Figure 13. Prototype architecture.

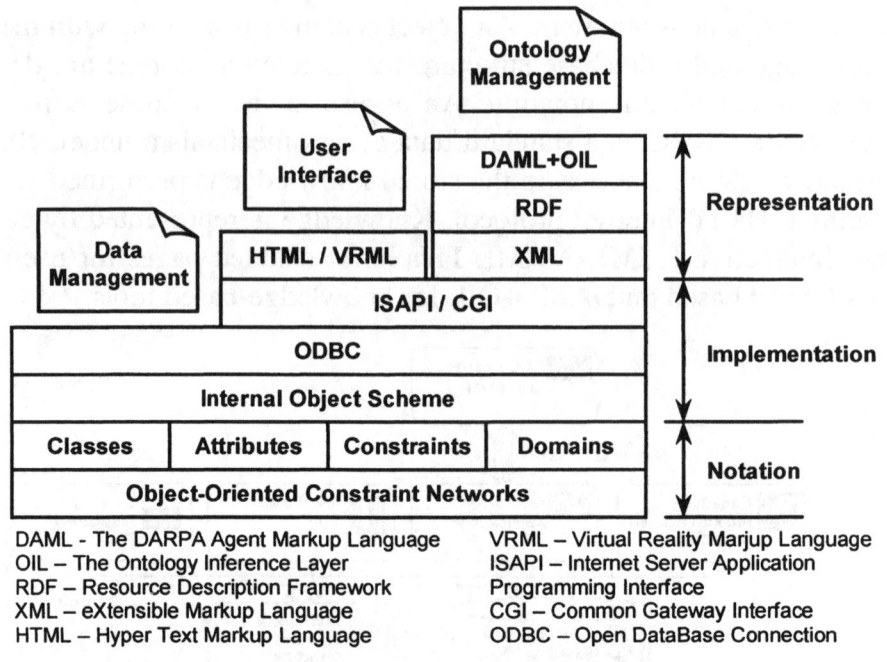

Figure 14. Standards of knowledge logistics information kernel.

For the agent community implementation it was necessary to choose a tool for the prototype development. Based on the analysis of the multi-agent systems development toolkits (e.g., FIPA-OS, 2002, ZEUS, 2002, JADE, 2002, Gorodetski et al., 2002) the following requirements to them for implementation of KL technology were formulated: (i) availability of the source code, (ii) possibility of standard agent functions and features extension (introduction of new functions specific for KL technology), (iii) possibility of external function calls (e.g., methods of registered COM/DCOM-objects, DLLs, etc.), (iv) possibility of configuration recovering after system faults, (v) availability of a name server storing information about registered agents, (vi) availability of message exchange mechanisms, (vii) availability of agent's knowledge repository creation and modification for storage of agent's knowledge, operation scenarios, and behaviour rules, (viii) availability of agent's "pro-activity" functions. Table 5 presents a comparison of some multi-agent systems development toolkits.

Table 5. Comparison of some multi-agent toolkits.

Toolkit Characteristic	FIPA-OS	Zeus	Jade	MAS DK
Developer	Nortel Networks	British Telecommunications Labs	S.p.A., University of Parma	SPIIRAS
Development environment	Java	Java	Java	C++, Java
Agents communication language	FIPA ACL	KQML, FIPA ACL	FIPA ACL	KQML
External applications calls	API	Java – applications	Java – applications	DLL
Agents interaction	P2P (using directory facilitator)	P2P, (using agent namespace)	P2P (using directory facilitator)	P2P (using namespace or broadcast), mediated
Agents hierarchy	Peer	Superior, subordinate, co-worker and peer	Peer	Peer (P)
Fault tolerance	N/A	N/A	N/A	N/A
Library of pre-defined negotiation protocols	-	+	-	-
Operating System	Windows; Unix	Windows; Solaris	Any OS with Java support	Windows

For the research prototype implementation the MAS DK environment developed in SPIIRAS was chosen as a toolkit for agents interaction modelling. Agents' scenarios use finite state automata starting when one or more predefined conditions are met (an agent receives a message from another agent or from the system). Each finite state automaton is implemented using the internal MAS DK language. It is represented by a set of conditional statements, messages to other agents, and agent's functions. Conditional statements are used to check error codes of problem-oriented functions, to compare values of different variables stored in the agents' repository, etc.

To extend agents' functions Microsoft Visual C++, Microsoft Access XP, Microsoft Access ODBC drivers, constraint satisfaction/propagation technology ILOG (Configurator 2.0, Solver 5.0, Concert Technology 1.0, Dispatcher 3.2) (ILOG, 2002), lexical reference system WordNet, etc. were used. It was motivated by a number of reasons including the following: (i) ILOG is a world leader in constraint satisfaction/propagation technologies and optimization algorithms that would enable a very efficient processing of large amounts of constraints, (ii) ODBC allows utilizing unified functions for accessing relational databases of different database management systems and enables modification of the DBMS used in the prototype, (iii) Microsoft Visual Studio 6.0 (Visual C++) allows accessing ILOG features, databases, writing efficient programs, and (iv) Microsoft Access XP enables rapid database design and creating simple database applications. Some intermediate and auxiliary forms for data preparation and results visualization were designed using Microsoft Access. For representation and interchange of agent's messages XML was used. Due to a large number of tools working with this format and specifications describing it the application of XML was useful and convenient.

During the system "KSNet" architecture development main system scenarios were designed using UML-based conceptual projects. Particularly, these projects include the agents' architecture, a list of actions performed by the agents in the different system scenarios, a list of messages for agent interactions. One of the developed projects is presented in Figure 15.

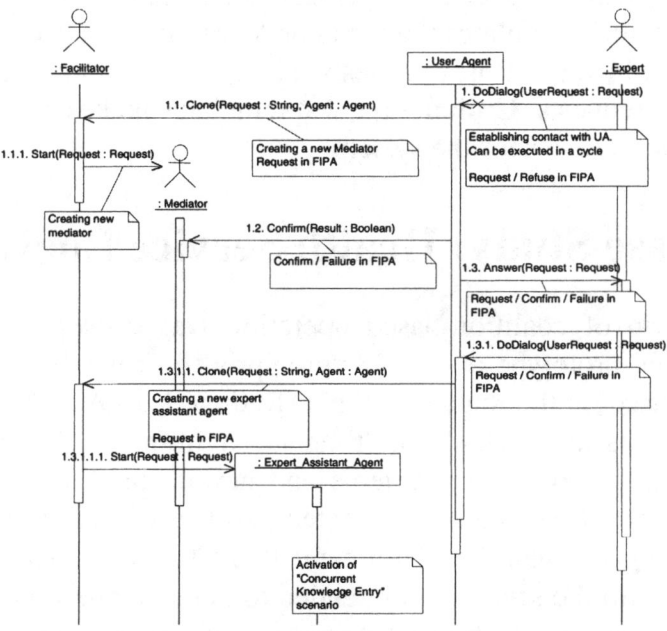

Figure 15. Knowledge entry process initialization by expert.

To start agent community operations the following operations were done. First, templates of all the agents were defined. A template includes agent's type, structure of knowledge repository, set of agent's messages for communication with other agents and a set of agent's work scenarios describing its behaviour. After that instances of agents were created. Each instance contains a unique agent's name and an address of its host. When the system starts only two agents are active: the facilitator and the monitoring agent.

They wait for new tasks and activate other agents in accordance with the prepared scenarios.

Some of possible scenarios of the user template-based request processing are presented in the following sections. Templates are dynamic software forms with fields for entering request terms, constraints and criteria specially developed for certain requests. In order to build such templates the machine learning technology is used to analyse a history of user requests, discovering common patterns and similar requests. Utilizing request templates makes recognition of the requests easier for the system.

6 Case Study: Health Service Logistics

The scenario of coalition-based operation was chosen as a case study for the prototype of the KSNet-approach. It is a hypothetical scenario based on the Sudanese Plain (Rathmell, 1999). The aim of the scenario is to provide a multi-agent environment, focusing on new aspects of coalition problems and new technologies demonstrating the ability of coalition-oriented agent services to function in an increasingly dynamic environment (CoAX, 2002). The experimentation with the scenario is intended to demonstrate how the developed KSNet-approach can be used for the health service logistics application.

The general problem considered in this case study designed for the system "KSNet" has the following formulation: "Define suppliers, transportation routes and schedules for building a hospital of given capacity at given location by given time". The following subproblems were selected:

- hospital related information (constraints on its structure, required quantities of components, required delivery schedules);

- available United Nations and friendly suppliers (constraints on suppliers' capabilities, capacities, locations) and ;
- available United Nations and friendly providers of transportation services (constraints on available types, routes, and time of delivery);
- geography and weather of the region (constraints on types, routes, and time of delivery, e.g. by air, by trucks, by off-road vehicles);
- political situation, e.g. who occupies the territory used for transportation, existence of military actions on the routes, etc. (additional constraints on routes of delivery).

As a result of the analysis of these problems the following modules were defined:

1. *Portable hospital allocation.* This subproblem is devoted to finding the most appropriate location for a hospital to be built considering such factors as locations of the disaster, water resources, nearby cities and towns, communications facilities (e.g., locations of airports, roads, etc.) and decision maker's choice and priorities.
2. *Routing problem.* This subproblem is devoted to finding the efficient ways of delivery of the hospital's components from suppliers considering such factors as communications facilities (e.g., locations of airports, roads, etc.), their conditions (e.g., good, damaged or destroyed roads), weather conditions (e.g., rains, storms, etc.) and decision maker's choice and priorities.
3. *Hospital configuration.* This subproblem is devoted to finding the efficient components for the hospital considering such factors as component suppliers, their capacities, prices, transportation time and costs and decision maker's choice and priorities.

Input data for user request input is prepared by an expert team using specially developed screen forms. Experts' tasks included definition of a list of possible hospital locations and a list of suppliers, a specification of dependencies between the weather and delivery types (routes), and analysis of hospital supplies delivery costs. Parts of ontologies corresponding to the described task were found in Internet's ontology libraries (Clin-Act, 2000, Cyc, 1998, NAICS, 2001, UNSPSC, 2001). The application ontology of this humanitarian task was built and a connection of the found sources was performed. Three wrappers were developed to process information about: (i) suppliers, (ii) transportation service, and (iii) weather conditions and prepared HTML forms for user request input.

One of the scenarios of agent community interaction during the user request processing is given in Figure 16. In the framework of the case study there were modelled agents' interactions for processing different user requests. Analysis of time distribution between members of the agent community during these scenarios processing has shown that a re-distribution of the tasks and adding new functions to the most unloaded agents can increase a rapidity of the system.

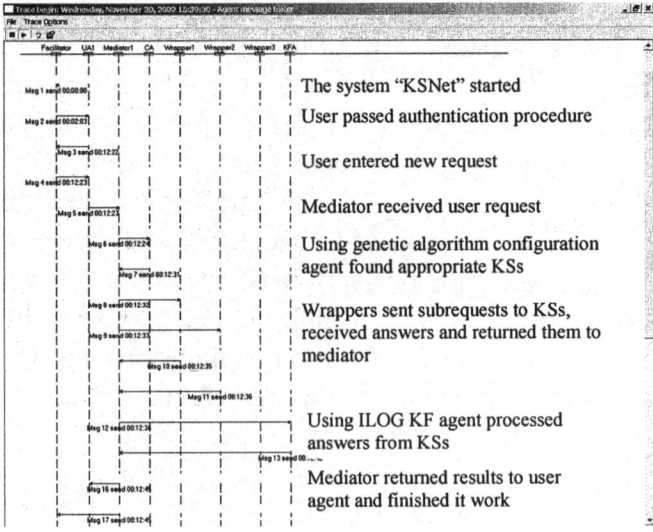

Figure 16. An example of the modelled agents' interaction.

7 Case Study: Virtual Supply Network

Generally, a Virtual Supply Network (VSN) can be represented as shown in Figure 17. Common VSN objective is to achieve "market-winning" fulfilment time with the least inventory risk for the product. Configuring deals with creating configuration solutions and selecting components and ways to configure these. In VSN each unit selects its direct suppliers ("first-level suppliers"). It is supposed that there are no central control units that could influence upon a choice of other units. In this case, a generic VSN pattern can be defined as a unit with its first-level suppliers (shown with dashed lines). This allows considering the generic VSN pattern as a configuration pattern.

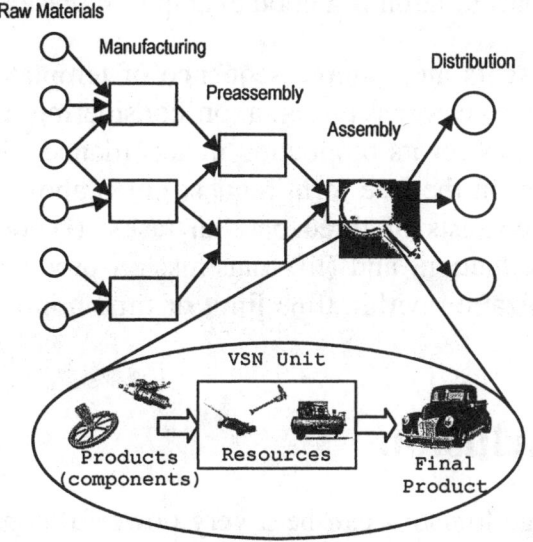

Figure 17. Structure of a generic VSN pattern.

This section is intended to demonstrate how the system "KSNet" may support the process of decision making by delivering obtained knowledge and generating possible solution(s) that can be taken into account by the user.

For illustrative purposes two major kinds of configuration tasks for business environment were selected: (i) complex system configuration task, and (ii) resource allocation task. The main idea of the system configuration task (a car in the example) is to obtain a feasible configuration of a system meeting specified requirements, with a system structure being known. The task of resource allocation assumes that there are some amount of work to be done and some facilities which can perform this work. Facilities can be plants with known capacities and such characteristics as production cost and time. The work consists of several operations (parallel and/or sequential) and each facility is capable to perform some of the operations. VSN configuration is a good example of these kinds of tasks.

Figure 18 presents an example sequence of templates for user request input and answer representation considering a configuration of a car and VSN for its production in accordance with user's preferences (based on the free form request given above). The production process consists of three parallel tasks: (i) body production, (ii) engine production, and (iii) transmission production. The goal is cost minimization within time limit or time minimization within cost limit.

8   Conclusion

The knowledge logistics can be a very powerful concept to enable collaboration between members of joint actions and operations. This concept can be applied in many other industrial, healthcare applications featuring large-scale dynamic virtual organizations with distributed operations, logistics operations addressing end-to-end rapid supply, markets via partnerships, etc.

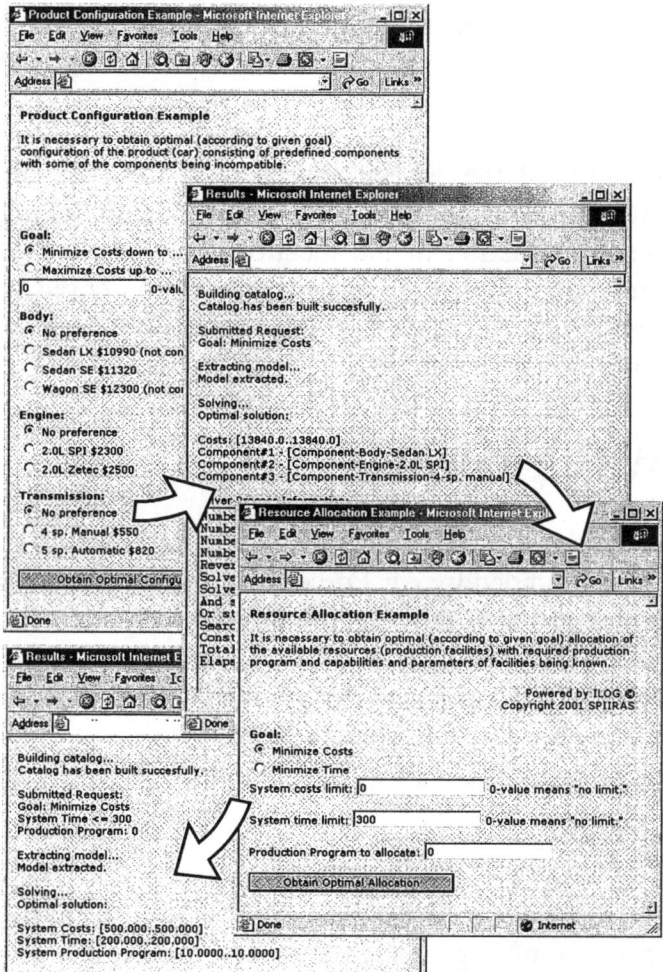

Figure 18. Example of user request templates.

Agent-based technology is a good basis for knowledge logistics since agents can operate in a distributed environment, independently from the user and apply ontologies to knowledge representation, sharing and exchange. The chapter presented an implementation of the agent community where agents cooperate, interact and negotiate with each other. Agent-based architecture increases scalability, efficiency and interoperability of the system "KSNet". Developed constraint-based contract net protocol allows obtaining more efficient negotiation results than the conventional contract net protocol.

An applicability of the KSNet-approach to the area of e-health (portable hospital configuration for a given location taking into account a current situation in the locations' region) and e-business demonstrates possibility of its usage for coalition-based operations support and organization of collaboration between global e-business environment members.

Acknowledgements

Some parts of the research were done as parts of the ISTC partner project # 1993P funded by Air Force Research Laboratory at Rome, NY, the project # 1.9 of the research program "Fundamental Basics of Information Technologies and Computer Systems" of the Russian Academy of Sciences, and the grant # 02-01-00284 of the Russian Foundation for Basic Research.

References

Adams, M.B., Deutsch, O.L., Hall, W.D., et al. (2000), "Closed-loop operation of large-scale enterprise: application of a decomposition approach," *Proceedings of DARPA ISO Symposium on Advanced in Enterprise Control (AEC)*, pp. 1-10.
AKT project web site (2002), http://www.aktors.org/.
Blythe, J., Kim, J., Ramachandran, S., and Gil, Y. (2001), "An integrated environment for knowledge acquisition," *Proceedings of International Conference on Intelligent User Interfaces*, pp. 13-20.
Cheah, Y. and Abidi, S.S.R. (2001), "Capturing tacit healthcare knowledge via an intelligent info-structure featuring knowledge

standardization and repair mechanisms," *Medicinska Informatika (Journal of the Croatian Society for Medical Informatics)*, vol. 5, pp. 49-52.

Clin-Act (Clinical Activity). The ON9.3 Library of Ontologies: Ontology Group of IP-CNR (a part of the Institute of Psychology of the Italian National Research Council (CNR)) (2000), http://saussure.irmkant.rm.cnr.it/onto/.

CoAX – Coalition Agents eXperiment. Coalition Research Home Page (2002), http://www.aiai.ed.ac.uk/project/coax/.

COGITO: E-Commerce with Guaiding Agents Based on Personalised Interaction Tools (2002), http://ipsi.fhg.de/~cogito.

Cunningham, M. (2001), "Buyer (and Seller). Be Aware," *DB2 Magazine*, vol. 6, is. 2, pp. 33-37.

DARPA Advanced Logistics Project web site (2002), http://www.darpa.mil/iso/alp.

Davis, R. and Smith, R.G. (1983), "Negotiation as a metaphor for distributed problem solving," *Artificial Intelligence*, vol. 20, no. 1, pp.63-109.

Dzbor, M., Paralic, J., and Paralic, M. (2000), "knowledge management in a distributed organisation," *Proceedings of BASYS '2000 - 4th IEEE/IFIP International Conference on Information Technology for Balanced Automation Systems in Manufacturing*, Kluwer Academic Publishers, London, ISBN 0-7923-7958-6, pp. 339-348.

FIPA-OS toolkit (2002), http://fipa-os.sourceforge.net/.

Foundation for Intelligent Physical Agents (FIPA) Documentation (2002), http://www.fipa.org.

Gorodetski, V., Karsayev, O., Kotenko, I., and Khabalov, A. (2002), "Software development kit for multi-agent systems design and implementation," Springer-Verlag Berlin Heidelberg, *Lecture Notes in Artificial Intelligence*, vol. 2296, pp. 121-130.

Howells, H., Davies, A., Macauley, B., and Zancanato, R. (1999), "Large scale knowledge based systems for airborne decision support," *Knowledge-Based Systems Journal*, vol. 12, pp. 215-222.

Hpkb-Upper-Level-Kernel-Latest: Upper Cyc/HPKB IKB Ontology with links to SENSUS, Version 1.4, February, 1998. Ontolingua Ontology Server (1998), http://www-ksl-svc.stanford.edu:5915.

Hummingbird web site (2002), http://www.hummingbird.com/products/eip/.

ILOG Corporate Web-site (2002), http://www.ilog.com.

Intelligent Agents. ISR Agent Research (2002), http://193.113.209.147/projects/agents.htm.

Jarrar, Y., Schiuma, G., and Zairi, M. (2002), "Defining organisational knowledge: a best practice perspective," *Proceedings of VI International Conference on "Quality Innovation Knowledge"*, pp. 486-496.

Java Agent DEvelopment Framework JADE (2002), http://sharon.cselt.it/projects/jade/

Kim, J., Ling, R., and Will, P. (1997), "Ontology engineering for active catalog," *Proceedings of AAAI Workshop on Using AI in Electronic Commerce, Virtual Organizations and Enterprise Knowledge Management to Reengineer the Corporation*, pp. 44-49.

Maedche, A. and Staab, S. (2000), "Discovering conceptual relations from text," in Horn, W. (ed.), *Proceedings of 14th European Conference on Artificial Intelligence (ECAI 2000)*, IOS Press, Amsterdam, pp. 321-325.

Nodine, M.H. and Unruh, A. (1997), "Facilitating open communicating in agent systems: the InfoSleuth infrastructure," *Technical Report MCC-INSL-056-97*, Microelectronics and Computer Technology Corporation.

North American Industry Classification System code, DAML Ontology Library, Stanford University, July (2001), http://opencyc.sourceforge.net/daml/naics.daml.

Ontolingua web site. Stanford University, Knowledge Systems Laboratory (2001), http://www-ksl-svc.stanford.edu:5915/&service =frame-editor.

Ontoprise: Semantics for the WEB (2003), http://www.ontoprise.de /products/index_html_en.

Park, J.Y., Gennari, J.H., and Musen, M.A. (1997), "Mappings for reuse in knowledge-based systems," *SMI Technical Report 97-0697*.

Payne, T., Paolucchi, M., Singh, R., and Sycara, K. (2002), "Communicating agents in open multi-agent systems," *Proceedings of First GSFC/JPL Workshop on Radical Agent Concepts (WRAC)*, pp. 365-371.

Pease, A., Chaudhri, V., Lehmann, F., and Farquhar, A. (2000), "Practical knowledge representation and the DARPA high performance knowledge bases project," in Cohn, A., Giunchiglia, F., and Selman, B. (eds.), *Proceedings of Conference on Knowledge Representation and Reasoning, KR-2000*, pp. 12-15.

Protégé-2000 Project web site. USA, Stanford Medical Informatics at the Stanford University School of Medicine (2000), http://protege.stanford.edu/index.shtml.

Rathmell, R.A. (1999), "A coalition force scenario "Binni – gateway to the golden bowl of Africa," in Tate, A. (ed.), *Proceedings on International Workshop on Knowledge-Based Planning for Coalition Forces*, pp. 115-125.

Rasmus, D.W. (2000), "Knowledge management: more than AI but less without it," *PC AI*, vol. 14, no. 2, (Knowledge Technology Inc., Phoenix, AZ), pp. 35-44.

Sandholm, T.W. and Lesser, V.R. (1995), "Issues in automated negotiation and electronic commerce: extending the contract net framework," in Huhns, M. and Singh, M. (eds.), *Readings in Agents*, Morgan Kaufmann Publishers, pp. 66-73.

Sandholm, T. (1999), "Distributed rational decision making," in Weiss, G. (ed.), *Multiagent Systems: a Modern Introduction to Distributed Artificial Intelligence*, MIT Press, pp. 201-258.

Smirnov, A., Pashkin, M., Chilov, N., and Levashova, T. (2002), "Distributed knowledge entry based on intelligent agents and virtual reality technologies," *Proceedings of 9th International Conference on Neural Information Processing; 4th Asia-Pacific Conference on Simulated Evolution and Learning; 2002 International Conference on Fuzzy Systems and Knowledge Discovery (ICONIP'02-SEAL'02-FSKD'02)*, vol. 2, pp. 437-441.

Smirnov, A., Pashkin, M., Chilov, N., Levashova, T., and Haritatos, F. (2003a), "Knowledge source network configuration approach to knowledge logistics," *International Journal of General Systems. Taylor & Francis Group*, vol. 32, no. 3, pp. 251-269.

Smirnov, A., Pashkin, M., Chilov, N., and Levashova, T. (2003b), "KSNet-approach to knowledge fusion from distributed sources," *Computing and Informatics*, vol. 22, pp. 105-142.

Smirnov, A., Pashkin, M., Chilov, N., and Levashova, T. (2003c), "Agent-based support of mass customization for corporate knowledge management," *Engineering Applications of Artificial Intelligence*, vol. 16, is. 4, pp. 349-364.

Smirnov, A., Pashkin, M., Chilov, N., Levashova, T., and Krizhanovsky, A. (2003d), "Ontology-driven knowledge logistics approach as constraint satisfaction problem," Springer-Verlag Berlin Heidelberg, *Lecture Notes in Computer Science*, vol. 2888, pp. 535-652.

Smith, R. (1980), "The contract net protocol: high-level communication and control in a distributed problem solver," *IEEE Transactions on Computers*, no 29(12), pp. 1104-1113.

Tecuci, G., Boicu, M., Marcu, D., et al. (2002), "Development and deployment of a disciple agent for center of gravity analysis," *Proceedings of Eighteenth National Conference on Artificial Intelligence and the Fourteenth Conference on Innovative Applications of Artificial Intelligence*, pp. 853-860.

Trellis project web site (2002), http://www.isi.edu/expect/projects/trellis/index.html.

The UNSPSC Code (Universal Standard Products and Services Classification Code), DAML Ontology Library, Stanford University, January (2001), http://www.ksl.stanford.edu/projects /DAML/UNSPSC.daml.

Vail, E.F. (1999), "Knowledge mapping: getting started with knowledge management," *Information Systems Management, Fall*, pp. 16-23.

Visser, P.R.S., Jones, D.M., Beer, M.D., Bench-Capon, T.J.M., Diaz, B.M., and Shave, M.J.R. (1999), "Resolving ontological heterogeneity in the KRAFT project," *Proceeding of the International Conference On Database and Expert System Applications, (DEXA-99)*, Springer-Verlag, *Lecture Notes in Computer Science*, vol. 1677, pp. 668-677.

Chapter 4

Architectural Styles and Patterns for Multi-Agent Systems

Manuel Kolp, T. Tung Do,
Stéphane Faulkner and T. T. Hang Hoang

A Multi-Agent System (MAS) is an organization of coordinated autonomous agents that interact in order to achieve common goals. Considering real world social organizations as an analogy (Zambonelli et al. 2000), this chapter proposes architectural styles and design patterns for MAS which adopt concepts from social theories. The styles are intended to represent a macro-level architecture of a MAS in terms of actor, goal and actor dependency and are evaluated with respect to software quality attributes. At a micro-level, social patterns give a finer-grain description of the MAS architecture and define how goals assigned to agents will be fulfilled. They are modeled within a conceptual framework analyzing them from five points of view: social, intentional, structural, communicational and dynamic. An e-business example illustrates our purpose.

1 Introduction

The characteristics and expectations of new application domains for the enterprise such as e-business, knowledge management, peer-to-peer computing or web services are deeply modifying software architecture engineering. Most of the system architecture designed for these kinds of application areas are now de facto concurrent and distributed. They tend to be open and dynamic, in that they exist in a changing organizational and operational environment where new components can be added, modified or removed at any time (Bass et al. 1998).

For these reasons – and more – Multi-Agent Systems (MAS) architectures are gaining popularity over traditional systems, including object-oriented ones (Zambonelli et al. 2003). MAS architectures do allow dynamic and evolving structures and components which can change at run-time to benefit from the capabilities of new system entities or replace obsolete ones.

Such architectures become rapidly complicated due to the ever-increasing complexity of these new business IT domains and their actors: as the expectations of users and business stakeholders change day after day, as the complexity of systems, information and communication technologies and organizations continually increases in today's dynamic environments, developers are expected to produce architectures that must handle more difficult and intricate requirements that were not taken into account ten years ago, making architectural design a central engineering issue in modern enterprise information system life-cycle (Shaw and Garland 1996).

An important technique that helps to manage this complexity when constructing and documenting such architectures is the reuse of design experience and knowledge. Thus, architectural styles and design patterns have become an attractive approach to reusing design knowledge (Pree 1994).

Architectural styles are intellectually manageable abstractions of system structure that describe how system components interact and work together (Shaw and Garland 1996).

Design patterns describe a problem commonly found in software designs and prescribe a flexible solution for the problem, so as to ease the reuse of that solution. This solution is repeatedly applied from one design to another, producing design structures that look quite similar across different applications (Gamma et al. 1995, Buschmann et al. 1996).

MAS architectures can be considered social structures composed of autonomous and proactive agents that interact and cooperate with each other to achieve common or private goals (Fox 1981, Zambonelli et al. 2000). This chapter will introspect socially based styles and patterns to construct such architectures and apply them on an e-business case study.

It proposes a set of generic architectural structures:

- At the architectural design level, organizational styles inspired from organization theory and strategic alliances will be used to design the overall MAS architecture. Styles from organization theory will describe the internal structure and design of the MAS architecture, while styles from strategic alliances will model the cooperation of independent architectural organizational entities that pursue shared goals (Kolp et al. 2001).
- At the detailed design level, social patterns drawn from research on cooperative and distributed architectures, will offer a more microscopic view of the social MAS architecture description. They will define the agents and the social dependencies that are necessary for the achievement of agent goals (Hayden et al. 1999).

The chapter is organized as follows. Section 2 introduces a macro level of organization-inspired architectural styles, and proposes an evaluation of architectural alternatives. Section 3 introduces a framework to introspect social design patterns for finer-grain design of an organizational architecture. An e-business case study will illustrate the use of styles and patterns proposed in the chapter. Finally, Section 4 summarizes the contributions of the chapter.

2   Organizational Architectural Styles

Software architectures describe a software system at a macroscopic level in terms of a manageable number of subsystems, components and modules inter-related through data and control dependencies (Bass et al. 1998).

System architectural design has been the focus of considerable research during the last fifteen years that has produced well-established architectural styles and frameworks for evaluating their effectiveness with respect to particular software qualities. Examples of styles are pipes-and-filters, event-based, layered, control loops and the like (Shaw and Garland 1996). Examples of software qualities include maintainability, modifiability, portability, etc (Bass et al. 1998). This section analyzes architectural styles for multi-agent software systems. Since the fundamental concepts of a Multi-Agent System (MAS) are intentional and social, rather than implementation-oriented, theories which study social structures can provide inspiration and insights (Fox 1981). But, what kind of social theory? There are theories that study group psychology, communities (virtual or otherwise) and social networks. Such theories study social structure as an emergent property of a social context. Instead, organizational theories, namely *Organization Theory* and *Strategic Alliances*, are interested in social structures that result from a design process. Organization Theory (e.g., (Mintzberg 1992, Scott 1998, Morabito et al. 1999)) describes the structure and design of an organization; *Strategic Alliances* (e.g., (Dussauge and Garrette 1999, Yoshino and Srinivasa Rangan 1995)) models the strategic collaborations of independent organizational stakeholders who pursue a set of agreed upon business goals. These theories propose organizational styles such as the structure-in-fives, the matrix, the chain of values, the bidding model, the joint venture, the arm's length model, the hierarchical contract, etc.

For instance, in organization theory, the **Structure-in-5** style considers an organization an aggregate of five sub-structures, as proposed by Minztberg (Mintzberg 1992). At the base level sits the *Operational Core* which carries out the basic tasks and procedures directly linked to the production of products and services (acquisition of inputs, transformation of inputs into outputs, distribution of outputs). At the top lies the *Strategic Apex* which makes executive decisions ensuring that the organization fulfils its mission in an effective way and defines the overall strategy of the organization in its environment. The *Middle Line* establishes a hierarchy of authority between the Strategic Apex and the Operational Core. It consists of managers responsible for supervising and coordinating the activities of the Operational Core. The *Technostructure* and the *Support* are separated from the main line of authority and influence the operating core only indirectly. The Technostructure serves the organization by making the work of others more effective, typically by standardizing work processes, outputs, and skills. It is also in charge of applying analytical procedures to adapt the organization to its operational environment. The Support provides specialized services, at various levels of the hierarchy, outside the basic operating work flow (e.g., legal counsel, R&D, payroll, cafeteria).

In strategic alliances, the **joint-venture** style involves agreement between two or more intra-industry partners to obtain the benefits of larger scale, partial investment and lower maintenance costs (Dussauge and Garrette 1999). A specific joint management actor coordinates tasks and manages the sharing of resources between partner actors. Each partner can manage and control itself on a local dimension and interact directly with other partners to exchange resources, such as data and knowledge. However, the strategic operation and coordination of such an organization, and its actors on a global dimension, are only ensured by the joint management actor in which the original actors possess equity participations.

An organizational style is a metaclass of organizational structures offering a set of design parameters to coordinate the assignment of organizational objectives and processes, thereby affecting how the organization itself functions (Fox 1981). Design parameters include, among others, goal and task assignments, standardization, supervision and control dependencies and strategy definitions.

These two organizational styles are described below. For further details, including organization case studies and formal specification, see (Do et al. 2003).

Figure 1 models the structure-in-5 style using the *i** strategic dependency model. *i** is a modeling framework for early requirements analysis (Yu 1995), which offers goal- and actor-based notions such as *actor, agent, role, position, goal, softgoal, task, resource, belief* and different kinds of social *dependency* between actors. It is a graph, where each node represents an *actor* and each link between two actors indicates that one actor depends on the other for some goal to be attained. A dependency describes an "agreement" (called *dependum*) between two actors: the *depender* and the *dependee*. The *depender* is the depending actor, and the *dependee,* the actor who is depended upon. The type of the dependency describes the nature of the agreement. *Goal* dependencies represent delegation of responsibility for fulfilling a goal; *softgoal* dependencies are similar to goal dependencies, but their fulfillment cannot be defined precisely (for instance, the appreciation is subjective or fulfillment is obtained only to a given extent); *task* dependencies are used in situations where the dependee is required to perform a given activity; and *resource* dependencies require the dependee to provide a resource to the depender.

Actors are represented as circles; dependums – goals, softgoals, tasks and resources – are respectively represented as ovals, clouds, hexagons and rectangles; dependencies have the form *depender* → *dependum* → *dependee*.

For instance in Figure 1, the *Technostructure, Middle Agency* and

Architectural Styles and Patterns for Multi-Agent Systems

Support actors depend on the *Apex* for strategic management. Since the goal *Strategic Management* does not have a precise description, it is represented as a softgoal (cloudy shape). The *Middle Agency* depends on the *Technostructure* and *Support* respectively through goal dependencies *Control* and *Logistics* represented as oval-shaped icons. The *Operational Core* is related to the *Technostructure* and *Support* actors through the *Standardize* task dependency and the *Non-operational Service* resource dependency, respectively.

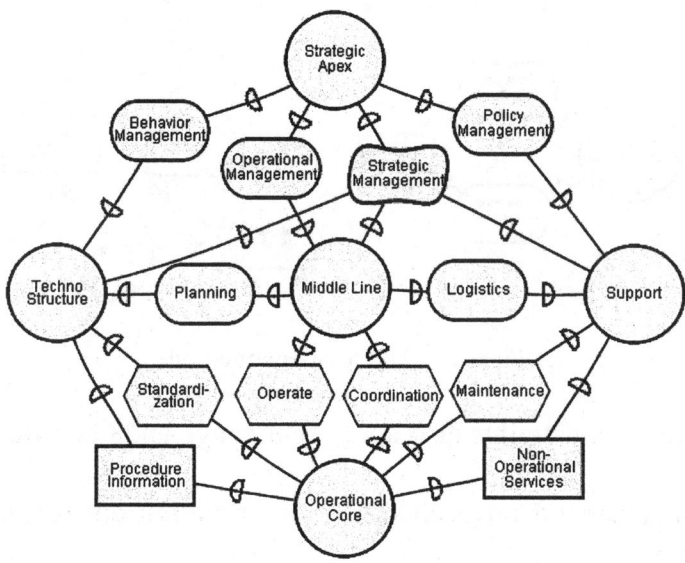

Figure 1. The Structure-in-5 Style

A number of constraints also apply (Mintzberg 1992):

- the dependencies between the *Strategic Apex* as depender and the *Technostructure*, *Middle Line* and *Support* as dependees must be of type goal
- a softgoal dependency models the strategic dependence of the *Technostructure*, *Middle Line* and *Support* on the *Strategic Apex*
- the relationships between the *Middle Line* and *Technostructure* and *Support* must be of goal dependencies

- the *Operational Core* relies on the *Technostructure* and *Support* through task and resource dependencies
- only task dependencies are permitted between the *Middle Line* (as depender or dependee) and the *Operational Core* (as dependee or depender).

Figure 2 models the joint-venture style using i*.

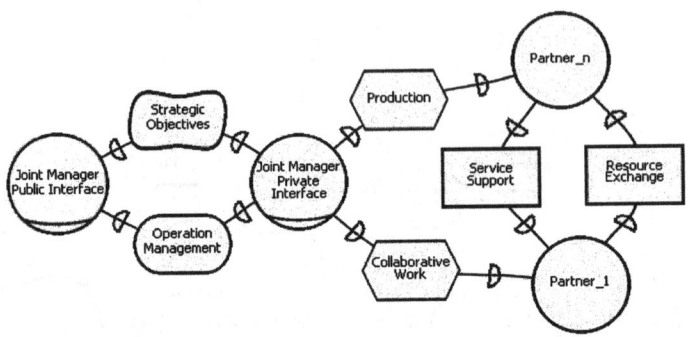

Figure 2. The Joint-venture Style

A number of constraints also apply (Dussauge and Garrette 1999):

- Partners depend on each other for providing and receiving resources.
- Operation coordination is ensured by the joint manager actor which depends on partners for the accomplishment of these assigned tasks.
- The joint manager actor must assume two roles: a private interface role to coordinate partners of the alliance and a public interface role to take strategic decisions, define policy for the private interface and represents the interests of the whole partnership with respect to external stakeholders.

2.1 Applying Organizational Styles

This section overviews the use of the structure-in-5 and joint-venture styles with the design of an architecture for a business-to-consumer (B2C) application called *E-Media*.

E-Media is a business-to-consumer system allowing on-line customers to buy different kinds of media items such as books, newspapers, magazines, audio CDs, videotapes and the like on the Internet. Customers can search the on-line store by either browsing the catalogue or using a search engine to query the database. *E-Media* also allows to process on-line orders, bills and delivery invoices, and keeps track of all web information of strategic importance for statistical analysis. The full case study is described in (Do et al. 2003)

Structure-in-5. Figure 3 suggests a possible assignment of system responsibilities for *E-Media* following the structure-in-5 style. It is decomposed into five principal components *Store Front, Coordinator, Billing Processor, Back Store* and *Decision Maker. Store Front* serves as the *Operational Core*. It interacts primarily with Customer and provides her with a usable front-end web application for consulting and shopping media items. *Back Store* constitutes the *Support* component. It manages the product database and communicates to the *Store Front* information on products selected by the user. It *stores* and *backs up* all web information from the *Store Front* about customers, products, sales, orders and bills to produce *statistical information* to the *Coordinator*. It provides the *Decision Maker* with *strategic information* (analyses, historical charts and sales reports).

The *Billing Processor* is in charge of handling orders and bills for the *Coordinator* and implementing the corresponding procedures for the *Store Front*. It also ensures the secure management of financial transactions for the *Decision Maker*. As the *Middle Line*, the *Coordinator* assumes the central position of the architecture. It ensures the coordination of *e-shopping* services provided by the *Operational Core* in-

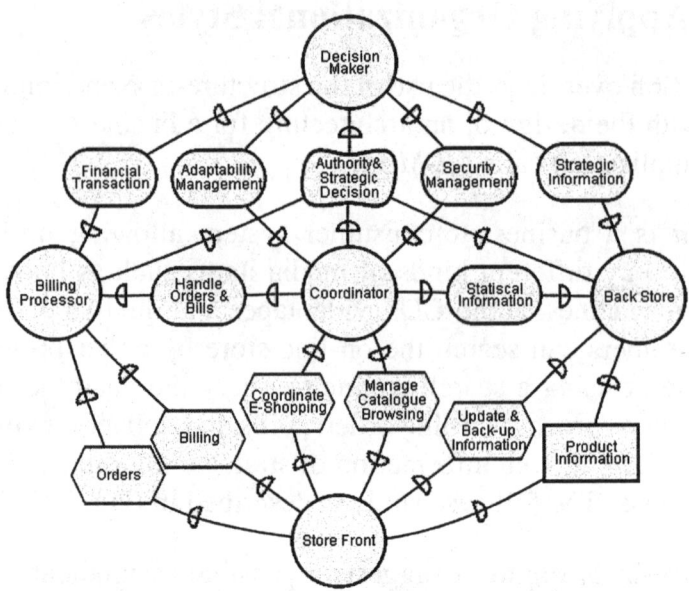

Figure 3. The E-media Architecture in Structure-in-5

cluding the management of conflicts between itself, the *Billing Processor*, the *Back Store* and the *Store Front*. To this end, it also handles and implements strategies to manage and prevent *security* gaps and *adaptability* issues. The *Decision Maker* assumes the *Strategic Apex* role. It defines the *Strategic Behavior* of the architecture ensuring that objectives and responsibilities delegated to the *Billing Processor*, *Coordinator* and *Back Store* are consistent with that global functionality.

Joint-venture. Following the joint-venture style, the *E-Media* architecture in Figure 4 is organized around a joint manager assuming two roles: the *E-store* role defines the *Customer Relationship Management* and the operational strategies of *E-Media*, i.e., *Sales* and *DB Management Strategy*. The *Back Store* deals with coordinativity supervising the other actors: a *Data Mining Processor* handling *Business Knowledge* processes, a *DataBase* storing the on-line catalogue and allowing *Catalogue Browsing*, a *Billing Processor* managing all *Financial Transactions* and a *Shopping Cart* implementing *E-Shopping* activities.

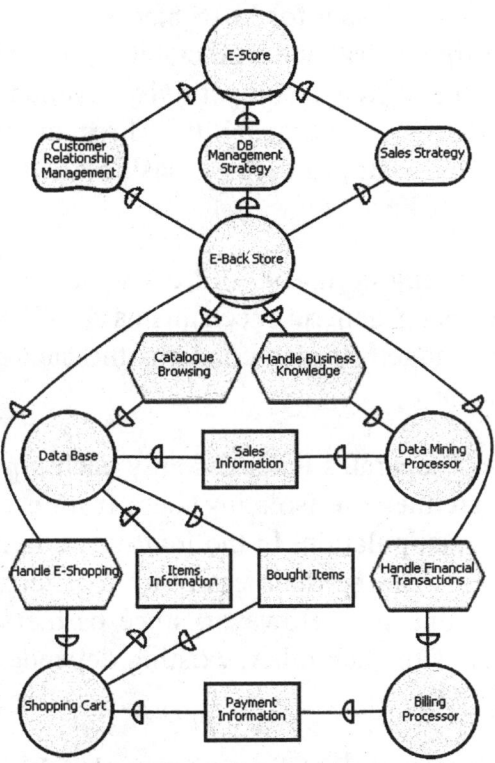

Figure 4. The E-media Architecture in Joint-venture

Each of these four last actors also interacts directly with each other to exchange data, information and knowledge: the *Shopping Cart* needs data and information about selected products from the *DataBase* and provides *Billing Processor* with financial information about purchased products and on-line customers. The *Data Mining Processor* gets sales data and information from the *Database* to produce business knowledge, i.e., historical charts sales reports and business forecasts.

2.2 Evaluation

Software quality attributes (Shaw and Garland 1996) (i.e., non-functional requirements describing how well the system accomplishes its functions) relevant for MAS have been studied in (Kolp et al. 2001). These are for instance predictability, security, adaptability, coordinability, cooperativity, competitivity, availability, fallibility-tolerance, modularity or aggregability. Three of them have been identified particularly strategic for e-business systems (Do et al. 2003, Kolp et al. 2001):

Adaptability. (Horling et al. 1999) deals with the way the system can be designed using generic mechanisms to allow web pages to be dynamically changed. It also concerns the catalogue update for inventory consistency.

The *structure-in-5* separates independently each typical component of the *E-Media* architecture isolating them from each other and allowing dynamic manipulation. In the *joint-venture*, manipulation of partner components can be done easily by registering new components to the joint manager. However, since partners can also communicate directly with each other, existing dependencies should be updated as well.

Security. (Bonati et al. 1998) Clients, exposed to the internet are, like servers, at risk in web applications. It is possible for web browsers and application servers to download or upload content and programs that could open up the client system to crackers and automated agents. JavaScript, Java applets, ActiveX controls, and plug-ins represent a certain risk to the system and the information it manages. Equally important are the procedures checking the consistency of data transactions.

In the *structure-in-5*, checks and control mechanisms can be integrated at different assuming redundancy from different perspectives. Contrary to the classical layered architecture (Shaw and Garland 1996), checks and controls are not restricted to adjacent levels. Besides, since the structure-in-5 permits to separate process (*Store Front*, *Billing Processor* and *Back Store*) from control (*Decision Maker* and *Monitor*), security and consistency of these two hierarchies can also be verified independently. The *joint-venture*, through its joint manager, proposes a central message server/controller. Exception mechanism, wiretapping supervising or monitoring can be supported by the joint manager to guarantee non-failability, reliability and completeness.

Availability. (Chess 1998) Network communication may not be very reliable causing sporadic loss of the server. There are data integrity concerns with the capability of the e-business system to do what needs to be done, as quickly and efficiently as possible in particular with the ability of the system to respond in time to client requests for its services.

The *structure-in-5* architecture prevents availability problems by differentiating process from control. Besides, contrary to the classical layered architecture (Shaw and Garland 1996), higher levels are more abstract than lower levels: lower levels only involve resources and task dependencies while higher ones propose intentional (goals and softgoals) relationships. In the *joint-venture*, the central position and role of the joint manager is a means for resolving conflicts between components and prevent availability issues. Through its joint manager, the architecture proposes a central message server/controller. Exception mechanisms, wiretapping supervising or monitoring can be centrally supported by to guarantee non-failability, reliability and completeness.

Table 1 summarizes the strengths and weaknesses of the two architectures with respect to the software quality attributes detailed above.

Table 1. Strengths and weaknesses of e-business organizational architectures

	Structure-in-5	Joint-venture
Security	+	+
Availability	+	+
Adaptability	+-	++

3 Social Patterns

A further step in the architectural design of MAS consists of specifying how the goals delegated to each actor are to be fulfilled (Kolp et al. 2001). For this step, designers can be guided by a catalogue of multi-agent patterns which offer a set of standard solutions. Considerable work has been done in software engineering for defining software patterns (see e.g., (Gamma et al. 1995, Buschmann et al. 1996)). Unfortunately, little emphasis has been put on social and intentional aspects. Moreover, the proposals of agent patterns that address these aspects (see e.g., (Aridor and Lange 1998, Deugo et al. 1999, Pree 1994)) are not intended to be used at a design level, but rather during implementation when low-level issues like agent communication, information gathering, or connection setup are addressed.

This section details the notion of social patterns according to five complementary dimensions as specified in (Do et al. 2003a): social, intentional, structural, communicational, and dynamic. As an illustration, it defines and studies a social pattern called *Booking*. Social patterns are design patterns focusing on social and intentional aspects that are recurrent in multi-agent and cooperative systems. In particular, the structures are inspired by the federated patterns introduced in (Hayden et al. 1999, Kolp et al. 2001). They are classified in two categories. The *Pair* patterns describe direct interactions between negotiating agents. Such patterns are the booking, the call-for-proposal, the subscription. The *Mediation* patterns feature intermediate agents that help other agents to obtain some agreement about an exchange of services. These are the monitor, the broker, the matchmaker, the mediator, the embassy, the wrapper.

3.1 Modeling Social Patterns

(Do et al. 2003a) proposes a conceptual framework based on five complementary modeling dimensions, to introspect social patterns. Each dimension reflects a particular aspect of a MAS architecture, as follows.

- The *social dimension* identifies the relevant agents in the system and their intentional interdependencies.
- The *intentional dimension* identifies and formalizes the services provided by agents to realize the intentions identified by the social dimension, independently of the plans that implement those services. This dimension answers the question: "What does each service do?"
- The *structural dimension* operationalizes the services identified by the intentional dimension in terms of agent-oriented concepts like beliefs, events, plans, and their relationships. This dimension answers the question: "How is each service operationalized?"
- The *communicational dimension* models the temporal exchange of events between agents.
- The *dynamic dimension* models the synchronization mechanisms between events and plans.

The social and the intentional dimensions are specific to MAS. The last three dimensions (structural, communicational, and dynamic) of the architecture are also relevant for traditional (non-agent) systems, but they have been adapted and extended them with agent-oriented concepts.

The rest of the section details the dimensions. Each of them will be illustrated through the Booking pattern.

3.1.1 Social Dimension

The social dimension specifies a number of agents interacting with each other and their intentional interdependencies using the i* social model.

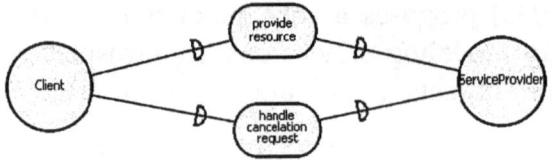

Figure 5. Social Diagram - Booking

Figure 5 shows the social-dimension diagram for the Booking pattern. The Client depends on the Service Provider to provide resources and cancel existing reservations.

3.1.2 Intentional Dimension

While the social dimension focuses on interdependencies between agents, the intentional view aims to model the rationale of an agent. In other words, it is concerned with the identification of *services* possessed by agents and available to achieve the intentions identified by the social dimension. Each service belongs to one agent. Service definitions can be formalized as intentions that describe the fulfillment condition of the service. The collection of services of an agent defines its behavior.

Table 2 lists several services of the Booking pattern with an informal definition.

The `FindPotentialSP` service allows a client to find service providers that can provide a requested resource. This service is fulfilled either by the `FindSP` or the `FindSPWithMM` services (the client finds potential service providers based on its own knowledge or via a matchmaker). The request is then sent to the potential service providers through the `SendReservationRequest` service. When receiving such a request, a service provider queries its database using the `QueryResourceAvailability` service and then answers the client through the `SendReservationDecision` service. Three alternative answers are possible: (1) the resource cannot be supplied; (2) the resource can be supplied but not for the moment: a waiting list option is proposed to the client; (3) the resource can be provided and an offer is made to the client. The client processes these answers with services `RecordSPRefusal`, `RecordWLProposal`, and `RecordOffer`, respectively. The service provider also records its negative answers, for later reminiscence, through its `RecordSPRefusal` service.

Table 2. Some services of the booking pattern

Service Name	Informal Definition	Agent
FindPotentialSP	Find service providers that can provide the requested resource	Client
FindSP	Find service providers that can provide the requested resource by the client's own knowledge	Client
FindSPWithMM	Find service providers that can provide the requested resource with the help of a matchmaker	Client
SendReservation Request	Send a booking request to the potential service providers	Client
QueryResource Availability	Query the database for information about the availability of the requested resource	Service Provider
SendReservation Decision	Send answer to client	Service Provider
RecordSPRefusal	Record a negative answer from the service provider	Client
RecordWLProposal	Record a proposal for a waiting list option	Client
RecordOffer	Record an offer for a resource	Client
RecordSPRefusal	Record a negative answer	Service Provider

3.1.3 Structural Dimension

Unlike the intentional dimension that answers the question "What does each service do?", the structural dimension answers the question "How is each service operationalized?". Services will be operationalized as *plans*, that is, sequences of actions.

As already said, a MAS is a social organization of intentional autonomous software entities called *agents*. The knowledge an agent has (about itself or about the environment to which it belongs) is stored in its *beliefs*. An agent can act in respond to the *events* it handles through its plans. A plan in turn, is used by the agent to read or modify its beliefs and send (or post) events to other agents (or to itself).

The structural dimension is modeled using a UML style class diagram extended for MAS engineering.

The required agent concepts extending the class diagram model are defined below. The structural dimension of the Booking pattern illustrates them.

Structural concepts Figure 6 depicts the concepts and their relationships, needed to build the structural dimension. Each concept defines a common template for classes of concrete MAS (for example, `Agent` in Figure 6 is a template for agent class `ServiceProvider` of Figure 7).

A **Belief** describes the knowledge that an agent has about itself and its environment. A belief is a tuple composed of a key and value fields.

Events describe stimuli, emitted by agents or automatically generated, in response to which the agents must take action. As shown by Figure 6, the structure of an event is composed of three parts: declaration of the attributes of the event, declaration of the methods to create the event, declaration of the beliefs and the condition used for an automatic event. The third part only appears for automatic events. Events can be described along three dimensions:

- *External / Internal* event: event that the agent sends to other agents / event that the agent posts to itself. This property is captured by the *scope* attribute.
- *Normal / BDI* event: An agent has some alternative plans in response to a BDI event and only one plan in response to a normal event. Whenever an event occurs, the agent initiates a plan to handle it. If the plan execution fails and if the event is a normal event then the event is said to have failed. If the event is a BDI event, a set of plans can be selected for execution and these are attempted in turn, in order to try to achieve successful plan execution. If all the plans in the set of selected plans have failed, the event is also said to have failed. The event type is captured by the *type* attribute.
- *Automatic / Not Automatic* event: an automatic event is automatically created when certain belief states arise. The *create when* statement specifies the logical condition which must arise for the event to be automatically created. The states of the beliefs that are defined by *use belief* is monitored to determine when to automatically create an event.

A **Plan** describes a sequence of actions that an agent can take when an event occurs. As shown by Figure 6, plans are structured in three parts: the Event part, the Belief part, and the Method part. The Event part declares events that the plan `handles` (i.e., events that trigger the execution of the plan) and events that the plan produces. The latter can be either `posted` (i.e., sent by an agent only to itself) or `sent` (i.e., sent to other agents). The Belief part declares beliefs that the plan `reads` and those that it `modifies`. The Method part describes the plan itself, that is, the actions performed when the plan is executed.

The **Agent** concept defines the behavior of an agent, including the type of events it posts and sends, the plans it uses to respond to the events it handles, and the beliefs it has as its knowledge.

The agent structure is composed of five parts: declaration of its attribute, declaration of events it posts (or sends) explicitly (i.e., without using its plans), declaration of its plans, declaration of its beliefs and declaration of its methods.

The beliefs of an agent can be of type *private*, *agent*, or *global* depending on how the agent can access beliefs data. A *private* access means that the agent can read and modify independent of all other agents, even those of the same agent class; an *agent* access means that the agent has shared read-only access to the belief, but only with other agents of the same class; a *global* access means that the agent has shared read-only access to the belief with all other agents.

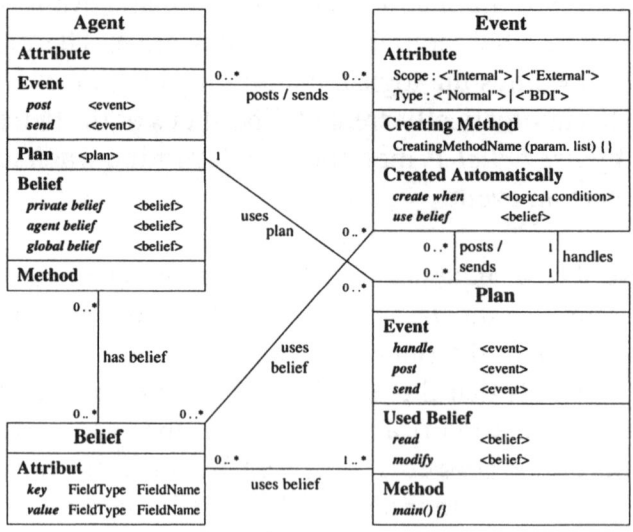

Figure 6. Structural Diagram Template

Booking Pattern Structural Model As an example, Figure 7 depicts the Booking pattern components. Due to lack of space, each construct described earlier is illustrated only through one Booking pattern component. Each of these components can be considered an instantiation of the (corresponding) template defined in the previous section.

ServiceProvider is one of the two agents composing the Booking pattern. It uses plans such as `QueryResourceAvailability` and `SendReservationDecision`. The plan name is also the name of the service that it operationalizes.

The global belief `ResourceType` is used to store the resource type and its descriptions (e.g., for an air ticket booking system, the resource type could be `economic / first class / business class`; for a hotel room booking system the resource type could be `single room / double room / ...`). This belief is declared as global since it will be used by both the client agent and the service provider agent. The other beliefs are declared as private since the service provider is the only agent that can manipulate them.

The constructor *method* allows to give a name to a service provider agent when created. This method may call other methods, for example `loadBK_SP()`, to initiate agent beliefs data.

SendReservationDecision is one of the Booking pattern plans the service provider uses to answer the client: the `SPRefusedExternal` event is sent when the answer is negative, `SPResourceProposal` when there is some resource available for the client, or `SPWLProposal` (Service Provider Waiting List Proposal) when the service provider is able to provide the type of resource asked by the client but not at the moment of the request since all these kind of resources are reserved. This plan is executed when the `AvailabilityQueried` event (containing the information about the availability of the resource type required by the client) occurs.

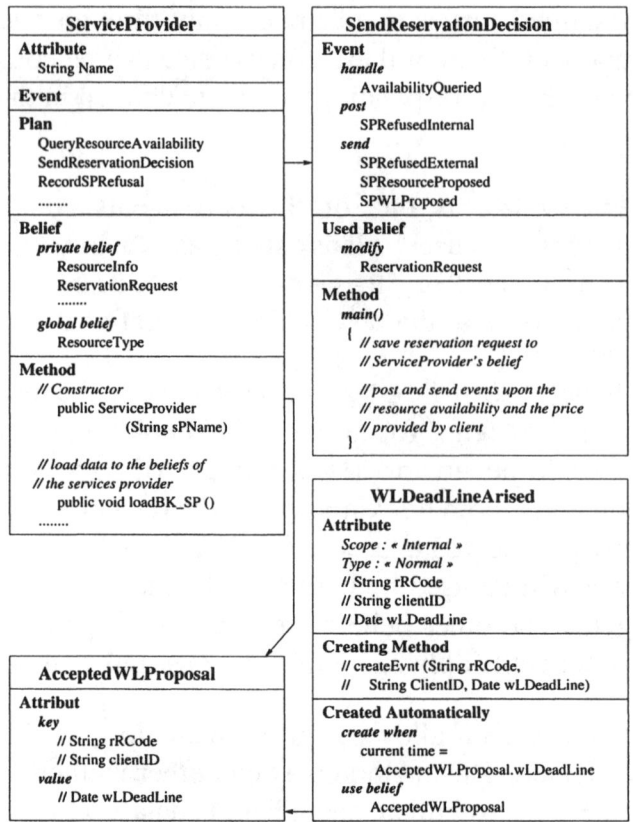

Figure 7. Structural Diagram - Booking

This plan also modifies ReservationRequest, the service provider's belief storing the client's reservation request before the service provider sends (or posts) his answer.

AcceptedWLProposal is one of the service provider's beliefs used to store the client's accepted waiting list proposal. The reservation request code rRCode and the clientID form the belief key. The reservationInfoCode attribute that contains the correspondent code of the resource type requested by the client, and the wLDeadLine that contains the time-out before which the service provider must send a *resource not available* message to the client if no resource is proposed, are declared as value fields.

WLDeadlineArised is an event that is posted automatically whenever the time-out wLDeadLine (of the AcceptedWL-Proposal belief) is reached. It will then invoke a plan to inform the client that the resource is not available.

3.1.4 Communication Dimension

Agents interact with each other by exchanging events. The communicational dimension models, in a temporal manner, events exchanged in the system. The sequence diagram model proposed in AUML (Bauer et al. 2001) is extended: *agent_name/Role:pattern_name* expresses the role (*pattern_name*) of the agent (*agent_name*) in the pattern; the arrows are labeled with the name of the exchanged events.

Figure 8 shows the sequence diagram for our Booking pattern. The client (customer1) sends a reservation request (ReservationRequestSent) containing the characteristics (place, room, etc.) of the resource it wishes to obtain from service providers. The service provider may alternatively answer with a denial (SPRefusedExternal), a waiting list proposal (SPWLProposed) or an approval, i.e, a resource proposal when there exists such a resource that satisfies the characteristics the client sent.

In the case of a waiting list proposal (SPWLProposed), when the client accepts it (AcceptedWLProposalSent), it sends a waiting list time-out (wLDeadLine) to the service provider. Before reaching the time-out, the service provider must send a refusal to the client, in the case it does not find an available slot in the waiting list (ResourceNotAvailableMessageSent), or propose a resource to the client. In the later case, the interaction continues as in the case that the resource proposal is sent to the client.

A resource that is not available becomes available when some client (`customer2` in Figure 8) cancels its reservation.

3.1.5 Dynamic Dimension

As described earlier, a plan can be invoked by an event that it handles and create new events. Relationships between plans and events can rapidly become complex. To cope with this problem, the synchronization and the relationships between plans and events are modeled with activity diagrams extended for agent-oriented systems. These diagrams specify the events that are created in parallel, the conditions under which an event is created, which plan handles which event, and so on.

An internal event is represented by a dashed arrow and an external

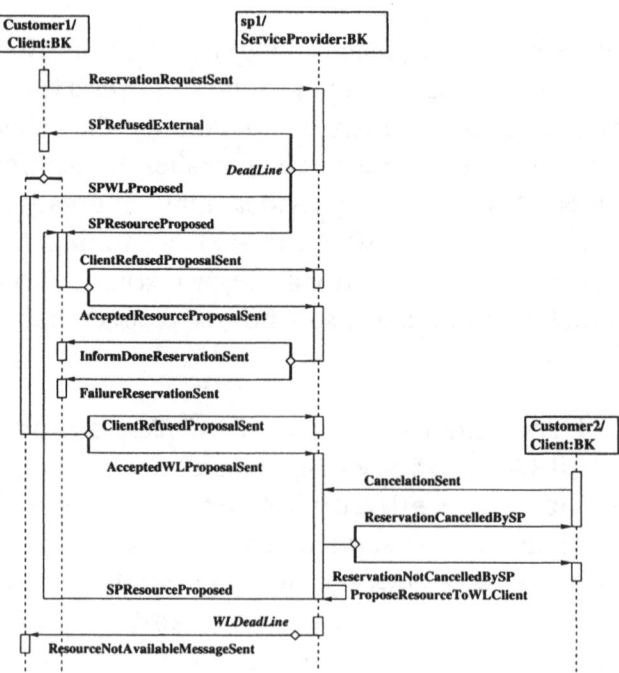

Figure 8. Communication Diagram - Booking

event by a solid arrow. As mentioned earlier, a BDI event may be handled by alternative plans. They are enclosed in a round-corner box. A plan is represented by a lozenge shape. Synchronization and branching are represented as usual.

Four activity diagrams actually model the dynamic dimension of the Booking pattern; due to lack of space, only one is presented, depicted by Figure 9. It models the flow of control from the emission of a reservation request to the reception by the client of the answer from the service provider (refusal, resource proposal, or waiting list proposal). Two swimlanes, one for each agent of the Booking pattern, compose the diagram.

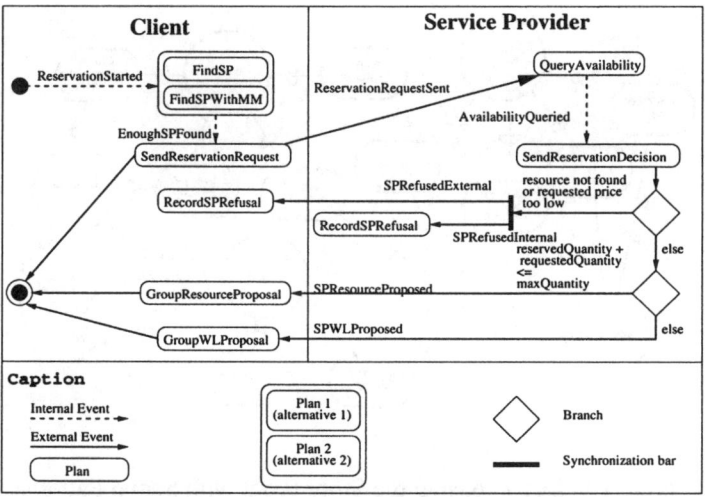

Figure 9. Dynamic Diagram - Booking

MaxPrice stores the maximum value that the client can afford to obtain a resource unit; quantity stores the number of resource units the client wishes to book; reservedQuantity and maxQuantity respectively store the actual quantity of resource units that are reserved and the maximum number of resource units that the service provider can provide.

At a lower level, each plan could also be modeled by an activity diagram for further detail if necessary.

3.2 Applying the Patterns

Figure 10 shows a possible use of the social patterns in the e-business system of Figure 3. In particular, it shows how to realize the dependencies *Manage catalogue browsing*, *Update Information* and *Product Information* from the point of view of the Store Front. The Store Front and the dependencies are decomposed into a combination of social patterns involving agents, pattern agents, subgoals and subtasks.

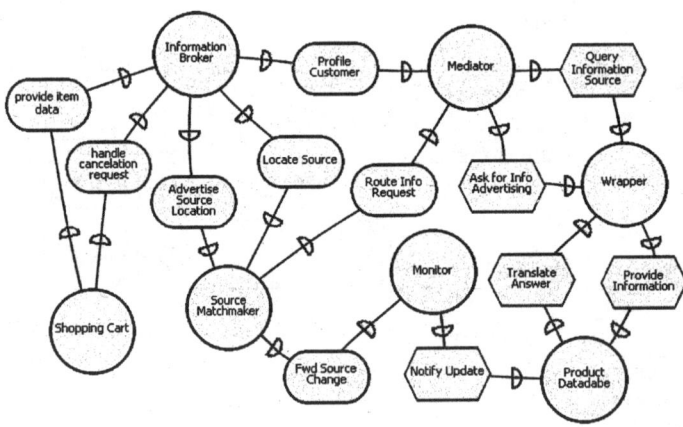

Figure 10. Decomposing the Store Front with Social Patterns

The booking pattern is applied between the *Shopping Cart* and the *Information Broker* to reserve available items. The broker pattern is applied to the *Information Broker*, which satisfies the Shopping Cart's requests of information by accessing the *Product Database*. The *Source Matchmaker* applies the matchmaker pattern to locate the appropriate source for the *Information Broker*, and the monitor pattern is used to check any possible change in the *Product Database*. Finally, the mediator pattern is applied to dispatch the interactions between the *Information Broker*, the *Source Matchmaker*, and the *Wrapper*, while the wrapper pattern makes the interaction between the *Information Broker* and the *Product Database* possible. Of course, other patterns can be applied. For instance, the call-for-proposal pattern could have been used to select a wrapper to which delegate the interaction with the *Product Database*, or the embassy to route the request of a wrapper to the *Product Database*.

4 Conclusion

Styles and patterns ease the task developers describing system architectures.

Organizational architecture styles and social design patterns must be used for designing multi-agent systems architectures. Since the fundamental concepts of this kind of systems are intentional and organizational, rather than implementation-oriented, they can be viewed as social organizations composed of autonomous and proactive agents that interact to achieve common or private goals.

This chapter has detailed and adapted the structure-in-5, a well-understood organizational style used by organization theorists and the joint-venture, an organizational style used to describe cooperative strategies in the business world.

The organizational styles constitute an architectural macro level. The micro level focuses on the notion of social design patterns for agents. Many existing patterns can be incorporated into system architecture, such as those identified in (Gamma et al. 1995, Buschmann et al. 1996). For agent inherent characteristics, patterns for distributed, and open architectures like the broker, matchmaker, embassy, mediator, wrapper, mediator are more appropriate (Hayden et al. 1999). They detail how goals and dependencies identified in an organizational style can be refined and achieved.

The chapter has introduced a design framework to formalize the *code of ethics* for social patterns – MAS design patterns inspired by social and intentional characteristics –, answering the question: what can one expect from a broker, mediator, embassy, etc.? The framework is used to:

- define social patterns and answer the above question according to five modeling dimensions: social, intentional, structural, communicational and dynamic.
- drive the design of the details of a MAS organizational architecture in terms of these social patterns.

References

Aridor, Y. and Lange, D.B. (1998), "Agent Design Patterns: Elements of Agent Application Design," *Proc. of the 2nd Int. Conf. on Autonomous Agents* (Agents'98), pp. 108-115, St Paul, Minneapolis, USA.

Bass, L., Clements, P., and Kazman, R. (1998), *Software Architecture in Practice*, Addison-Wesley.

Bauer, B., Muller, J.P., and Odell, J. (2001) "Agent UML: A Formalism for Specifying Multiagent Interaction," *Proc. of the 1st Int. Workshop on Agent-Oriented Software Engineering* (AOSE'00), pp. 121-140, Limerick, Ireland.

Bonati, P.A., Kraus, S., Salinas, J., and Subrahmanian, V.S. (1998), "Data-security in heterogeneous agent systems," *Cooperative Information Agents*, pp. 290-305. Springer Verlag.

Buschmann, F., Meunier, R., Rohnert, H., Sommerlad, P., and Stal, M. (1996), *Pattern-Oriented Software Architecture – a System of Patterns*. John Wiley & Sons.

Chess, D. M. (1998), "Security Issues in Mobile Code Systems," in Vigna, G. (ed.), *Mobile Agents and Security*, pp. 1-14. Springer-Verlag.

Deugo, D., Oppacher, F., Kuester, J., and Otte, I.V. (1999), "Patterns as a Means for Intelligent Software Engineering," *Proc. of the Int. Conf. of Artificial Intelligence* (IC-AI'99), vol. II, pp. 605-611, CSRA.

Do, T.T., Faulkner, S., and Kolp, M. (2003), "Organizational Multi-Agent Architectures for Information Systems," *Proc. of the 5th Int. Conf. on Enterprise Information Systems*, (ICEIS'03), vol. IV, pp. 89-96, Angers, France.

Do, T.T., Kolp, M., and Pirotte, A. (2003a), "Social Patterns for Designing Multi-Agent Systems," *Proc. of the 15th Int. Conf. on Software Engineering and Knowledge Engineering* (SEKE'03), pp. 103-110, San Fransisco, USA.

Dussauge, P. and Garrette, B. (1999), *Cooperative Strategy: Competing Successfully Through Strategic Alliances*, Wiley and Sons.

Fox, M. (1981), "An Organizational View of Distributed systems," *Transactions on Systems, Man and Cybernetics*, 11(1), pp. 70-80.

Gamma, E., Helm, R., Johnson, J., and Vlissides, J. (1995), *Design Patterns: Elements of Reusable Object-Oriented Software*, Addison-Wesley.

Hayden, S., Carrick, C., and Yang, Q. (1999), "Architectural Design Patterns for Multiagent Coordination," *Proc. of the 3rd Int. Conf. on Agent Systems*, pp. 412-413, Seattle, USA.

Horling, B., Lesser, V., Vincent, R., Bazzan, A., and Xuan, P. (1999), "Diagnosis as an integral part of multi-agent adaptability," *Technical Report UM-CS-1999-003*, University of Massachusetts, USA.

Kolp, M., Giorgini, P., and Mylopoulos, J. (2001), "A Goal-Based Organizational Perspective on Multi-Agents Architectures," *Proc. of the 8th Int. Workshop on Intelligent Agents: Agent Theories, Architectures, and Languages* (ATAL'01), pp. 128-140, Seattle, USA.

Mintzberg, H. (1992), *Structure in Fives: Designing Effective Organizations*, Prentice-Hall.

Morabito, J., Sack, I., and Bhate, A. (1999), *Organization Modeling: Innovative Architectures for the 21st Century*, Prentice Hall.

Pree, W. (1994), *Design Pattern for Object Oriented Development*, Addison Wesley.

Scott, W.R. (1998), *Organizations: Rational, Natural, and Open Systems*, Prentice Hall.

Shaw, M. and Garlan, D. (1996), *Software Architecture: Perspectives on an Emerging Discipline*, Prentice Hall.

Yoshino, M.Y. and Srinivasa Rangan, U. (1995), *Strategic Alliances: an Entrepreneurial Approach to Globalization*, Harvard Business School Press.

Yu, E. (1995), *Modeling Strategic Relationships for Process Reengineering*, PhD thesis, University of Toronto, Department of Computer Science, Canada.

Zambonelli, F., Jennings, N.R., and Wooldridge, M. (2000), "Organizational abstractions for the analysis and design of multi-agent systems," *Proc. of the 1st Intern. Workshop on Agent-Oriented Software Engineering*, pp. 127-141, Limerick, Ireland.

Zambonelli, F., Jennings, N.R., and Wooldridge, M. (2003), "Developing multiagent systems: the Gaia Methodology," *ACM Trans on Software Engineering and Methodology*, 12(3), pp. 317-370.

Chapter 5

Design and Behavior of a Massive Organization of Agents

Alain Cardon

1 Introduction

Researches in the domain of multi-agent systems have currently a very important development. The plastic, dynamic and distributed characters of the these systems permit several advantages: in simulation, they can easily express the behavior of real systems for which equational models are insufficient and, in effective working, they allow to produce some typically adaptive behaviors. Multi-agent systems are used in varied domains where they present an interesting alternative to the classical approaches. But there exists a limitation to their utilization: one doesn't know how easily conceive or cleverly control the behavior of multi-agent systems of very large size, containing about ten or hundred thousand agent (Huhns & Singh 1998).

We are interested in systems whose organization is really complex and that are typically adaptive in their environment (Le Moigne 1990, Cardon 2000). Such systems are, for example, the ones under the living systems paradigm, as autonomous robots with intentions, or simulators of ecosystem of very large size developing scalability, or again systems generating sense with an artificial consciousness. We think that massive multi-agent systems (MMAS) are especially adequate to represent such complex phenomena. Agents, in these systems, are components at the conception level that have a simple structure, in order to be constructed easily and generated by automatic reproduction (Cardon - Vacher 2000). They have, structurally, a very limited knowledge and they can reconsider their goals according to the knowledge they manipulate with their communications. MMAS are composed of hundred to thousands of weak agents.

In massive organization, agents have been created while using a specific method of agentification that permitted to reify the multiple functionalities of which we wanted to initially endow the system. Agents endeavour to reach their goals while organizing themselves at the best, while using their capacities of communication via their acquaintance networks, to form some multiple groups: they exchange messages between them according to the language of which they are endowed. They achieve, by their actions and their pro-actions, the activation of the organization in the whole. The system has a general behavior determined, on one hand, by the action of interface agents dedicated to the action on the system environment and, on the other hand, by the activity of the strictly internal agents in the system.

The question is then to define, to conceive and principally to control the behavioral characters of the organization of agents, and this during its working. We are going to present two examples before presenting a solution for massive multi-agent systems.

The first example of well-controlled system composed of very numerous simple elements is the case of the particulate systems. It is about physical system composed of very numerous elementary particles. One knows how to perfectly define, in the classical mechanics, the notion of trajectory of a particle, expressed with an equational form. When the particles are without interaction, the behavior of the system is then entirely deterministic and given by all the trajectories of particles, that are well calculable. But when particles are in interaction and when these interactions are permanent, the system has a complex behavior that is with difficulty definable with equations. We are leaded to adopt a statistical representation of the behavior of the system, whose states are neither anymore steady nor easily predictable (Prigogine 1996).

The second example of reference, in the domain of the computer science this time, is the one of large object oriented systems where the control is made during the realization and the running of the system. It is current to conceive and to construct systems having thousands of objects. Object oriented systems are seen as extended and constraint software interpretation of particulate systems, but with permanent interactions. An object is a more complex entity that an elementary particle, that interact with other objects, whose interactions are entirely and definitely controlled in the diagrams of interactions, at the time of the construction and before working (Muller 1997). In that way, one conceives object oriented systems according to a very controlled gait so that their working is entirely controllable.

Finally, multi-agent systems (MAS) can be seen like a conceptual rupture with object oriented systems in the sense that agents are more complex than object and are especially autonomous and proactive (Wooldridge & al. 1994). Interactions between agents, that are fundamental in the working of a multi-agent system, cannot be specified completely at the time of the construction. These interactions will evolve in time and we get, by some aspects, an agent organization similar a dynamic and unsteady particulate system. So, the massive MAS will be considered as very dynamic and complex system with a strong character of autonomy, according to the actions and communications between agents.

But a multi-agent system has the particularity to unify in its architecture the knowledge and the processing of the knowledge. The processing of this distributed knowledge will be expressed as the state of a morphological space, depicting actions of agents in a geometrical way. This expression of re-organization of MAS will be computed using a specific organization of agents, the morphological organization. Then, we could use the analysis of this morphology to modify in real-time the actions in MAS, realizing the self-adaptability of the system on a systemic loop.

There is a rupture between the classical mechanics of the steady systems and the one of systems evolving far of the equilibrium. There is also a rupture between the classical software systems foreseen entirely and constructed to achieve some precise tasks and the unsteady software systems, adjusting to their environment of use and even to their own characters of working. There is a rupture between the ideal mechanics that produces things of use entirely mastered and the living systems, natural or artificial, that evolve in their length and for their account.

We successively present the characters of the unsteady particulate systems, the ones of the control in the design and working of object oriented systems, and massive multi-agent systems. We propose an incremental agentification method with self-learning, and a means to have a control of the system by itself, in the behavior of its internal organization of agents, by the definition of a morphological space of which we specify characters and the agent interpretation at a time. We propose a notion of entropy for the system, and finally a state equation expressing the behavior.

2 The Systems with Particles

The first example of a system with a complex behavior will be chosen in physics. Let's consider an enclosed system formed of a large number of particles. Every particle is a certain ideal elementary material entity. It is, in the setting of the classic mechanics, entirely definite by its position p and its moment $q = mV$ product of its mass by its speed. The knowledge of the particle comes down to those of position and moment therefore. Equations of the movement, that are the very classical equations of Newton, give as solution the position and the moment of a particle from its initial conditions, taking into account force to which is submitted the particle. The core notion, in classic mechanics, is the one of trajectory that defines the exact situation of all material particles during the time.

In the equations of calculation of the trajectory, the variable expressing time plays a reversible role. We can foresee the position and the moment of the particle thus to all ulterior and previous instants of the time, while changing t into - t. We can determine the state and the place where are the particles at each moment of the future or of the past therefore. We characterize a particulate system in which the behavior of its particles is given by equations in which the time is reversible, a *deterministic* system.

With this approach of particulate systems based on the determination of trajectories, one goes in fact to interest to states that will be especially steady: the so-called *equilibrium* states where characters of the system don't change even though some particles are in movement. For such a system, an equilibrium state is a state that is necessarily reached while leaving from initial conditions valued in a certain domain: if the initial conditions change a little in this domain, it is always this same equilibrium state that will be reached. This equilibrium state will be therefore meaningful of the system. This case is in fact ideal and doesn't seem to correspond to the situation of the living systems and not for all of the particulate ones.

The space permitting to represent the behavior of the particulate system is the so-called *space of phases*, whose measurements are N times the three classic coordinates of space and the quantity of movement, if there are N particles. One represents the trajectory of the whole of particles that is the system, like a point in this space.

But one can also represent in the space of phases a set of points of which each is not anymore the precision of trajectories of all particles given one instant, but the probability of the of particle in this point. One represents a cloud of points of which each is the density probability of the density of particles for coordinates of space and moment thus. It is the Gibbs model.

One gets two ways for the representation of the state of the system therefore: one by the individual trajectories of all particles and one other by the probability of the particles in each point of the phase space. And it could have seemed obvious that these two representations provided the same description and that they were equivalents. But this is not the case for all particulate systems because every system cannot be characterized steady how.

Let's consider a system formed of a large set of such material particles while knowing that these particles can now have collisions and that these collisions are permanent: they are not temporary but they characterize the working of the system. Trajectories of particles are then not independent and the problem, considering the important number of particles becomes very complex because one is not able to mathematically solve the equations of trajectories!

2.1 The Unsteady Systems

In fact, when the system is composed of particles in permanent interactions, it is impossible to determine its global behavior from the only trajectories of the particles. In this case, that to be-to-say whose movements are correlated, H. Poincaré showed (Poincaré 1893) that the equations of the movement are not anymore integrable, that they cannot produce a solution. This result, clearly negative, was a long time ignored because it puts out a fundamental problem: it is impossible to based on the individual trajectory particle notion to describe the behavior of a system composed of numerous particles in permanent interactions (in fact more than two).

The systems composed of particles in permanent interactions are categorized as *unsteady* systems (Prigogine 1996). It will remain then, to define and to predict the states of the system, a probabilistic description of the position of the particle in the space of phases, while using models with suitable operators.

But the distinction between the deterministic behavior provided at the level of individual trajectories of particles and the global probabilistic behavior goes farther. Ilya Prigogine (Prigogine ref. mentioned) showed that for systems constituted of numerous particles in interaction, the probabilistic description produces states that the individual trajectory notion cannot provide. The probabilistic description, by its global character, contains additional information that the individual trajectory notion doesn't have. He has shown in fact that there is not anymore, for the unsteady systems, equivalence between the two descriptions, the individual based on trajectories and the global based on probabilities. Only the probabilistic description permits to represent the non-deterministic characters of the behavior of such an unsteady system.

In fact, the unsteady systems have a behavior that takes into account the arrow of time. The time, for their working, is oriented toward the future, it can't be consider as reversible and their behavior can't be described by equations in which the time plays a symmetrical role, like the equations of the classical mechanics.

The unsteady systems that are in fact very currents in the real world, have a particular behavior. For a steady system, all the fluctuations can be considered as accidental and few meaningful for its behavior, because the system comes quickly and necessarily after a certain number of oscillations, to its equilibrium state. The behavior, and the control, of unsteady systems are quite different. Fluctuations are not accidental, but play a central role. The state of stabilization, that is necessarily fragile, depends of fluctuations very strongly and cannot be foreseen from the only initial conditions.

We have necessary to take into account the initial fluctuations that destabilized it, the organized state of this initial state, the state of its entities in interrelations and their capacities of communication. For a physical system composed of particles in interaction, the behavior won't be determined anymore by the individual equations of particles but by other equations in the space of phases, that are global and defined in probability: the behavior is not totally controllable. The problem of the prediction of the future state of the system is then more complex.

2.2 Operators of Determination of the Behavior for an Unsteady System

The classical equations of the movement specify the trajectories of particles while defining speeds and accelerations and while joining force to the acceleration. The central value of this formulation is then the Hamiltonian H (p, q) that represents the energy of the system, which is dependent of the positions of particles, noted p and of their moments noted q.

According to H. Poincaré (Poincaré op. mentioned), the dynamic systems are expressed with an extended Hamiltonian:

$$H = H_0 (p) + \lambda V (q),$$

where $H = H_0 (p)$ is an integrable Hamiltonian describing the kinetic energy of the system and $\lambda V (q)$ is a term that represents the potential energy and interactions between particles.

H. Poincaré pointed out that in general interactions could not be removed of equations by change of variables and that the system was then no integrable: it became impossible to calculate trajectories. The fact of no integrability is owed to resonance between particles. A degree of liberty is mathematically expressed by a frequency. Every point of the space of phases is characterized by a linear combination of frequencies that annuls for certain values of these frequencies. The calculation of trajectories room this linear combination as denominator and we get the so-called points of resonance thus, where the trajectory is not defined because the terms become infinite: the dynamic system is then no integrable. Every interaction generates others relations developed by transitivity with other particles, creating couplings and dragging a flux of relations. These couplings, putting in action numerous particles, are unsteady because they alter themselves by association and transformation of local couplings. The system has a fluctuating behavior, with variations in its organization expressed by numerous bifurcations in its global space state.

The classical description represents the characters of particles by variables: the knowledge of particles is expressed with the variables on the equations, whose values represent the observable states directly. It is a simple and simplifying vision of the reality. We need in fact to mediate this direct interpretation stance while introducing an operator's notion. An operator expresses the behavior of a system from a certain change of state and some of its values, called the proper values, correspond to the effectively observed characters.

The statistical approach of the behavior of a particulate system consists in a distribution of probability ρ (p, q, t) on the system, taking into account the states of all particles that compose it. The operator of evolution is the Liouville's operator and the equation of evolution of the system is then:

$$i \frac{\partial \rho}{\partial t} = L \rho$$

The formal solution is $\rho(t) = \rho_0 e^{(-i L t)}$. The notion of trajectory is then not primitive but deducted of the solution of the Liouville's equation using this operator.

Liouville's operator is about a hermitic operator that provides some real proper values in Hilbert space where it takes its solutions. This formulation produces a deterministic behavior therefore for the system and I. Prigogine proposed (in Prigogine op. mentioned), while spreading the space of Hilbert, to associate to every proper value a complex oscillating term of the form:

$$L_n = \omega_n - i\,\gamma_n$$

permitting to break the temporal symmetry in the equation of evolution. Thus, for t > 0 (the future) the contribution of the probability decreases whereas for t < 0 (the past) it increases: the knowledge of the past is bigger than the one of the future.

I. Prigogine developed a model based on the strong coupling of particles and no more on their trajectories. He has defined a probabilistic representation of systems far of the equilibrium, a dynamics integrating thermodynamics for the unsteady systems. He introduced an approach therefore by operators in classical mechanics, as it is the case in quantum mechanics. Let's recall that the quantum mechanics is based on the notion of wave function associated to particles and on the absence of concept of trajectory.

The unsteady systems in fact seem to self-organize themselves from multiple forks defining changes of order in their organization. Their behavior go through by multiple forks showing a caesura that is spatial either temporal, in their behavioral evolution. Their behavior is not predictable with certainty.

This capacity of self-organization is very precisely the one that is going to characterize massive agent organizations. The way one supervises the behavior of the agent organizations cannot be the one very classical and very ideal of the computed systems whose behavior is specified entirely in advance, located in variables. They are not systems that solve very precise problems of which one validates, still in advance, the reach of the solutions. And we are able to introduce some self-control in the agents that is impossible with particles we can't restrain without deterioration.

Today, one deals with two types of computed systems:
- systems with a very regular behavior, that solve very precisely some well-posed problems,
- unsteady systems, that self-organize themselves in relation to environmental stimuli and that adjust their behavior by the internal self-control of their organization.

Let's recall that the instability doesn't necessarily lead to a chaotic behavior for systems and in particular for computed systems. Computed systems have a kind of behavior where they adjust in very free ways to environmental stimuli, by the transformations of coupling between their components. It is quite what is searched for in social autonomous robotics.

But it remains to find the mean to define and to make control of the behavior of such computed systems. If the object approach permits a total control, it permits to achieve that of systems with an only regular behavior. It is therefore necessary to find new mean for the control of massive multi-agent systems that are not object oriented systems.

3 The Object Approach: a Very Controlled Process of Construction and Run of Systems

We can consider three ways to construct a system:
1. the functional way considers the system as composed of functions, and therefore of variables in interactions. It is about finding these functions, to represent them and to put them into relation,

2. the way of object considers the system as composed of objects encapsulating data and behavior. It is about finding these objects, that will be permanent, and to describe their relations (their links), that will be fixed. Objects speed up automatically, sending messages that trigger methods,
3. the third way will be the one of agent by which one considers the system as composed of many autonomous and cooperative agents. These agents operate on a world of entities, that are not agents, and of which their cooperative, negotiated, evolutionary behavior and their common actions permit to endow the behavior of the system with adaptability.

It is about three successive stages in the history of the construction of systems.

3.1 The Object Approach and the Software Engineering

The most common approach today to construct a complex software that one wants to entirely control, from its design to its exploitation, is the object approach, with a gait of construction leading from the specifications of the system to its realization and validation. We precise the basic and permanent entities, which are the objects. The system will be their aggregation and will assure their functioning and communications in very precise, controlled and synchronized way. Let's note that we don't consider some active objects such mobile objects on networks associated to processes and that are like actors.

Such approach is not neutral. It is located in the particular domain of the well-posed problem resolution and it places the architectural description of the system to a low enough granularity level. This level is very close by the operationality that systematically avoids the treatment of the systems no merely broken up as, for example, are the adaptive systems (Cardon 2000).

The centre of interest of the Software Engineering is located in the programming of large size (Booch 1994). The goal is clearly to rationalize the developments of applications leading to codes of very large size achieved by teams managed in hierarchical order. The way to conceive the system will be in agreement with the characters of the produced system: rationality and control. From a clear problem considered as very complicated because its solution amounts to set in relation very numerous elements, one starts a gait of design while leaning on the definition of every object, the whole composing the system by aggregation (Cardon & al. 2001).

The Software Engineering using an object approach permits to clearly define, in a step of analysis, the elements of the system and their relations. We are able to develop comfortably these elements later, in the step of coding with programs while always refer to these perfectly definite and permanent structural elements, and whose relations have been well specified once for all. In fact, the communications between objects, the sending of messages triggering the activation of methods, are in advance adjusted and the graph of communications between classes is fixed. Once the classes are defined and the objects are instanced, the system can only answer to typical messages and can activate its methods, according to a predefined order of sequencing that the dynamic diagrams have adjusted very well.

3.2 Objects and Object-Oriented Design of Systems

The basic concept, and that will be kept of way all along the different phases of realization of a system in the object approach is, evidently, the object. An object is a concept or an abstraction or again the corresponding of certain physical thing that one endows of a precise definition without ambiguity and of strictly definite limits in its behavior (Rumbaught 1997).

An object is therefore an entity that exists in the time and space of the computed system and whose behavior is essentially reactive. The reactions to solicitations that the object recognizes are fixed, and it has a behavior stated in a contractual way (Meyer 1987). It is an element of design, of programming and of action answering with exactness and reliability to foreseen in advance solicitations.

Object
> *An object is a concept, an abstraction, or the corresponding of something of the real world that we endow of a precise definition without ambiguity and of behavioral limits strictly defined. This is an entity having a permanent identity, a state and a behavior at the same time precise and stable.*

An object has a duplicate role. On the one hand, it must represent like a sign represents something in the system in construction (Peirce 1984). On the other hand, it must permit the convenient realization that is the design and the coding, while inserting itself in the good modules. This duplicate role is very ambitious, because it comes to unify at a time by only one completely definite entity, the abstract thing having a basic organizational role in the system, and also the very concrete and operational entity that will be the element of the programmed system. The goal is to pass from the first of these categories, the stated concept, the most reasonably to the other, the object, and as naturally as possible, while controlling gaits of teams that conceive rationally and achieve the system together.

The object can be discerned in an object oriented system likes a multi-facet particle whose trajectory is completely defined: its behavior is fixed once and for all the time. The behavior of the system is then easily expressible: it is given by the whole of behaviors of all objects that we can consider one by one. And we clarify this behavior in advance, in the famous diagrams of classes, of spreading and of dynamic with the UML method (Muller 1997).

3.3 Limits of the Object Approach

The gait of object modelling is found on strong properties of systems that we have to construct: these systems must be functionally well divided into parts and their behavior must be regular and stable. That is with these properties that we determine the basic components for the system, the objects, so that their aggregations constitute the system in whole. We can conversely consider first the global system whose a functional decomposition is going to permit to find, finally, all the basic objects. In the two cases, the system is the compositional sum of its elements, with regard to its structure and especially to its behavior.

The behavior of the object oriented system is the result of the completely mastered arrangement of the behaviors of the parts (the packages), of sub-systems (the families of objects), while going until the objects. Such a behavior is therefore typically deterministic in the sense that it produces, while functioning, what we know that it must produce and that is fixed in the step of design.

Such object oriented systems are in fact *closed systems* (von Bertalanffy 1973). A closed system is defined from its own functionality and no information from the environment can modify its structure, its components. Information is like a well-used tool, not a particular component linking with the environment and generating the behavior of the objects. The system is defined and built out of the environment it is used. Then, on can define a closed system in the following way (von Bertalanffy op. referred):

Closed System
A closed system is made of functional components each of them having a well-defined behavior and the general functionality results of the composition of the peculiar fixed functions generated by its components.

It is clear that this perspective of construction inherits the technician gait of realization of systems in a functional approach and notably of the Forrester's systemic (Thiel 1998). It is perfectly effective to achieve systems with very precise functional characters, as for example systems of real time control, or systems of calculation of trajectories and piloting. The strength of the object paradigm is, in these domains, incontestable.

But the objects becomes some active objects and free themselves of their rigid structuring (Wooldridge 1994, Guessoum - Briot 1999), what puts of another problems of consistency and coherence in systems. The agent's way comes closer.

4 Massive Multi-Agent Systems

Today one takes an interest to some systems that have no strictly regular behavior, that must develop original actions or even creative' ones like autonomous robots. The perspectives asked by the Situated Artificial Intelligence (Steels & al. 1995) and the New Artificial Intelligence (Meyer 1996) are not anymore those definite for the regular and steady systems, whose behavior is specified of definitive manner in the step of design. Such systems are not some clever solvers for a well-set and closed problem class: they are systems that must adjust to very varied environments, thanks to their capacities of structural modifications and theirs multiple original actions. These capacities of modifications and actions will result from the self-promoter faculty of their plastic internal structure.

This means that the organization of components constituting such systems is capable of deep enough re-structuring and that the system works like rather an operational closure of antipoetic systems (Varela 1989). Indeed, the organization of the system components cannot be steady or regular and the research of permanent links definitely fixed will be only very local and temporary. The structure of components will be only generative for the behavior of the system and, besides, the system will be a few more that the sum of its definite components stated in the step of construction (Morin 1986).

The modelling of such system not enters really in the setting, become too narrow, of the analysis and design with objects: their analysis is located above the one of objects that are and remain very close entities of coding. It will be necessary to operate a design at the behavioral level, permitting the variation, the transformation of the components and of their relations, especially permitting to put the system in situation of action in its environment, by the emergence of states adapted to these situations.

We will set up the architecture of these systems on the agent's paradigm and this in a strictly multi-agent point of view.

4.1 Agents

There is, in Computer Science, two ways to define agents (Wooldridge & Jennings 1994). The way characterized as "weak" which considers the agent like a software or material entity that is:

- autonomous,
- with some social faculties,
- reactive and pro-active, to be-to-say that is capable to redefine its goals and to take initiatives without any external solicitation.

The autonomy comes to consider this kind of agent like a process, or a thread that is like an entity capable to execute a lot of instructions on a processor. The social faculties amount to introduce in its behavior a notion of complement by involvement to a more global plan of agent group. The use of weak agent type drags the existence of groups, with realization of common goals, therefore necessarily. The notion of pro-activity is fundamental and sums up to introduce in the agent the notion of intention and modification of its own goals, that is to keep at distance from the automatic reaction message - method of the objects. Characterized thus, the agent is an entity that spreads and enriches in a behavioral way the notion of object or active object.

The second way to consider agents, characterized as "strong" way, considers that the agent has in addition of the three previous characters, related mental forms like those of humans, as valuable knowledge in fact, beliefs, intentions and even emotions. That is an approach that tempts to represent entities possessing a certain consciousness, but with structure made of classical components. The agent is then itself a very complex system and the notion of multi-agent system founding on such agents is, today again, very difficult to achieve in computation (Rao - Georgeff 1991).

A very general definition of an agent can be the one proposed by J. Ferber (in Ferber 1995):

Agent
> *One calls an agent a real or abstract entity capable to act on itself and on its environment, that generates a partial representation of this environment, that can communicate with other agents in case of a multi-agent universe, and whose behavior is the consequence of its observations, knowledge and interactions with other agents.*

This definition puts agent's notion like an action entity defined at the construction step of a system and operating in the setting of an open problem to solve.

In every case, general agent's structure will be the one as expressed in figure 1 below, with four parts more or less developed according to the specificity of the agent:

- a module for knowledge,
- a module for communication,
- a module for behavior,

- a module for action.

Let's notice that the behavior module of such an agent is generally given by an automaton or an Augmented Transition Network (Thorne & al. 1968, Bobrow & Fraser 1968) which are therefore strictly rational and deterministic.

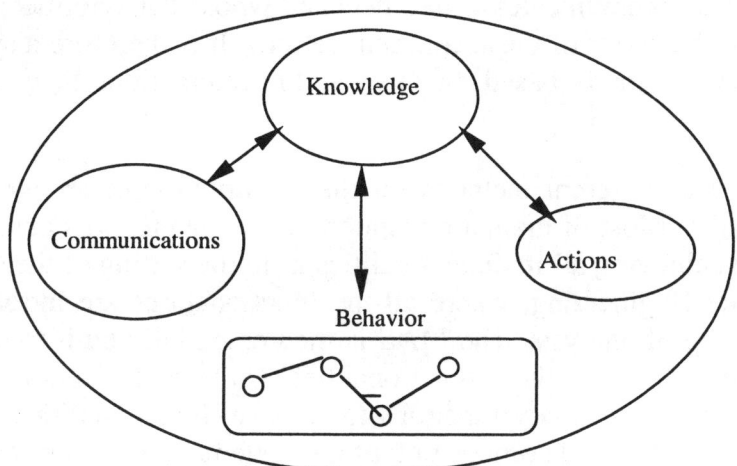

Figure 1. The general structure of an agent

A multi-agent system (MAS) is constituted of a set of agent organizations and is situated in an environment composed of many objects that are not agents that are essentially reactive in a permanent way. This system communicates with its environment by the action of specific agents so-called the interfacing agents. The agents of the MAS use objects of their world as well as actions of the other agents achieve some various actions and unite their actions to define some collective behaviors.

The efficient, visible behavior of MAS will essentially be achieved by the behavior of the agents and will be constructed therefore of distributed manner. This is in the agents, and essentially in the agents that will be distributed the characters of action, the effects the system in whole produces on the environment (Axelrod 1997).

4.2 Nondeterminism and Instability in Massive Multi-Agent Systems

We are going to specify a method of construction of a massive multi-agent organization. Every agent will be considered like a simple enough software entity, that to be-to-say it will be a weak agent that communicates with others and whose behavior takes into account the result of these communications. It is therefore a system whose behavior is based on the variable interactions between its agents.

There exist different methods for the construction of an agent organization. Most of them are inspired of the manner we construct a system achieving all its functionalities as in the setting of the object Software Engineering, where all the functionalities are introduced at the step of analysis. The MAS is then essentially built to satisfy the best as possible these functionalities, while associating to every agent some elementary functions to achieve. Functionalities of the system are the good composition of these elementary functions.

We will choose granularity at a time finer with regard to functionalities, and typically plurivoc, that to be-to-say based on the redundancy and the plurality of characters. We set that the system will have its functionalities distributed in no steady and no consistent agents: at every precise functionality we will associate groups of agents whose cooperative actions should permit to achieve the concerned function, but also its opposite, inverse, contrary, near and similar functions.... This agentification process leads to a certain redundancy and large diversity with regard to faculties of agents. That will be necessary to permit a complex behavior in an organizational way and also to make operate the system strictly by emergence. Groups of agents won't be in anything functional but rather versatile, ambiguous and they will be able to make groups emerging using communications between them, in reifying certain functions rather than others.

Such a multi-agent system is seen therefore like a collective of agents that has social characters of aggregation and regrouping, and a very strong restructuring capability. One will be able to distinguish two parts in the system:

1. one part composed of interfacing agents, communicating with the environment, taking in charge the information coming from the outside and also managing actions on the environment,
2. another internal part with agents taking information from objects of their world and of the interfacing agents, acting and exchanging messages between them.

These two organizations are necessarily strongly interactive and the global organization, by communication between agents, has integrating processes permitting the cooperation and the meaningful emergence of groups (C.f. Fig 2).

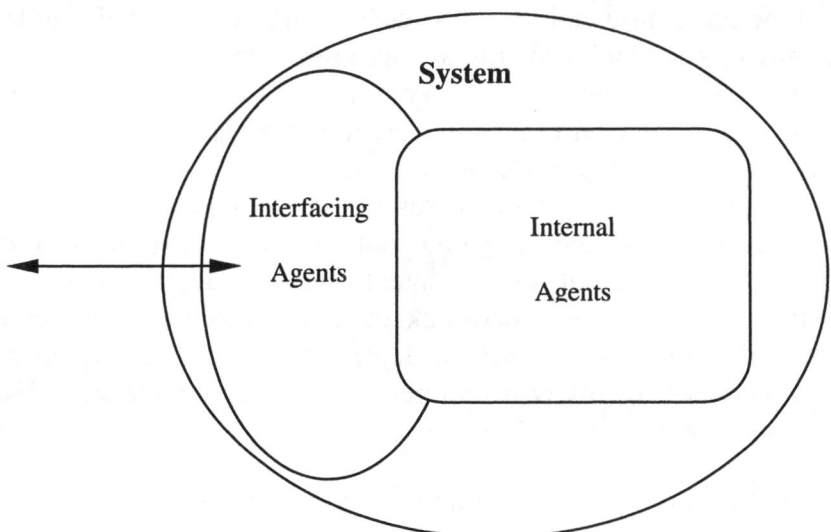

Figure 2. The two main organizations of agents

We claim that a MMAS won't have a deterministic behavior. For each stimulus coming from the environment, agents will be activated while filtering it. Because the agents are constructed using a plurivoc manner, similar, nearly similar and opposite agents ... will be activated for this input. And these agents transitively will activate some others, by their communications.

Each of these agents, is pro-active, it is a process or at least a thread. All these processes will have the same priority to be processed or a priority distributing in few groups of priorities. The scheduler of the system won't be able to order these processes and should choose the one that will be activated at random (in the case of a mono-processor computer). The access to the shared variables, in the case of competitor's process on a distributed system, will undergo the exclusion in critical section, what will lead to satisfy some process took at random rather than others, that go to activate some agents before some others.

However, the activations of agents are not commutative. These activations are contextual and, for each agent, activation depends of the previous activation of the agents known by the acquaintances. We will obtain a running of the system, steps by steps of execution of processes that won't be reversible to the computing sense of the term. The functioning pulls at random for certain options of treatment, like the choice of processes or accesses to the shared variables that will generate the global state of the system undergo with bifurcations. Some states of the agent organization, and while solving the problem of the dead-lock of activations and communications, will be very different and depend on whether a sequence of action between agents will be operated before or after the activity of a some agent.

One will also note that for certain *stimuli* very close by at the signification of information they express, behaviors of the system, after a certain number of steps of work, will be very different: some close by inputs produce distant states therefore. The working of such a system is then no deterministic and sensitive to the initial conditions: it is therefore about an unsteady system. We will propose a means to reduce this instability, while produce the system as self-adaptive, by analogy with the living systems.

4.3 An Agentification Method for the Massive Multi-Agent Systems

In the step of design of a multi-agent system, we need to first define basic categories that belong to the agents, so that the system will have some precise but distributed functionality. It is therefore necessary to define *a priori* all the categories permitting the agents to operate, that is the world where spreads out agents. It is also necessary to specify the limit of the system, what distinguishes itself of its environment and allowing the perception and knowledge of its environment. Finally, it is necessary to define the means of actions and the possible behaviors of the system, its operative functionalities according to different situations. We set that all these categories exist on the form of ontology's, of knowledge systems with multiple returns and that it will be possible to consider them by their minimal features, without having recourse to complex formulations, like rules and meta-rules (Lenat & al.1990, Pitrat 1993).

We set the hypothesis that it is possible to construct a massive multi-agent system from a ***generic organization*** the so-called initial agents. This organization will be defined with few specialization, conceived with very redundant and plurivoc semantic traits, while using some essential categories and while systematically taking into account the opposite and declensions characters of the categories.

It is clear that such a multi-agent system is adjusted the treatment of problems for which we can find numerous complex categories. And these categories have the property they are not reducible a unique hierarchical ontology, otherwise the problem could be treated easily while following other more classic way that are perfectly effective.

Class of problems using massive agent organization
Systems for which a design using massive agent organization is adequate are those founding on ontology, where the determination of all concepts give rise to multiple returns leading from each to the whole, and for which the global behavior can be estimated only by dynamic projections of the ontology called emergences.

They are systems that are complex in an organizational way indeed and that we can describe only in a partial way. So, the classical systems the users want to entirely control are out of this scope.

Numerous agentification methods (Drogoul & Collinot 1998, Ferber – Gutknecht 1998, Demazeau & Muller 1991), distinguish three levels of abstraction in a multi-agent system:

1. the individual agent level,
2. the level of interactions between agents,
3. the level of the organization of the system.

The system will be constructed from these three levels, reifying them in agent organization and using the notion of role. An agent has three roles thus:

- the agent's functional role,
- the relational role,
- the organizational role.

The system distributes these three roles in the activity of the agents that hold them according to their actions and after negotiations with their acquaintances. That defines some groups in which agents have some complementary roles thus, groups that are well structured and coherent.

We propose a different approach. Initially, every entity that can be an entity of action in the system is considered at a very fine level of granularity (a functional entity for robots, a functional entity for an ecosystem, a semantic entity for a system of knowledge of language...) and will be an agent. One can say therefore, following the position of G. Booch (Booch 1994) who declared that to conceive an object oriented system one must consider that all recognized things are objects, that every thing of functional and active character is a weak agent. This agent is few specialized but being able to possibly become. What won't be functional, evolutionary, autonomous, will be an object of the agent world, a permanent thing. While referring to functionalities of very fine granularity, one will get a lot of agents thus, but of weak structural complexity. We will call them the *fundamental agents*.

The roles that these agents will be able to hold, in relation to their functionalities, will be categorized and will be represented systematically by specific agents (role of defender for a system with soccer agents, role of flight for an afraid autonomous robot, role of predator for an agent reifying a living organism in an ecosystem...). These agents should communicate between them, generate others agents ruin some agents and also communicate with the fundamental agents and associate at last to form groups with meaningful behavior. We will call them the *role agent*s.

Behavior of groups will be, in this organization and by the fact of its working, the co-operation of fundamental agents but surrounded with many role agents. These groups will form themselves by the collaboration, the communication and the concordance of determined or few determined actions of fundamental agents on which operate role agents entrusting them, determining them by their activity, giving them an operational sense. This approach is therefore related with notions of so-called *pregnance* flux investing *saliences* of the morphogenetic theory of R. Thom (Thom 1972, Cardon 2000). Role agents will have local specificity (bound to a certain fundamental agent type), either grouped (action of determination on groups of agents), either finally organizational (action of aggregation and cohesion at the level of the different aggregation groups).

In the design step, for each fundamental agent or agent with a role corresponding a feature of a certain category, we will associate, bound by their acquaintances, contrary, opposite, dual, similar, close by... categories reified in a lot of specific agents. It will be possible to define fuzzy closeness between categories and therefore fuzzy gaps between agents (operating at the social behavior level), as for features that they reify (Cardon- Durand 1998).

We get a system containing very numerous weak agents, that is a massively agentified system, where the whole of agents presents very large behavioral possibilities. Such a system is opposed to a functional one in which the roles are assigned strictly in the construction therefore. Here, roles must invest themselves, change, and group themselves producing the behavior of the system and changing the fundamental agents. The advantage of this agentification, in which roles are reified in agents, is double:

- the behavior of each fundamental agent with an active role is explicitly, that is in a visible manner, wrapped with an aggregation of role agents that invested its behavior, by struggles and negotiations between them. And these roles, reified by multiple agents of whom states of activities are variable, represent some functional tendencies always, with dominance, contradiction, systematic no-uniformity, and especially no strict functionalities.
- the fundamental agent groups are formed from intersections of heaps of fundamental agent with their role agents that is with common role agents, assuring the aggregation concretely. The notion of group of agents is thus concrete and expressed by heaps of agents cooperating and communicating together. These heaps are at a time made with fundamental agents and their dominant role agents.

We are going to develop the system in an incremental way and especially no functional. It is about increasing the system whereas it is in functioning or has functioned either and has modified its agent organizations. This leads to a methodology of construction inspired of the one proposed by J.P. Muller, in another domain (Muller 1998).

This generating organization possesses the property to generate a minimal, partially functional only, plurivoc and succinct behavior for the system. These minimal properties are, for example for the behavioral system of an autonomous robot (Dorigo & al. 1998):

- spatial and temporal apprehension of the surrounding space: simple vision of code-bar, reduced plans of circulation, GMT time,
- need to reload its energy: merely to know the way to go to the refill station when it is necessary using a fixed and guided path,
- capacity to distinguish and to identify features in the environment: merely to analyze the vertical and horizontal features defining zones of different shade upon the background,
- capacity to memorize some relative facts in the environment: capacity of a simple memorization, using linear scripts or generating automata,
- need of action and communication: elementary fundamental tendencies to act, to move in a coherent manner (Cardon-Guessoum 2000),
- capacity to reason on representations and to rebuild these representations: faculty to infer on the elementary knowledge and linear scripts.

All retained categories, on the form of set of semantic traits associated altogether and taking into account the fundamental characters and of the functional roles with a metric to appraise gaps, are agentified in one or several initial organizations of agents. And after that, we are going to increase them by an incremental process of agentification.

The steps of the construction of the system are then the following:

1. Definition of the fundamental categories C_f of an ideal reference system, for example the system of intentional representation of the environment for an autonomous robot capable of social acts. The C_f set is very complex in a combinatorial way and is not approved for a development element by element. These categories specify at a time the basic entities, the functional components and the roles they can hold locally or in groups, or in the organization.
2. Definition of an initial family F_i of categories, corresponding to some of the previous ones, strictly included in C_f. We choose to construct the system from a reasonable reduction of C_f. F_i is very approved for development whereas C_f is not.
   Setting up O_A the initial agent organization corresponding to the F_i and made of a first organization of agents reifying the categories. To every category took in account, we have a set of corresponding agents (fundamental agents and their role partners) filtering the feature, its variations (interpretations), dual, contrary, opposite, semantically close by ... categories. We introduce the maximal liberty degrees in the system. We define also an initial organization reifying "objects" of the domain (what is not an agent but a resource or a thing).
3. The O_A organization constitutes, while functioning, a dynamic system $S = O_A \times T$. We get an application named *Effect* of F_i in S then:

   $$\textit{Effect}: \{ F_i \} \rightarrow S,$$

   where the S is the space of observation of the behavior of the system, which is well observable.
   This application links an abstract space composed of families of categories reifying semantic traits in numerous agents with a set of observable situations of the system in use, expressing its behavior. We shall propose in the following a means to observe this behavior effectively, that is to define the *Effect* function. We

can note thus in the image of the application *Effect* (F_i) some permanencies or regularities, some anomalies or dysfunctions. These observations induce to modify some agents, some families of agents, to add new ones, to define some intermediate families between the existing ones, to either add or tamper semantic traits, to enrich new categories in the system or to remove it via agents.

In fact, an agent's addition tampers O_A in its whole, by the communications between agent's acquaintances. The current family F_c (initially F_c is F_i) of reified categories in agents is the result of these modifications:

$F_c \leftarrow (F_c$ modified by reconstruction$)$
$O_A \leftarrow (O_A$ according to the modification of F_c,$)$
the sign $\leftarrow$ meaning "becomes".

4. We can define in the behavior of agents of O_A some avoidable sets that are to reduce the organizational degree of the system or, on the contrary, to release constraints, while playing on indicators of agent groups, network of acquaintances or on the decision structures of agents. We can even introduce sexual genetic in the F_c families (Cardon 2000, Cardon- Vacher 2000).

$F_c \leftarrow (F_c$ modified by local and genetic change of agents of $O_A)$.

5. The construction of the dynamic system leaves therefore from a precise initial state, organized from the F_i categories. We develop the system by successive modifications - insertions of agents, in reifying new categories, with modification of the organizational degree of O_A. Agents are introduced during the working of the system.

6. We compare the system of agents O_A, definite by its current family F_c of categories, with the frame of reference, definite from the general set of categories C_f. If there is not a good adequacy, one returns to the step 3. Otherwise, the system is considered as adequate to its referring and the agentification is finished for the builder.

Let us note that modifications are made after some steps of working of the system, introducing a new element of no regularity: the agentification will depend of the moments of insertions and of the orders of the insertions. This is a process of construction where the builder adapts and forms the system while taking into account of its working. It is clear that the working tampers the current organization, by creation of new agents of internal manner into the system (with cloning or reproduction) and by change of agents. The builder improves, in a process of learning by insertion of agents, the behavior to the system. Insertions are alike a behavioral learning, no direct and no directed.

Thus, with this method, it is possible to test, from a generating organization possessing the minimal valued characters, if this one is regularly increasing, by successive additions, while preserving the consistency of the system.

The important point is then the control of such an organization of agents.

5 Analysis of the Behavior of a Massive Agent Organization: the Control Problem

We need now to determine the behavioral characters of a massive agent organization. The problem will be solvable at the computable level because agents are active software entities. It will be possible, at the conceptual level and at the implementing level, to follow-up agent's organizations in progress by means of evaluation of the meaning of the computations. Let's note that it is impossible to achieve such an observation for particulate systems without change them. Our proposition is therefore similar to the utilization of control fields in physics but without modifying it, except if one wishes that.

5.1 The Characterization of an Agent Organization

The agent that we considered until are simple enough and has three main characters only (Jennings and al. 1998):

- they are some social entities. Agents communicate between them while using a very determined, precise and no ambiguous communication language. They exchanging knowledge, but with the ontological and relative characters to an act of language entirely definite in advance (Searle 1969).
- they are reactive entities. Agents have a perception of their environment, of the state of this environment and of the modifications of this state, leading them to adopt some behavior, that to be-to-say to involve some predefined actions. They are constructed to involve these actions in a reasonable delay, coherent with the dynamics of the environment.
- they are some pro-active entities, that can involve actions without being necessarily driven by explicit external solicitations. Agents have a behavior endowed of some autonomy of decision and action therefore.

Such agents are fundamentally rational ones. They act, at the individual level, strictly as their intern structure permits their actions. Their behavior is therefore, in this sense, perfectly deterministic: they can perfectly be specified. But the fact that they are very numerous, pro-active, that each of them is represented by a process, that their degree of behavioral liberty is relatively large, that they can alter and generate other agents, lead us to consider their organization like a complex system, whose global behavior is not easily represented and whose an exact prediction of the behavior seems very delicate.

We can specify the three general characters that define agents while expressing properties that they necessarily must own. These properties are more precise than the general characters. They are those stated by Krogh (Krogh 1995), to which we add (point 6) the notion of dependence. An agent possesses the six following properties:

- it is self-centred, that is that it has some own goals (reactivity and pro-activity),
- it is self-motivated (pro-activity),
- it is interacting (sociality),
- it is heterogeneous, that to be-to-say formed of no homogeneous parts (reactivity, pro-activity and sociality)
- it is persistent (sociality),
- it is dependent of certain another one (sociality).

The main problem is then, while leaving from every agent's individual activity, to define the activity of the agents in whole, that to be-to-say to specify the significance and the behavior at the global level of the agent organization. We know that this organization, when it is a massive one, form an unsteady and no deterministic system. A statistical approach of the behavior of the system seems delicate because agents are very cognitive elements that are strongly correlated in a cognitive way. It is difficult, in this case, to define probabilities for a medium behavior.

The main problem
The main problems in the manipulation of a large set of agents are the representation and the control of the behavior of the organization for all the agents.

Every agent's behavior is provided by its behavioral module that is generally an automaton (or an ATN). Initially, the module of knowledge arranges certain pre-definite knowledge and its goals are specified at the construction step. The agent is in some state, it acts on objects of its world and can send messages to other agents. The receipt of a message for an agent can change the knowledge of the recipient (with structural modifications and rewriting of code in the knowledge sub-system), can tamper its ATN (by change of states and / or steps, and therefore of the structure of the ATN itself) and even can change its protocols of communication. These changes can base themselves on random choices, while using a randomly function in the host system where agent is computed.

5.2 The Morphological Space, the Correspondent of the Space of Phases for the MMAS

We are going to represent the behavior of a set of agents, in a way independent of the problems the organization solves and therefore without the semantic of the system. This semantic will be added to this representation, to increase it thereafter.

We set a form as an element of R^n provided of a metric. A dynamic space is the product of a topological metric space by the time, supplying transitions between current state and the next states. Let's notice that agents are considered likes some simple rational entities: it is possible to associate at each of them a precise notion of state.

State of an agent
An agent's state is the meaningful characters that permit to entirely describe its current situation in its organization at every instant, this state being an element of R^n.

We wish to express each of these characters with an element of R. That is possible because, let's recall it, an agent is rational and deterministic. Thus, an agent's state will be expressed as a point of R^n if there are n characters that define it. The problem is to determine these characters independently of the semantics.

Activity map (Lesage 2000)
The activity map of an agent organization is a representation of the temporal evolution of the significant characters of every agent's state.

If we want to represent an activity map by geometric forms evolving in the time, we need to represent every agent by a vector. Let's look for what are the meaningful characters of every agent's behavior. Let's notice that an object doesn't have such characters, because it is permanent and doesn't undergo some structural evolution: at most, we can define for it the number of time each of its methods has been triggered.

The agent's intrinsic characters will be defined from the three following categories (C.f. Fig. 3):

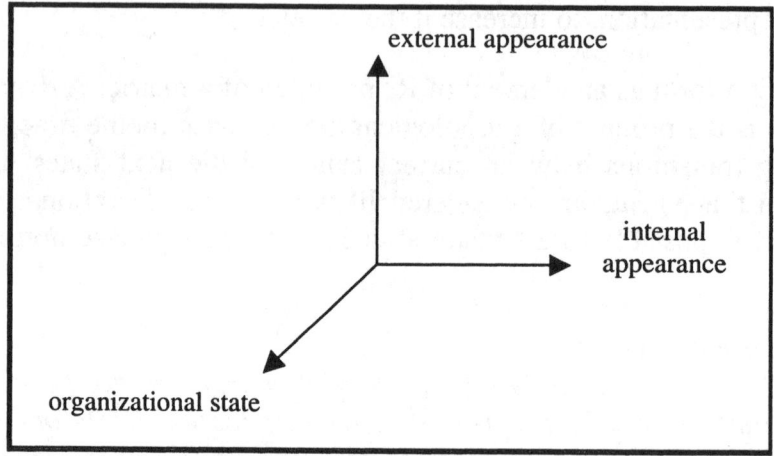

Figure 3. The three categories of characters of an agent

1. the appearance of the agent from its outside, that is for the other agents. This category characterizes the agent's situation in relation to its environment,
2. the agent's internal appearance facing of the reach of its goals. It is about its state in relation to goals it must reach again, the relation to functions that it must necessarily assure,

3. the agent's state at the level of its own working, that is its internal dynamic. This is the measure of the quality of its organization and the variation of this one along the time.

We go, while taking into account of these three characters of general categories, to propose the ten dimensions for the space of representation (C.f. Fig. 3).

1. The agent's first category is its *external appearance* for its immediate environment, that to be-to-say its actor's own situation in relation to its acquaintances and its environment. One keeps notions of:
   - *supremacy*: it is the measure of the fact the agent is either located in position by force for agents recognized by its acquaintances, that it has or no many allies and that its enemies are or no powerful,
   - *independence*: it is the measure of the agent's autonomy in its actions, specifying if it is necessary or no to find allies to reach its goals, if its actions are submitted in relation to actions of other agents, if its knowledge structure is included or not in those of some other agents,
   - *persistence*: it is the measure of the agent's longevity, of its life that can be possibly very brief or strong long.
2. The second character is its *internal aspect* that is its structural state in relation to its assigned functions. One keeps notions of:
   - *easiness*: it is the measure of the fact the agent reached its current state with difficulties, with more or less resistances. For example, if its behavior is represented by an ATN, this character measures the easiness of transition clearing, the rear returns. This indication specifies also the support, or the resistance, met by agent from other agents to reach its goals.
   - *speed*: it is the speed taken by the agent to reach its goals that it fixes itself either that is given at construction. It is for example the speed of get over the states of its behavioral automaton when the agent possesses one.

3. The third character concerns its organizational internal states, that is its own functioning understood like a process of information exchange between its modules and that forms its structure. One keeps the following notions:
   - *intensity of the internal activation flux*: it is the measure of the quantity of information exchanged between its intern components permitting it to lead to a visible activation from the outside in groups to which it belongs.
   - *complexification*: it is the measure of its structural transformation generated by certain dysfunctions or certain difficult situations, and having required transforming some elements of its structure. This measure determines its structural evolution, so is a simplification or a complexification.
   - *communicating frequency*: it is the measure, out of the semantic, of communicational relation between the agent and the others, representative in fact of a measure of its linkage in the organization.
   - *organizational gap*: it is an assessment taking in account the operationality of the agent's structure that permitted it to achieve some social actions leading to put itself an appreciation of the distance between its own state and the global state of the environment (there is not a supervisor in the system!). It is the appreciation of the adequacy of its "set in situation" in its world. This character is fundamental and precise the relation between the agent, understood like an autonomous entity, and the organization, composed of all agents, that is between each part of the organization and the whole (Morin op. mentioned).

Each of these characters can be represented itself by a measure in R, or by a suitable function. It is necessary to consider that the multi-agent system, as we already noted it, is immersed in its environment: it is in continuous communication with its environment. It exchanges information therefore with this one and interface agents achieve this action. We must have to add a typical character therefore because the system is *opened* in its environment, while importing and exporting information. We will define a character measuring the intensity of this transport of information, that represents the quantity and the *transport of information* exchanged with the environment thus. Note that agents of interfacings assure this measure that appears situated at the level of the system itself, normally, but these agents communicate evidently with the others about this external information. The measure specifies all agents' implications therefore in the environment of the system.

These ten characters, grouped in three general categories, can be represented in R^{10} and permit to associate to every agent what one will call its *aspect vector* thus. The agent's structure will be increased, merely, to be able to produce, to each moment, these ten indications, which are only numerical. We will call *organizational dimension* the dimension of the morphological space.

External aspect of the agent	Supremacy
	Independence
	Persistence
Internal aspect of the agent	Easiness
	Velocity
Organizational state of the agent	Intensity of the internal flux
	Complexification
	Communicating frequency
	Organizational gap
Opening of the system	Transport of information

Figure 4. The ten characters of behavioral aspect of an agent

Morphological space
It is the sub-space of R^{10} presenting in a vector the characters of activity of each agent, according to notions of supremacy, independence, persistence, easiness, velocity, intensity of the internal flux, complexifi-

cation, communicational frequency, organizational gap and transport of information.

This means that an agent for whom these ten fundamental characters are given is perfectly defined as for its obvious behavior in its environment, in the organization, that to be-to-say in the domain of application where it is operational.

We could also represent the state of the system in a space $R^{10.N}$ if there are N agents, like a particular point in a space of phases for agents. We will prefer to use R^{10} for space of description of the agent behavior, what is more compliant to our objectives. We will interest to the form of the cloud of points representing behavioral state of agents thus, as well as to the distortion of this form. On the other hand, the number of agents varies during the working and it is desirable to preserve a permanent dimension in the space of representation.

Central hypothesis
  The morphological space has, for an organization of agents, the role of the space of phases of a particulate system.

One can also use a reduced form of the morphological space, taking in account that the three measurements of space corresponding to the three general categories: external aspect, internal aspect and own organization. We get a morphological space said reduced space where are represented the general features of agents only. We also can use a projection of R^{10} in R^k, with k <10, and interest themselves to certain peculiar aspects of agents.

It is well obvious that some types of agents, and in particular the strong agents as the BDI one (Rao & al. op. mentioned) have capacities giving importance to some of these characters, as the organizational gap. But we say that each agent, whatever it is, can at least of concise manner, provide a measure of all these ten characters.

The activity map of an agent organization will be then a cloud of points in R^{10} corresponding to the ten aspectual characters of the agent behavior. These characters are expressed from its only structure and activity. It is now possible to specify the notion of representation of an agent organization: the notion of landscape of agents.

Landscape of agents
A landscape of agents is a geometric expression of an activity map.

This representation leans on the essentially geometric characters of the cloud of points, while possibly using notions of algebraic topology, that is simplicial complex (Giavitto 1998). It will be in fact achieved by a specific agent organization. The geometric elements that we are going to consider are elements of the dynamic system R^{10} x T (Thom 1972), that measure the importance of behaviors of agents and of agent groups, for a while. The conformation of a form of this space can be represented by a polyhedron. We can express the fact that this form has, for example, prominent parts, ravines or trays, lengthening or dense parts and discontinuities, that permits to represent the behaviors, the manner of which the activities of the agents have produce the behavior of the system, enough faithfully. We go to use, to describe these concepts, some particular agents we call the *morphological agents*.

5.3 The Organization of Morphological Agents Assuring the Representation of the Aspectual Organization

We are going to manipulate two organizations of agents therefore: one organization of agents that defines, by their actions in group, the functionalities of the system, and another representing the activation of these agents in terms of the morphological space. We will call the first organization the *aspectual organization* with the aspectual agents that gives the aspects of functionalities and of operationality of the system, and the second the *morphological organization* with morphological agents that expresses the form of the first organization in a geometrical way (C.f. Fig 5).

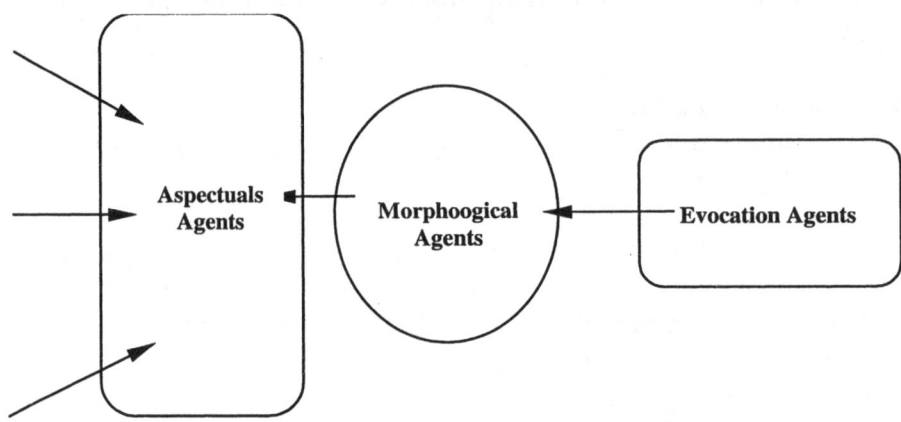

Figure 5. The three agent organizations

The morphological reified space

> *The reification of the morphological space corresponds to the definition of a specific agent organization that dynamically expresses with agents all the characters of the morphological space: that is the reified morphological space.*

We need to represent the characters of the aspectual agents activation with an organization of agents rather than operate with the set of points of the morphological space, for the two following reasons:

- the notion of point, or of pinpoint particle, in the morphological space is not adapted to the characters of our survey. It is another kind than agent's concept who stands typically in an equation setting. We don't try here to produce an operator for the change of state of the agent organization, but we want follow in real time the changes of this organization, and at the computable level. We prefer therefore the notion of dynamic attraction basins, changing in the time, in which the same attraction basin absorbs close by agents like groups absorb lonesome agents, and this while the system is working. Thus, the morphological space will be formed of organizations of agents reifying basins of attraction and expressing, according to their depth and to the variation of this depth, the aspectual agent density in the considered region of the morphological space.

- the geometric indications appearing in the morphological space will have to be used immediately, to react with the aspectual agents, and only an organization of agents can interact well with another. We wish in fact to strongly couple the aspectual agents and those representing their characters, the morphological agents, to give the system as self-controllable.

A morphological agent will have for function to represent, according to its type and role, the local activations and the modification of activations of the aspectual agents, according to the ten characters of the morphological space (C.f. Fig 6).

Structurally, a morphological agent is a weak agent, whose behavior component is an automaton with four states (C.f. Fig. 7), to take into account the four states of the decision according to L. Sfez (Sfez 1992) and while following what we already used otherwise (Cardon - Durand 1997). These four states are:

1. *initialization*: the morphological agent enters in activity,
2. *deliberation*: the agent looks for in its allied by acquaintances to reach its goals,
3. *decision*: the agent tempts to master its objectors to reinforce itself in the groups it belongs,
4. *action*: the agent can act, having allied agents and having put stronger aside it's opposing. Its action essentially comes to lo-

calize some aspectual agents in a certain well of potential that it represents and of which it provides a conformation (depth, width, number and density of aspectual inside).

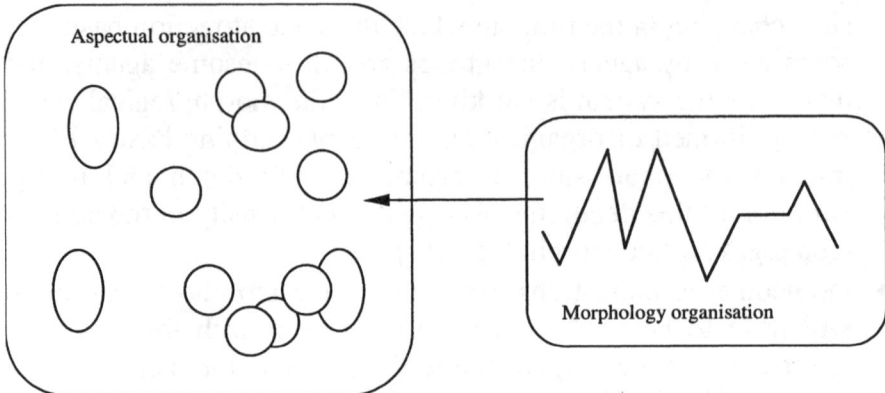

Figure 6. Action of the morphological agents producing a representation of the aspectual organization

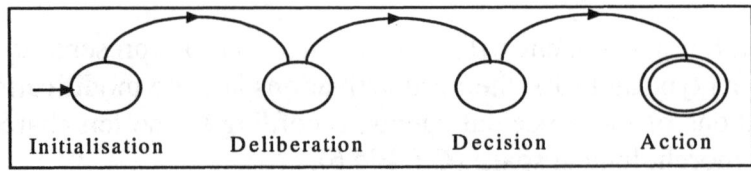

Figure 7. Structure of the behavioral component of a morphological agent

A morphological agent acts while essentially taking information in behavior of aspectual agents, with the measures of the ten fundamental characters of these agents. It places in fact each aspectual agent in a meaningful ball of R^{10}. This process is technically complex, because it binds some aspectual agents to a morphological agent considered like a changing potential well, *via* pre-morphological agents that scrutinizes characters of every aspectual agent and direct it toward the morphological agent that is created for it, or that is reinforced (Lesage 2000). A reinforced morphological agent has tendency to absorb its neighbours that are less strong and a weak morphological agent has a tendency to associate with the comparable and close by morphological agents to form a stronger agent.

Morphological agents are going to achieve "in-line" a behavioral analysis of the aspectual agent organization. They operate according to theirs strictly rational goals, tempt to clear typical forms from the set of characters expressed by the aspectual agent landscape, while assembling in groups (in groups of potential wells), that is while constituting so-called *coherent groups* (Cardon 2000). A coherent group is a structured and meaningful aggregation of morphological agents. This is a heap made of close by morphological agents, for a certain metrics on the morphological agent space, valued by an internal measure to the agent while appraising proximity with its acquaintances.

Chreods are reified emergences in a set of morphological agents (that is a set of groups of potential wells each holding some aspectual agents). These sets endowed of geometric signification (density, form, development) on morphological agents. They are represented by particular polyhedrons of which topological forms have an applicable significance to represent the way whose aspectual agents group themselves, according to their behavior (Giavitto op. mentioned). The landscape of morphological agents (a landscape on a landscape!) admits different readings that will be done by a third organization of agents: the evocation agents.

5.4 Characters of Coherent Groups

Organizations of aspectual and morphological agents are in permanent interaction. Chreods, exhibited by morphological agents, are going to be able to influence the activity of morphological agents, while activating for example more strongly those that are compliant in appearing in coherent groups and while possibly giving back recessive those that are isolated. While interpreting in an organizational way these actions, one can say that the system is endowed of a mechanism of permanent features formation, of leanings in its mechanism or re-organization, and independently of the semantic characters of its aspectual agents. They are organizational tendencies, strictly internal and depending only of the behavior of the morphological organization.

A coherent group is formed of a set of close at hand morphological agents, according to the metric of the morphological space. It has the form of a polyhedron. Every point of this polyhedron is a singular morphological agent that represents a group of aspectual agents that have significance, semantics for the domain of application of the system. We can recover this semantics in the coherent group, if we take into account the features that the aspectual agents reified.

Interpretation of a coherent group
The interpretation of a coherent group the fact to take into account the semantic characters of the aspectual agent that composes it, added to the characters of its geometric form and of the evolution of this one.

The question is to give significance to the coherent group, while taking into account of its geometric form and the aspectual agent semantics that it identifies. One can characterize coherent groups in a geometric way with their centred aspect (axes of symmetries), the stretched out, partitioning, the joint points, the flat form or expanded, regular or not... and to carry these characters on the whole of the semantic features of its aspectual agents. These characters show the way the semantic features express themselves, how they are organized between them, how they complete themselves or are opposed, reinforce or inhibit them. And the most interesting analysis will be the follow-up of the coherent groups distortion in the time, expressing the way whose system express its features from its agents, how it makes the significance emerging in its organization.

5.5 Evocation Agents and Self-Adaptability of the System

The system in its whole, in the communicational movement of its agents, is going to draw, in the morphological organization, some coherent groups from the aspectual activations. The analysis of the morphology of the aspectual agent landscape, and particularly of the coherent groups, produces an active picture of the organization to every instant providing its organizational state. It is well about producing characters precisely defining the state of the system, like an operator as the one of Liouville provides the state of a particulate system at every instant of the time. The knowledge of this state permits to say for what way the system goes, what is effectively its global evolutionary action. But this knowledge is not direct; agents can't be analyzed one by one. It is no more statistical but, in our way, it is a geometric and global analyze, done in real time, during the working of the aspectual organization.

A third agent organization, after aspectual and morphological agents, is going to take in consideration the state of the landscape of morphological agents, to achieve an analysis of the morphology of the system, interpreting coherent groups (C.f. Fig. 6). It is about representing the sense of the activation of the aspectual organization, from its characters of aspect expressed by morphological agents into coherent groups. This organization of agents, the *evocation agents*, is going to provide an interpretation of that has been expressed by the geometric and semantics information coming from the landscape of morphological agents (C.f. Fig 8). This organization operates above the aspectual agent landscape that detains the semantics of the actions.

Evocation agents have a similar structure to the aspectual ones, while being very cognitive. They are going to represent the significance of the activation of agent organizations, interpreting the activity of the aspectual agents of interfacing, the aspectual agent behavior and its analysis with the morphological agents.

And the organization of evocation agents is going to permit to make a lot more that this analysis, because the system is a computable one. It is going to permit to achieve a control of the aspectual agent organization, by the action of the evocation agents while using the morphological analysis. The system is going to be able to take account of the significance of its morphology in progress, to memorize it organizationally, that is to take account implicitly of it in its future activation, in its future commitments. The system can function like a *self-organizational memory*.

Evocation agents will be able to act thus on the global aspectual activation consistency. They will be able to do choices and to take decisions for the global behavior of the aspectual and morphological organizations. They will be able to take account of the past actions, of the tendencies of aspectual agents, while keeping strategies of inhibition of action for certain aspectual or morphological agents of interfacing, controlling so the general line of organizational emergence achieved in the system, to all levels. Let's notice that these strategic actions will be indirect in relation to every agent's behavior, permitting to constitute a system with emergence of organizational sense, with intrinsic characters of no-stability and learning by regular structural distortion (Cardon 2000). The systemic loop is now curled and the system controls itself by self-adaptability.

Self-adaptive system
> *A self-adaptive system is a system composed of three organizations of elements: a first organization of proactive elements that generates the elementary functionalities, a second organization that expresses the geometric characters of the re-organizational actions of the previous one and a third organization that values*

> *the behavior and semantics of the first one using the second and that can, from this analysis, influences the behavior of the first organization continuously.*

It is clear that all the living organisms can be represented according to this characterization, the last two organizations doing the self-control and the regulation of movements done in the first. It is also well obvious that only some massive agent organizations are adequate to the computable representation of such systems.

Finally, the action of evocation agents, by the fact that they tamper the active aspectual agents, achieves to each moment in a continuous way, some control and a kind of memorization of the behavior of the system (C.f. Fig. 8). This looped action permits to control the stability of the system, to advise it in the loop without end, of production of states compliant to its capacities of reorganization and adequate to its "setting in situation" in its environment. The working of the system amounts to a continuous succession of self-controlled emergent states.

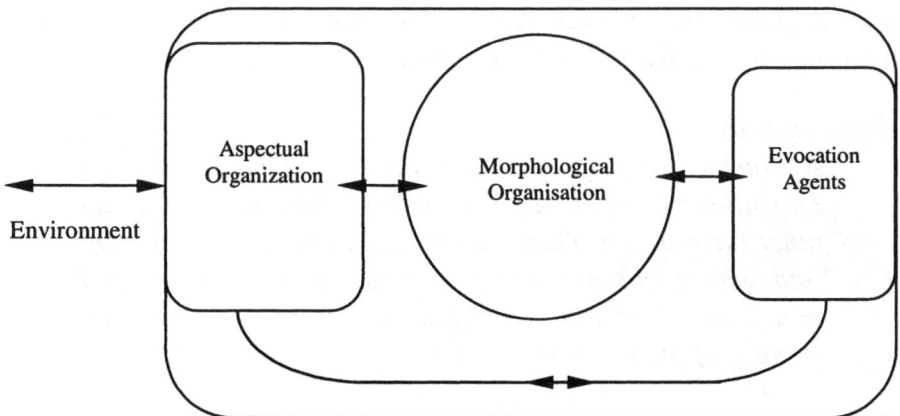

Figure 8. The systemic loop of a massive multi-agent system: the three different organizations of agents

For the system, the complexity of the interfacing and aspectual agents drags the complexity of all other organizations of agents. The system can generate an elementary organizational emergence thus: the organizational significance of the prominent form exhibited. It can also generate a very fine emergence: for example the emergence of sentences from intentions expressing linguistic tendencies (Lesage 2000).

6 Entropy and Equation of Trajectory of MMAS

A MMAS has a behavior like an unstable system. We are looking for an entropy notion and one equation predicting the state of such a system. We can consider the case of the so-called *open systems* (Von Bertalanffy 1973). The notion of open system comes from the von Bertalanffy researches and presents an alternative at the notion of mechanical classical closed system we shown about the object oriented systems. An open system operates with a strong relation with its environment and can't be defined without information or material coming from this environment.

Open system
An open system is a system made of numerous polymorph components in continuous inter-relation, that must have a continuous exchange with its environment and whose components died and can be regenerated modifying their functionalities, by the fact of the functioning of the system.

The construction and functioning of such a system need to take into account the information coming from its environment. Massive MAS produced with the incremental method of agentification is an example of such artificial open system.

6.1 Entropy

The second principle of thermodynamics specifies that a closed system always lead towards states of equal or less elevated order that is to bigger probability states. Entropy is the measure of this directed evolution. The entropy of a closed system is therefore still increasing or null. The famous Clausius equation expressing this principle is:

$$dS \geq 0, \text{ where S is the entropy}$$

An open system, on the other hand, cannot satisfy this principle. Its evolution permits it indeed to reach states of order that is more elevated and of less probability. I. Prigogine (Prigogine 1965) defined a new notion of entropy for these systems, using in fact two kinds of entropy:

$$dS = d_eS + d_iS$$
$$d_eS \leq 0 \text{ or } d_eS \geq 0$$
$$d_iS \geq 0$$

where d_eS is the variation of entropy coming from the environment, by input of energy or information into the system, and d_iS is the internal entropy variation, by the irreversible internal process activation. The external entropy can be negative, when the external contribution comes to create an order in the system. The internal entropy is, classically, still positive or null.

It is well obvious that a massive multi-agent system, composed of multiple processes in competition, takes information in its environment, to transform it in knowledge into the agents. The external entropy will be the measure of these informational contributions transformed into knowledge that is finally in code. The internal entropy will measure the deterioration of the working of agents, that is the fact that process treatments drag the non-determinism and increase the reproduction and cloning of agents. Thus, the system will be able to organize itself toward an increasing order, to the sense of manipulated knowledge, while using to best the external information contribution. It satisfies in that way the notion of Prigogine entropy.

6.2 Equation of Trajectory: a Reduction with Regard to the Morphological Analysis

The notion of equation of general trajectory of a MMAS will be an interpretation of the von Bertalanffy equation for the open systems (von Bertalanffy op. mentioned).

We have to set into relation in a temporal evolution equation what the system produces with what permits or inhibits this production. This is a multi-agent system and what it produced can be considered as a kind of activated knowledge. This knowledge is distributed on various forms in agents and serve of foundation to their activity. The system produces this knowledge like an emergence, on the form of distinguished characters and is constructed by agent's organizations.

Let's name K_i the i^{eme} kind of knowledge that the system can produce or manipulate. Let's recall that agentification has been achieved from ontology's, that the knowledge is in the system in a finished number of characters and that each peculiar knowledge admits qualifications and links toward other knowledge.

In the working of the MMAS, we interest therefore to the emergence of some knowledge noted K_i, either:

$$\frac{\partial K_i}{\partial t},$$

that represents the quantity of K_i expressed during the MMAS running.

Two characters will represent what make easier and inhibit this emergence of knowledge:

1. the movement of generation of this knowledge, expressed by the action of agents, its transport by its activation in the agents,
2. the ratio of production of this knowledge, in relation to what inhibits or permits it.

We define T_i the transport of the knowledge K_i that is what concerns the movement of displacement of this knowledge in the system, by the activation of agents. It is about good of a kind of movement, because the agents manipulating K_i are activated and these activations, by communications between them, drags the manipulation of K_i in others agents. Thus, K_i is activated in a distributed and dynamic way, like a transport of matter in a physical system.

One will keep, while basing on characters of external aspect of the agents, that the T_i transport is in fact a function depending of the communicational frequency between agents, of the organizational gap and of the three characters of external aspect of agents: supremacy, independence, persistence.

We define the ratio of production P_i of the knowledge K_i as the quantity of K_i manipulated in the activity of agents. It is about a concept expressing the efficient generation of the knowledge from internal knowledge system into the agents. From external information, K_i is activated in some agents, and the organizational movement of the system permits other generations of K_i.

We will keep that this ratio of production is a function taking in account the internal aspect characters of the agents: easiness, velocity, intensity of the internal flux, complexification.

We get so the following equation of production of a knowledge K_i by emergence, according to T_i and P_i, the so-called *equation of evolution* of the massive multi-agent system:

$$\frac{\partial K_i}{\partial t} = T_i + P_i$$

for the i^{nto} manipulated knowledge

This is the state equation expressing the variation of the quantity of generated knowledge in relation to values of transport T_i and of ratio of production P_i.

This equation is similar to the von Bertalanffy's one about the open systems. It denotes a precise knowledge, the i^{eme}, while representing that conducted to an emergence for this peculiar knowledge. The massive multi-agents system is then represented by as many equations that we can have kinds of knowledge, while observing that these kinds of knowledge are not independent. The equation of state characterizes a multi-agent system thus like an open system, stabilizing far of the equilibrium on no steady states and where the global organization influences on the local behaviors of agents.

6.3 Validity of the State Equations

Let's notice whereas it is difficult to show that one can get some stable states for a system governed by equations. Always each knowledge is linked with some others, and the behavior of a distinct knowledge is not very significant (Searle 69). This is the limit of the equational approach. Only the morphological approach allows the organizational representation of the manipulated knowledge.

The morphological organization, analyzed by the organization of evocation, is the calculable realization of an operator of the type of the Liouville's one describing the evolution of the aspectual organization, while expressing modifications and organizational and semantics tendencies of the aspectual organization. The difference between this real time agent computable determination of the behavior of a massive agent organization and one equation of state that put the existence of a behavior is typical of the difference between computer science and mathematics, between efficient computable and existential formalization.

6.4 Degraded Forms of the State Equation

The state equation admits two limit cases, where one recovers the classic systems behavior. When the T_i transport is missing, when communicational activity of agents, to the sense of transport of this knowledge, is non-existent, the working of the system is like the one of a simple neural network. Agents don't have anymore to generate the K_i knowledge and are reduced to neurons, or isolated reactive functional components.

When the ratio of production P_i is missing, the system is no more generating of knowledge, agents don't play the role of knowledge generator anymore but are merely functional activators for this knowledge. There are not anymore innovative characters in the system and this one like an object oriented system. There is always communication between agents, but without their proaction they don't generate innovative behavior leading to the production of new knowledge or new aspects of this knowledge.

7 Conclusion

We have presented massive multi-agent systems like unsteady software systems whose behavior develops some emergent characters. These systems have an evolution, they are made to learn and to adjust themselves to the situation on environment. Their design is not a usual way producing all the specifications and a strong validation. They aren't problem solvers. They are systems functioning with a behavior oriented in the time following their continuous development.

Intentions to the reorganization, in such systems, use laws of the non-equilibrium of aspectual agent organizations and are achieved by the habits acquired by the system to function as it ordinarily functions. The same working tampers agents, creates some of new by cloning or by reproduction. But tendencies draw themselves that are lines of habit of the behavior of agent organizations, according to the control operated by evocation agents on the aspectual organization and that lead the system to generate such a peculiar state rather than another one. Thus, the system has implicit intentions to enter into certain ways of emergence, and open out these intentions under explicit form, for example behavioral or linguistic. Such a system is an organizational memory that operates by continuous implicit learning while self-controlling its re-organizations. It is self-adaptive by nature, not being able to be otherwise.

The number of agents is not stationary, agents are not steady, communications between them are variable, the working is at a time appreciable to the initial conditions and no deterministic. It seems therefore absolutely vain want a priori to specify and to control such a system in its whole, as the Software Engineering teaches us to make it for the complicated systems ones. It is necessary to look for another way of design and the incrementally agentification by learning we propose is such a way.

Such a system, of which the behavior is oriented in the time, learns and evolves by the fact of its self re-organization. This is not anymore a system understood and conceived in a finish fashion and delivered for a technical use at the benefit of its users: it is, in the software domain, an artefact of a real living system.

References

Axelrod, R. (1997), *The Complexity of Cooperation: Agent-Based Model of Competition and Cooperation*, Princeton University Press.

von Bertalanffy, L. (1973), *Théorie générale des systèmes*, Bordas, Paris.

Bobrow, D.G. and Fraser, J.B. (1969), "An augmented state transition network analysis procedure," *Proc. Int. Joint Conf. on AI*, vol. I, p. 557.

Booch, G. (1994), *Analyse et Conception Orientées Objet*, Addison Wesley.

Borodzicz, E., Aragones, J., and Pidgeon, N. (1993), "Risk communication in crisis: meaning and culture in emeregency response organisations," *European Conference on Technology & Experience in Safety Analysis and Risk Management*, Rome, Italy.

Brodie, M. and Ceri, S. (1992), "On Intelligent and Co-operative Information Systems: a workshop summary," *Int. Journal of Intelligent and Co-operative Information Systems*, vol. 1, no 2., pp. 249-289.

Brooks, R. (1999), *Cambrian Intelligence, the Early History of the New AI*, The MIT Press.

Bussmann, S. and Demazeau, Y. (1994), "An agent model combining reactive and cognitive capabilities," *Proc of IEEE International Conference on Intelligent Robots and Systems, IROS'94*, München.

Cardon, A. and Durand, S. (1997), "A model of crisis management system including mental representations," *Proceedings of the AAAI Spring Symposium*, Stanford University, California, USA, 23-26 March.

Cardon, A. and Guessoum, Z. (2000), "Systèmes multi-agents adaptatifs," *8ème Journées Francophones sur l'Intelligence Ar-*

tificielle Distribuée et les Systèmes Multi-Agents, p. 100, 116, ed. Hermès, Saint-Jean-la-Vêtre, France, 2-5 October.

Cardon, A. and Lesage, F. (1998), "Toward adaptive information systems: considering concern and intentionality," *Proc. KAW'98*, Banff, Canada.

Cardon, A., Vacher, J.P., and Galinho, T. (2000), "Genetic algorithm using multi-objective in a multi-agent system," *Robotic and Autonomous Systems*, 33 (2-3), pp. 179-190, Elsevier.

Cardon, A. and Dabancourt, C. (2001), *Initiation à l'algorithmique objet*, Eyrolles, Paris.

Cardon, A. (1999), *Conscience artificielle et systèmes adaptatifs*, Eyrolles, Paris.

Clergue, G. (1997), *L'apprentissage de la complexité*, Hermès, Paris.

Cohen et al. (1995), *Simulating Organizations, Computational Models of Institutions and Groups*, AAAI Press, The MIT Press, California, USA.

Davis, R. (1982), "Report on the Workshop on Distributed Artificial Intelligence," *SIGART Newsletter*, no. 80, pp. 13-23.

Demazeau, Y. and Muller, J.P. (1991), *Decentralised AI 2*, North-Holland.

Dorigo, M. and Colombetti (1998), *Robot Shaping. An Experiment in Behavior Engineering*, Intelligent Robotics and Autonomous Agents series, vol. 2. MIT Press.

Drogoul, A. and Collinot, A. (1998), "Applying an agent-oriented methodology to the design of artificial organisations: a case study in robotic soccer," *Autonomous Agents and Multi-agent Systems*, 1(1).

Ferber, J. and Gutknecht, O. (1998), "A meta-model for the design and analysis of organizations in multi-agents systems," *Proc. of the 3rd International Conference on Multi-Agent Systems, ICMAS'98*, pp.128–135, IEEE Computer Society.

Ferber, J. (1995), *Les Systèmes Multi-Agents*, InterEdition, Paris.

Giavitto, J.-L. (1998), *Dossier d'habilitation à diriger des recherches*, LRI URA CNRS 410, Université de Paris-Sud, Orsay.

Guessoum, Z. and Briot, J.P. (1999), "From active objects to autonomous agents," *IEEE Concurrency*, 7(3): 68-76, Nov.

Huns, M.N. and Singh, M.P. (1998), *Readings in Agents*, Morgan Kaufmann Publishers Inc, Sa, Francisco.

Kieras, D. and Polson, P.G. (1985), "An approach to the formal analysis of user complexity," *Int. J. of Man-Machine Studies*, vol. 22, pp. 365-394.

Krogh, C. (1995), "The Rights of Agents," *Intelligent Agents II, IJCAI'95 Workshop (ATAL)*, pp. 1-16, Springer Berlin.

Kuokka, D. and Harada, L. (1995), "On using KQML for matchmaking," *Proc. of First International Conference on Multiagent Systems*, Menlo Park, CA., AAAI Press, USA.

Lapierre, J.W. (1992), *L'Analyse des Systemes*, Syros.

Le Moigne, J.-L. (1990), *La Modélisation des Systèmes Complexes*, Dunod, Paris.

Lenat, D. and Guha, R.V. (1990), *Building Large Knowledge-Based Systems, Representation and Inference in the Cyc Project*, Addison Wesley Publishing Co.

Meyer, J.-A. (1996), "Artificial life and the animat approach to artificial intelligence," in Boden (ed.), *Artificial Intelligence*, Academic Press.

Morin, E. (1986), *La méthode, Tome 3 : La connaisance de la connaissance*, Essais, Seuil.

Muller, P.A. (1997), *Modélisation objet avec UML*, Eyrolles Paris.

Muller, J.P. (1998), "Vers une méthodologie de conception de systèmes multiagents de résolution de problèmes par émergence," *JFIADSMA'98*, Hermes.

Pitrat, J. (1993), "L'Intelligence Artificielle: au-delà de l'intelligence humaine," in *Le cerveau: la machine-pensée*, DRT, D. de Béchillon, L'harmattan.

Poincaré, H. (1893), *Les méthodes nouvelles de la mécanique céleste*, Gauthier Villars Paris.

Prigogine, I. (1982), *Physique, temps et devenir*, Masson, Paris.

Prigogine, I. (1996), *La fin des certitudes*, Ed. Odile Jacob Paris.

Rao, A.S. and Georgeff, H.P. (1991), "Modeling agents within a

BDI-architecture," in Fikes, R. and Sandevll, E. (eds.), *Proc. of the 2rd International Conf. on Principles of Knowledge Representation and Reasoning KR'91*, pp. 473-484, Cambridge Mass, Morgan Kaufmann.

Rao, A.S. (1996), "Decision procedures for propositional linear-time belief-desire-intention logics," in Wooldridge, Müller, Tambe (eds.), *Intelligence Agents II*, Springer.

Searle, J.R. (1969), *Speechs Acts*, Cambridge University Press.

Steels, L. and Brooks, R. (1995), *The Artificial Life Route to Artificial Intelligence, Building Embodied Situated Agents*, Lawrence Arlbaum Associates.

Sfez, L. (1992), *Critique de la décision*, Presses de la fondation nationale des sciences politiques.

Thom, R. (1972), *Stabilité structurelle et morphogénèse*, W.A. Benjamin, INC, Reading, Massachusetts, USA.

Thorne, J., Bratley, P., and Dewar, H. (1968), "The syntactic analysis on English by machine," in Michie, D. (ed.), *Machine Intelligence*, vol. 3, NY Elsevier.

Varela, F. (1989), *Autonomie et connaissance, Essai sur le vivant*, Seuil, Paris.

Weis, G. (1999), *Multiagent Systems: a Modern Approach to Distributes Artificial Intelligence*, MIT Press.

Wooldridge, M. and Jennings, N.R. (1994), *Agent Theories, Architectures and Languages: a Survey*; Lectures Notes in A.I., 890, Springer Verlag.

Chapter 6

Developing Agent-Based Applications with JADE

F. Bergenti, A. Poggi, G. Rimassa, P. Turci and M. Tomaiuolo

JADE (Java Agent Development Framework) is an "open source" FIPA-compliant software environment to build agent systems. JADE offers an agent middleware to implement efficient FIPA2000 compliant multi-agent systems and supports their development through the availability of a predefined programmable agent model, an ontology development support, and a set of management and testing tools. This chapter describes JADE and its use in three international projects to develop applications in the fields of: corporate memory management, integration of fixed and mobile networks, and integration of Web services.

1 Introduction

Agents are one of the most promising information technologies (Genesereth & Ketchpel, 1994; Wooldridge & Jennings, 1995; Jennings, 2001; Wooldridge, 2002); however, agent-based technologies cannot keep their promises, and will not become widespread, until there are standards to support agent interoperability and adequate environments for the development of agent systems.

In these last decade, different projects and organizations worked on the standardization of agent-based technologies (see, for example,

KSE (Patil et al., 1992), OMG (Milojicic et al., 1998) and FIPA (FIPA, 1997)).

The main important results seems be achieved by FIPA. FIPA (Foundation for Intelligent Physical Agents) is an international non-profit association of companies and organizations sharing the effort to produce specifications for generic agent technologies. FIPA does not promote a technology for just a single application domain but a set of general technologies for different application areas that developers can integrate to make complex systems with a high degree of interoperability. The standardization work of FIPA is centered on the definition of the Agent Communication Language (ACL), but also specifies the key agents necessary for the management of an agent system and the shared ontology to be used for the interaction between two systems.

In this chapter, we present JADE (Java Agent DEvelopment framework), one of most known and used software framework to write agent applications in compliance with the last FIPA specifications for interoperable intelligent multi-agent systems. Moreover, the chapter presents three examples of the use of JADE to develop systems in the fields of: ubiquitous networks, corporate memory management, Web services access and composition. The next section describes the main features of JADE. Sections three, four and five respectively present three JADE based projects: LEAP, CoMMA and Agenticities. LEAP addressed the need for open infrastructures and services that support dynamic, mobile enterprises and extended JADE to support mobile and wireless applications. CoMMA realized a multi-agent system to help users in the management of an organization corporate memory and in particular to facilitate the creation, dissemination, transmission and reuse of knowledge in the organization. Agentcities created an on-line, distributed testbed to explore and validate the potential of agent technology for future dynamic service environments. Finally the last section summarizes the advantages of using JADE to develop agent applications, its limits and gives a short comparison with other similar software frameworks.

2 JADE

JADE (Java Agent DEvelopment framework) is a software framework to aid the development of agent applications in compliance with the FIPA specifications for interoperable intelligent multi-agent systems. JADE is an "open source" project, and the complete system can be downloaded from JADE Home Page (JADE, 1999; Bellifemine et al, 2001).

The JADE system can be described from two different points of view. On the one hand, JADE is a middleware for FIPA-compliant Multi Agent Systems, supporting application agents whenever they need to exploit some feature covered by the FIPA standard specification (message passing, agent life-cycle management, etc.). On the other hand, JADE is a Java framework for developing FIPA-compliant agent applications, making FIPA standard assets available to the programmer through object oriented abstractions.

2.1 Platform Architecture

JADE agent platform architecture tries to offer flexible and efficient messaging, transparently choosing the best transport available and leveraging state-of-the-art distributed object technology embedded within the Java runtime environment. While appearing as a single entity to the outside world, a JADE agent platform is itself a distributed system, since it can be split over several hosts with one of them acting as a front end where AMS and DF agents are placed. A JADE system comprises one or more Agent Containers, each living in a separate Java Virtual Machine and delivering run-time environment support to some JADE agents. The JADE middleware tries to provide efficient and flexible messaging services to user applications. An agent platform must contain a component called Agent Communication Channel (ACC for short), whose task is to transparently provide a Message Transport Service (MTS for short), relying upon one or more FIPA compliant Message Transport Protocols (MTPs), e.g., IIOP, HTTP and WAP protocols.

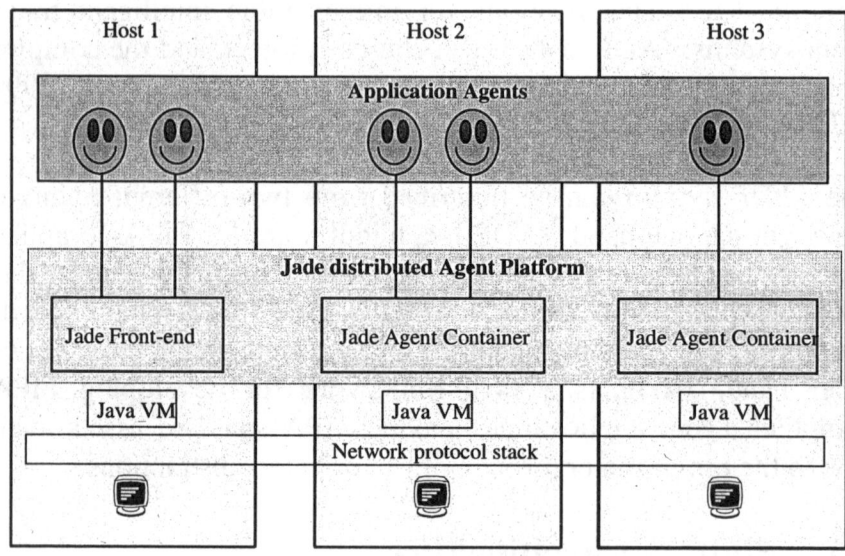

Figure 1. Jade agent platform architecture.

JADE distinguishes between inter-platform messaging (the sender and the receiver agents live on different platforms) and intra-platform messaging (the two interacting agents are within the same platform). While inter-platform messaging has to comply with FIPA specifications, intra-platform message delivery is strictly a JADE issue, so a more convenient proprietary transport can be exploited. JADE usually uses Java RMI for intra-platform communications, but support different transport protocols for intra-platform

Developing Agent-Based Applications with JADE

messaging to allow the distribution of the platform on devices providing different transport protocols.

Since JADE is a distributed agent platform, the ACC component is split in different parts, running on the different agent containers that make up the platform. The major features of JADE ACC are: i) multiple MTPs, deployed as plug-ins on multiple containers; ii) one hop message routing for outgoing and incoming messages, and iii) protocol independent address caching.

The general JADE messaging framework allows to deploy new transport ports during normal platform operation: the JADE administrator can add a new protocol to any agent container, simply logging in the management GUI and providing the Java class that implements the MTP.

An agent platform can now have any number of addresses, scattered around different hosts. Message routing support is needed to manage this rather general topology; the ACC provides a routing service that is guaranteed to require at most one hop. When a message reaches the platform through one of the available external communication ports, the ACC looks up the receiver agent ID to retrieve the agent container where it must dispatch the incoming ACL message. If the agent lives within the same container, the ACC uses an optimized local call; otherwise it relies on Java RMI.

When an agent wants to send a message to another, living on a different platform, it asks its local ACC for delivery service. The ACC reads the address list of the agent ID of the message recipient and tries all the addresses until one of them succeeds; for a specific address, the ACC discovers which MTP has to be used (FIPA addresses are URLs, so they contain a part that identifies the protocol) and checks to see whether that MTP is installed on the current agent container. If so, the locally available MTP is used, otherwise the ACC routes the message to a suitable container using a table that stores the deployment location of each MTP in the agent platform.

The JADE messaging subsystem also has an address caching feature that allows direct communication between agents, without unnecessary table lookups: intra-platform addresses and standard FIPA addresses are cached on each container exactly in the same way: on cache hits, the messaging subsystem does not even need to know whether the receiver is local, intra-platform or inter-platform. The cache is updated according to an optimistic attitude (i.e., if a cached address becomes stale the message delivery operation fails with an exception and the cached item is refreshed) and the cache replacement policy is the usual Least Recently Used one.

The JADE ACC can also be deployed on its own, without a complete agent container. This is meant to enable users to build and deploy agent level gateways and firewalls: a standalone ACC lives within a JVM that can route and filter ACL messages but cannot host FIPA agents.

2.2 Agent Architecture

An agent is defined as a collection of behaviours that are scheduled and executed to carry on agent duties. Behaviours represent logical threads of a software agent implementation. According to Active Object design pattern, every JADE agent runs in its own Java thread, satisfying autonomy property; instead, to limit the threads required to run an agent platform, all agent behaviours are executed cooperatively within a single Java thread. So, JADE uses a thread-per-agent execution model with cooperative intra-agent scheduling.

JADE agents schedule their behaviour with a "cooperative scheduling on top of the stack," in which all behaviours are run from a single stack frame (on top of the stack) and a behaviour runs until it returns from its main function and cannot be preempted by other behaviours (cooperative scheduling).

JADE agent architecture model is an effort to provide fine-grained parallelism on coarser grained hardware. A likewise, stack based execution model is followed by Illinois Concert runtime system (Karamcheti et al., 1996) for parallel object oriented languages. Concert executes concurrent method calls optimistically on the stack, reverting to real thread spawning only when the method is about to block, saving the context for the current call only when forced to.

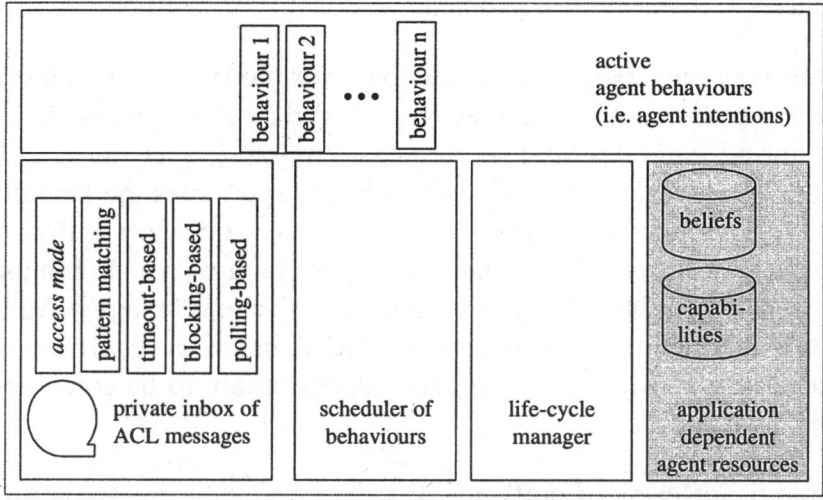

Figure 2. JADE agent architecture.

JADE thread-per-agent model can deal alone with the most common situations involving only agents: this is because every JADE agent owns a single message queue from which ACL messages are retrieved. Having multiple threads but a single mailbox would bring no benefit in message dispatching. On the other hand, when writing agent wrappers for non-agent software, there can be many interesting events from the environment beyond ACL message arrivals. Therefore, application developers are free to choose whatever concurrency model they feel is needed for their particular wrapper agent; ordinary Java threading is still possible from within an agent behaviour.

The developer implementing an agent must extend Agent class and implement agent-specific tasks by writing one or more Behaviour subclasses. User defined agents inherit from their superclass the capability of registering and deregistering with their platform and a basic set of methods (e.g. send and receive ACL messages, use standard interaction protocols, register with several domains). Moreover, user agents inherit from their Agent superclass some methods to manage the agent behaviours.

JADE contains ready made behaviours for the most common tasks in agent programming, such as sending and receiving messages and structuring complex tasks as aggregations of simpler ones. JADE recursive aggregation of behaviour objects resembles the technique used for graphical user interfaces, where every interface widget can be a leaf of a tree whose intermediate nodes are special container widgets, with rendering and children management features. An important distinction, however, exists: JADE behaviours reify execution tasks, so task scheduling and suspension are to be considered, too.

Thinking in terms of software patterns, if Composite is the main structural pattern used for JADE behaviours, on the behavioural side we have Chain of Responsibility: agent scheduling directly affects only top-level nodes of the behaviour tree, but every composite behaviour is responsible for its children scheduling within its time frame.

3 LEAP

LEAP (Lightweight Extensible Agent Platform) is an agent platform to allow the seamless deployment of agents on all Java-enabled, connected devices ranging from cellular phones to enterprise servers (LEAP, 2000). It is the result of an international project funded by the European Commission (LEAP, 2000). The project started in January 2000 and ended in June 2002.

In order to meet its goal, the LEAP project decided to start the development of its platform from JADE. In fact, the main result of the project is a new kernel for JADE that allows running legacy JADE agents on handy devices without any modification, provided that the device offers sufficient resources and processing power. When running on a device with no severe resource constraints, LEAP provides the same functionality as the original kernel of JADE. Summarizing, the following are the most important characteristics of LEAP:

1. It runs seamlessly on desktop PCs and on handy devices with limited resources, such as Palm Pilots and Java-enabled phones;
2. It adapts its functionality to the available resources in terms of memory, processing power, screen, etc.;
3. It guarantees connectivity to handy devices via wireless networks, like TCP/IP over GSM and GPRS, and IEEE 802.11 wireless LANs.

Figure 3 shows some pictures of LEAP running on different devices, with different connectivity.

The design of LEAP makes it sufficiently lightweight to execute on a handy device, but also sufficiently flexible and open to be a first-class choice for enterprise servers. LEAP can be deployed according to a set of profiles that identify the functionality available on each particular device. The basic profile supports only the functionality that FIPA requires and it suits the smallest device that we address, i.e., a cellular phone. The full-featured profile provides the functionality of an agent platform designed to run on desktop computers and it copes well with any device with sufficient memory and processing power. The choice of implementing LEAP as a lightweight and extensible kernel for JADE allows using the services that JADE offers across all profiles.

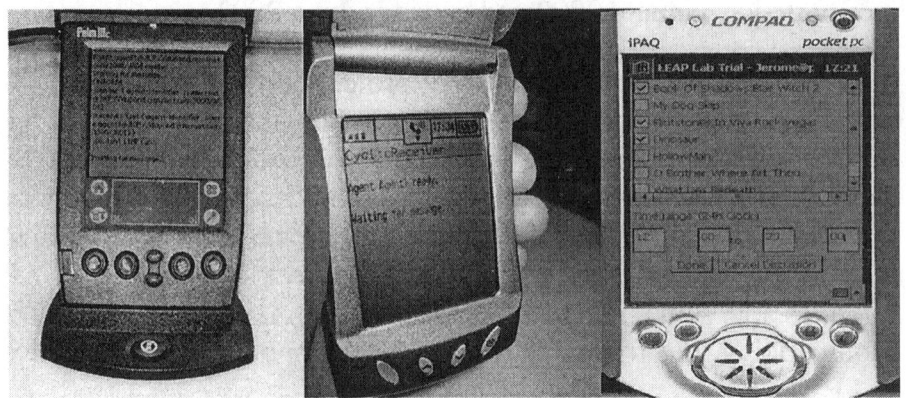

Figure 3. Three devices running LEAP.

In its current implementation, the basic profile provides the subset of the APIs of JADE that do not deal with the behaviour abstraction, and it does not integrate the run-time agent based management tools provided by JADE platform. Such limitations save memory and allow running LEAP on the current implementation of the KVM for Palm Pilots, which reserves only 200Kbytes of heap memory for Java applications. On the contrary, the full-featured profile integrates all tools that JADE provides and, from the developer point of view, it offers the same functionality and the same APIs of JADE. All profiles are instantiations of the FIPA abstract architecture and agents running on platforms configured with different profiles can interoperate. LEAP allows different locations of the same platform to be deployed according to different profiles.

The implementation of LEAP is basically different from that of JADE because the latter was not implemented taking into account the limitations of handy devices. The introduction of the concept of profile required a substantial redesign of the internals of JADE, but we succeeded in it without changing the APIs and agents developed for JADE can still run on LEAP. As a consequence, the community of JADE users can run existing applications on handy devices without any modification, provided that the device offers sufficient resources. Reasonably, such applications will need to spread the computation effort across the fixed infrastructure and the handy device. As applications are already decomposed into agents, the load balancing of the tasks is as simple as allocating agents to network nodes on the fixed and on the mobile network. These deployment issues are transparent to the developer as LEAP implements location transparency.

LEAP is not only meant for handy devices, it is also intended to support enterprise servers and we cannot constrain its functionality when running on devices with no limitations on resources. To achieve this, we decided to go for the worst case, i.e., J2ME CLDC (Sun Microsystems, 2000), and we implemented an extensible architecture over a layer of adaptation. Such a layer is capable of matching the classes available on J2ME CLDC with the ones available on the others Java 2 editions. This allows using the classes that express the maximum functionality where available and others with restricted functionality otherwise, without changing the source code of the agents.

3.1 LEAP Architecture

LEAP is naturally distributed as it is composed of a number of locations that provide the run-time resources that agents need to execute. Mimicking the nomenclature of JADE, we say that locations contain agents and we call them agent containers, or simply containers. We impose no restrictions on the way containers can be deployed across the nodes of the network, but the best way of deploying the platform is having one container running in one Java virtual machine for every node. It is worth noting that splitting the platform into several containers distributed on different devices is not mandatory. According to the context and to the requirements of a given application, it is possible to concentrate the whole platform into a single container or to distribute it across the nodes of a network. For example, if the application consists simply of a personal assistant supporting the user in managing the information on her palmtop, probably the best deployment choice is to have a single-container platform running on the palmtop. On the contrary, given an application where each user in a team is assisted by an agent running on her mobile phone and all such agents interoperate, the choice of a distributed platform composed of one main container and several peripheral containers on the users' phones can be the best solution.

The nodes of the network can be of any kind, from cellular phones to enterprise servers, and they can access indifferently any transport mechanism for which a proper handler in the platform is available. At the moment, only handlers based on TCP/IP are available. Such handlers have been designed to support TCP/IP both over wired and wireless connections, such as GSM, GPRS and IEEE 802.11, exploiting the possibility of using full-duplex connections.

The choice of spreading the platform across the network poses some problems for agents running on other platforms to communicate with agents running on LEAP. Following the design of JADE, we solved this problem introducing a privileged container, called main container, to allow agents on other platforms to see the platform as a whole. The main container is unique and it acts as a front-end for the platform. It maintains platform-wide information and provides platform-wide services. This container must be always reachable by all active containers in the platform. A LEAP platform is composed of a single main container and a set of peripheral containers, allowing a high modularity by running lightweight peripheral containers on handy devices. The main container provides FIPA mandatory services, i.e., the AMS and the DF, and it is mandatory to preserve FIPA compliancy. The amount of resources needed by the AMS and by the DF in their current implementation suggests that the main container should run on a full-featured computer.

4 CoMMA

CoMMA (Corporate Memory Management through Agents) is a FIPA compliant multi-agent system for the management of a corporate memory, implemented by using JADE (Gandon *et al.* 2002). It is the result of an international project funded by the European Commission (CoMMA, 2000). The project started in January 2000 and ended in Jannuary 2002. The CoMMA system was completely implemented and used in different companies to offer a helping service for enhancing the insertion of new employees and as a support system for technology monitoring.

The innovative aspect of the system is the integration of several emerging technologies that were generally used separately in the former information retrieval and management systems. These technologies are: agent technology, knowledge modeling, XML technology, information retrieval techniques and machine learning techniques (Euzenat, 1996; Corby & Dieng, 1997; Berney & Ferneley, 1999; Rabarijaona et al, 1999).

The multi-agent approach, relying on loosely-coupled software components, is naturally prone to facilitate integration of different and heterogeneous technologies in one system. The CoMMA developers decided, therefore, to use agents for wrapping information repositories defining the corporate memory, for the retrieval of information, for enhancing scaling, flexibility and extensibility of the corporate memory and to adapt the system interface to the users.

One of the points that makes CoMMA system different from the majority of the former multi-agent information systems is that agents are not only used for the retrieval of information, but also for the insertion of new information in the corporate memory.

The use of JADE increases system modularity and flexibility. The separation between the software platform infrastructure managing agent life-cycle, distribution and communication and the software implementing agent tasks decouples modifications in these two parts. The behavior based agent model offered by JADE allows to separate the software code realizing the different tasks of the agents; therefore, the modification of a task or the introduction of new tasks usually do not cause the modification of other parts of agent code.

Moreover, given that the main complexity of the CoMMA system is given by the interaction between the different types of agents cooperating in the different tasks of the system, the availability in JADE of a FIPA ACL library for agent communication and a set of predefined behaviours for the management of FIPA communication protocols reduces a lot the cost of realizing the multi-agent system

4.1 CoMMA Architecture

The system aims at helping users in the management of an organization corporate memory and in particular at facilitating the creation, dissemination, transmission and reuse of knowledge in an organization. The services offered by the CoMMA system are the result of three main tasks: insertion of XML annotations of new or updated documents, search of existing documents, and autonomous document delivery in a push fashion to provide her/him with information about new interesting documents (Figure 4 shows a schematic view of the CoMMA multi-agent system).

These tasks are performed through the cooperation among different kinds of agents that can be divided in four sub-societies: document and annotation management; ontology (enterprise and user models) management; user management; agent interconnection and matchmaking.

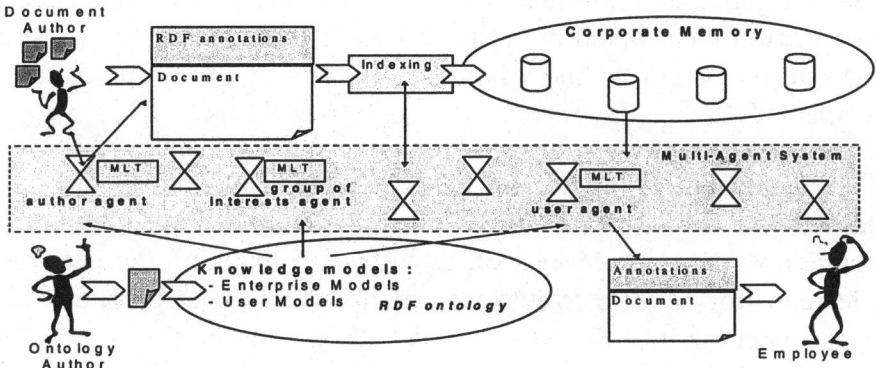

Figure 4. Schematic view of the CoMMA multi-agent system.

The agents belonging to the document dedicated sub-society are concerned with the exploitation of the documents and annotations composing the corporate memory, they will search and retrieve the references matching the query of the user with the help of the ontological agents. A hierarchical organization for the document subsociety has been chosen since separates the task of maintaining document repositories from the task of intelligent interface towards the other agents of the system.

The agents belonging to the ontology dedicated sub-society are concerned with the management of the ontological aspects of the information retrieval activity, especially the queries about the hierarchy of concepts and the different views. The ontology repository, composed of RDF schema forms, maintains a set of concepts and their relationships. Documents of the community are annotated using these ontologies and ontologies are used to search documents into the corporate memory and to navigate into it. In particular, the CoMMA ontology describes the documents maintained in the organization corporate memory and the enterprise model describes the structure of the organization ruling, for example, the access to the different type of documents of the corporate memory. A replicated organization for the ontology sub-society has been chosen since ontologies shared by users should be quite stable and most of the queries will need the whole ontology to apply inference algorithms.

The agents belonging to the user dedicated sub-society are concerned with the interface, the monitoring, the assistance and the adaptation to the user. Moreover, they have to maintain the user profile repository and distribute information about user profile to the agents needing it.

Finally the agents belonging to the interconnection dedicated sub-society are in charge of the matchmaking of the other agents based upon their respective needs.

5 Agentcities

Agentcities is a network of FIPA compliant agent platforms that constitute a distributed environment to demonstrate the potential of autonomous agents. It started on the second half of 2001 as a research project funded by the European Commission (Agentcities, 2001).

One of the aims of the project is the development of a network architecture to allow the integration of platforms based on different technologies and models. It provides basic white pages and yellow pages services to allow the dynamic discovery of the hosted agents and the services they offer.

An important outcome is the exploitation of the capability of agent-based applications to adapt to rapidly evolving environments. This is particularly appropriate to dynamic societies where agents act as buyers and sellers negotiating their goods and services, and composing simple services offered by different providers into new compound services.

To allow the integration of different applications and technologies in open environments, high level communication technologies are needed. The project largely relies on semantic languages, ontologies and protocols in compliance with the FIPA standards.

5.1 Network

The Agentcities network grows around a backbone of 14 agent platforms, mostly hosted in Europe. These platforms are deployed as a testbed, hosting the services and the prototype applications developed during the lifetime of the project. The backbone is an important resource for other organizations, even external to the project, that can connect their own agent-based services, making the network really open and continuously evolving.

Currently, the Agentcities network counts 160 registered platforms. The platforms are based on more than a dozen of heterogeneous technologies, including Zeus (Nwana et al., 1998), FIPA-OS (Poslad et al., 1999), Comtec (Comptec, 1999), AAP (AAP, 1999), Opal (Purvis et al., 2002). More than 2/3 of them are based on JADE and its derived technologies, as LEAP (LEAP, 2000) and BlueJADE (BlueJADE, 2001).

Figure 5. The Agentcities backbone.

5.2 Service Composition

The main rationale for using agents is their ability to adapt to rapidly evolving environments and yet being able to achieve their goals. In many cases, this can only be accomplished by collaborating with other agents and leveraging on the services provided by cooperating agents. This is particularly true when the desired goal is the creation of a new service to be provided to the community, as this scenario often calls for the composition of a number of simple services that are required to create the desired compound service.

The Event Organizer is an agent-based prototype application showing the results that can be achieved using the services provided by the Agentcities project. It allows a conference chair to organize an event, booking all needed venues and arranging all needed services, and then sell the tickets for the new event.

Using the web interface of the Event Organizer, its user can list a set of needed services, fixing desired constraints on each individual service and among different services. The global goal is then split into sub-goals, assigned to skilled solver agents.

The Event Organizer uses the marketplace infrastructure deployed on the Agentcities network to search for relevant venues. These are matched against cross-service constraints and, if found, a proper solution is proposed to the user as a list of services that allow the arrangement of the event. These services are then negotiated on the marketplace with their providers and a list of contracts is returned to the user. Finally, when the new event is successfully organized, the tickets for it can be sold, once again using the marketplace infrastructure.

The process requires the cooperation of a number of partners. Each of them can exploit the directory services to dynamically discover the location of the others. The Event Organizer directly interacts with a Trade House to search for venues and negotiate selected services. Other agent-based applications, as the Venue Finder and the SME Access, are responsible to offer goods on the Trade House and to negotiate them on behalf of their users. A Banking Service takes care of managing the banking accounts of the involved partners, securing all requests against tampering and eavesdropping. An Auction House is used to create auctions and sell the tickets of the new event.

The interesting part of the game is that these tickets are available for other agent-based applications. In fact an Evening Organizer, that helps its user to arrange an evening out, for example booking a restaurant and buying the tickets for a concert, can discover the new event and bid for some tickets on the Auction House.

6 Conclusions

In this chapter, we presented JADE (Java Agent DEvelopment framework), a software framework to aid the development of agent applications in compliance with the FIPA2000 specifications for interoperable intelligent multi-agent systems.

JADE is written in Java language and comprises various Java packages, giving application programmers both ready-made pieces of functionality and abstract interfaces for custom, application dependent tasks. Java was the chosen programming language because of its many attractive features, which are particularly geared towards object-oriented programming in distributed heterogeneous environments.

Starting from the FIPA assumption that only the external behaviour of system components should be specified, leaving the implementation details and internal architectures to agent developers, we produced a very general but primitive agent model that can serve as a useful basis to implement, for example, reactive or BDI architectures. In addition, the behaviour abstraction of our agent model permits an easy integration of external software. For example, we created a JessBehaviour that makes it possible to use JESS (Friedman-Hill, 1998) as agent reasoning engine. In comparison to the agent development tools introduced in the previous section, JADE offers a more efficient implementation and a more general agent model. Such an agent model is more "primitive" than the agent models offered, for example, by RETSINA (Sycara et al., 1996), however, the overhead given by such sophisticated agent models might not be justified for agents that have to perform some simple tasks. In addition, sophisticated agent models such as BDI and reactive architectures, as previously mentioned, can be implemented on the top of our "primitive" agents model.

JADE offers a FIPA ACL library and a set of predefined behaviours for the management of the different FIPA communication protocols. The presence of these components should reduce a lot the cost of realizing multi-agent systems given that interaction between agents is usually the most important feature (or at lest one of the most important features) of this kind of system.

JADE is not the only FIPA-compliant software development system: different systems have been realized or are under development (see, for example AAP (AAP, 1999), ASL (Kerr et al., 1998), Beegent (Kawamura et al., 1999), Comtec (Comtec, 1999), FIPA-OS (Poslad, 1999), Opal (Purvis et al., 2002), Zeus (Nwana et al., 1998)). However, JADE seems the most appreciate and used; in fact, in the agent platforms network, developed inside the agent-cities initiative (Agentcities, 2001) and involving more than 160 nodes, more than 2/3 are realized on the top of JADE or on the top of derived platforms, i.e., BlueJADE (BlueJADE, 2001) and LEAP (LEAP, 2000).

Acknowledgments

Thanks to all the people that contributed to development of JADE and to all the partners of the EC projects introduced in the paper. This work is partially supported by Telecom Italia Laboratories, Torino and by the European Commission through the contracts IST-1999-12217 - CoMMA - Corporate Memory Management through Agents and IST-1999-10211 – LEAP – Lightweight Extensible Agent Platform and IST-2000-28385 - Agentcities.RTD.

References

AAP (1999), AAP Project Home Page," available at http://sf.us.agentcities.net/aap.

Agentcities, (2001), "Agentcities Project Home Page," available at http://www.agentcities.org.

Bellifemine, F., Poggi, A., Rimassa, G. (2001), "Developing multi agent systems with a FIPA-compliant agent framework," *Software Practice & Experience*, 31:103-128.

Berney, B., Ferneley, E., (1999), "CASMIR: information retrieval based on collaborative user profiling," in *Proc of PAAM99*, pp. 41-56, London, U.K.

BlueJADE, (2001), "BlueJADE Project Home Page," available at https://sourceforge.net/projects/bluejade.

CoMMA, (2000), "CoMMA Project Home Page," available at http://www.ii.atosgroup.com/sophia/comma/HomePage.htm.

Comtec (1999), "Comtec Project Home Page," available at http://ias.comtec.co.jp/ap.

Corby, O., Dieng, R., (1997), "A CommonKADS expertise model Web server," in *Proc. Fifth Int. Symp. on the Management of Industrial and Corporate Knowledge*, Compigne.

Euzenat, J., (1996), "Corporate memory through cooperative creation of knowledge base systems and hyper-documents," in *Proc. of Knowledge Acquisition Workshop (KAW'96)*, Banff, Canada.

FIPA, (1997), "FIPA Organization Home Page," available at http://www.fipa.org.

Gandon, F., Poggi, A., Rimassa, G., and Turci, P. (2002), "Multi-agents corporate memory management system," *Applied Artificial Intelligence*, 9-10 (22): 699-720.

Genesereth, M.R.., Ketchpel, S.P., (1994), "Software agents," *Communications of ACM*, 37(7):48-53.

Friedman-Hill, E.J., (1998), "Java Expert System Shell, available at http://herzberg.ca.sandia.gov/jess.

JADE, (1999), "JADE Project Home Page," available at http://jade.cselt.it.

Jennings, N.R., (2001), "An agent-based approach for building complex software systems," *Communications of the ACM*, 44(4):35–41.

Karamcheti, V., Plevyak, J., Chien, A., (1996), "Runtime mechanisms for efficient dynamic multithreading," *Journal of Parallel and Distributed Computing*, 37:21-40.

Kawamura, T., Yoshioka, N., Hasegawa, T., Ohsuga, A., Honiden, S., (1999), "Bee-gent: bonding and encapsulation enhancement agent framework for development of distributed systems," *Proc. of the 6th Asia-Pacific Software Engineering Conference*, Takamatsu, Japan.

Kerr, D., O'Sullivan, D., Evans, R., Richardson, R., Somers, F., (1998), "Experiences using intelligent agent technologies as a unifying approach to network and service management," in *Proc. of IS&N 98*, Antwerp, Belgium.

LEAP, (2000), "LEAP Project Home Page," available at http://leap.crm-paris.com.

Milojicic, D., Breugst, M., Busse, I., Campbell, J., Covaci, S., Friedman, B., Kosaka, K., Lange, D., Ono, K., Oshima, M., Tham, C., Virdhagriswaran, S., White J. (1998), "MASIF - the OMG mobile agent system interoperability facility," *Proc. 2nd Int. Workshop Mobile Agents*, pp. 50-67, Springer-Verlag, Berlin, Germany.

Nwana, H.S., Ndumu, D.T., Lee, L.C., (1998), "ZEUS: an advanced tool-kit for engineering distributed multi-agent systems," *Proc of PAAM98*, pp. 377-391, London, U.K.

Patil, R.S., Fikes, R.E., Patel-Scheneider, P.F., McKay, D., Finin, T., Gruber, T., Neches, R., (1992), "The DARPA knowledge sharing effort: progress report," *Proc. Third Conf. on Principles of Knowledge Representation and Reasoning*, pp 103-114. Cambridge, MA.

Poslad, S., Buckle, P. and Hadingham, R. (1999), "The FIPA-OS Agent Platform: Open Source for Open Standards," available at http://fipa-os.sourceforge.net.

Purvis, M., Cranefield, S., Nowostawski, M., Ward, R., Carter, D., and Oliveira, M.A., (2002), "Agentcities interaction using the Opal platform," *Proc. of the Workshop – Agentcities: Research in Large-Scale Open Agents Environments, AAMAS 2002*, Bologna, Italy.

Rabarijaona, A., Dieng, R., Corby, O., (1999), "Building a XML-based corporate memory," *Proc. European Knowledge Acquisition Workshop (EKAW'99)*, LNAI 1621, Springer-Verlag, Berlin, Germany.

Somacher, M., Tomaiuolo, M., Turci, P. (2002), "Goal delegation in multiagent system," *Proc. of the AIIA 2002*, Siena, Italy.

Sun Microsystems, (2000), "Java 2 Platform Micro Edition (J2ME) Technology for Creating Mobile Devices," available at http://java.sun.com.

Sycara, K., Pannu, A., Williamson, M., Zeng, D., (1996), "Distributed intelligent agents," *IEEE Expert*, 11(6):36-46.

Wooldridge, M., Jennings, N.R., (1995), "Intelligent agents: theory and practice," *The Knowledge Engineering Review*, 10(2):115-152.

Wooldridge, M., (2002), *An Introduction to Multiagent Systems*, John Wiley & Sons Ltd., Indianapolis, IN.

Chapter 7

A Collective Can Do Better

N.D. Monekosso and P. Remagnino

Can we devise simple solutions to complex problems? Is it possible to do so by making use of elemental modules, which when collaborating create emerging intelligence? The answer is yes. No complex mathematical models are required. Nature offers a variety of techniques that lend themselves well to solving complex problems making use of simpler atomic entities. Insects, for instance, as individuals are very simple, but in a collective they are powerful systems able to solve very complex tasks. This chapter describes how the insect world can inspire engineers and computer scientists to devise simple solutions to complex problems. After all the simplest solution is always the best.[1]

1 Introduction

Nature demonstrates that complex behaviour can emerge from the interaction of simpler living creatures. Ants, bees, wasps are well known examples of insect species that through cooperation build organised societies. Can this emerging intelligence be reproduced to solve complex problems, usually solved with complex mathematical models? It is possible when simple elemental modules are devised to interact intelligently. Intelligence can indeed emerge either directly through direct communication between modules dedicated to sub-

---

[1]One should not increase, beyond what is necessary, the number of entities required to explain anything, Occam's Razor, and nature is as simple as it gets.

problems, or indirectly by means of a common space available for all modules to share their experience in order to achieve better performance as a whole system.

Nature offers good examples. Ants and bees in the real world solve difficult problems, not as individuals but as a collective, providing a holistic solution through simple communication and cooperation. In nature communication among insect species occurs through the secretion of chemicals, each one having a different purpose, to convey a different type of message. Individual insects secrete minute quantities of pheromone to exchange information with insects of the same species (Hölldobler and Wilson 1995). Other insects detect the information and join in the effort to accomplish a task, for example finding the shortest path to a food source.

Computer scientists have workied on models of intelligence of varying degree of complexity for decades now. The idea of storing information in a common repository in not uncommon. Initial models of the human mind were mimicked by rule-based systems, where emerging but predictable functionality was built on rules and meta-rules. Learning was added later on, allowing, for instance, for real time generation of rules and meta-rules (Giarratano 1998). In the early 80s blackboard systems were introduced, where knowledge sources operate on data and information stored on the blackboard, a common repository. No direct communication between knowledge sources was allowed only through the blackboard system. Distributed blackboard systems were also devised to create a pervasive layer over an array of processors capable of spawning a large number of processes. A flavour of the 90s was the agent paradigm, introduced to make smaller, modular and more portable individual units, capable of solving specific tasks and collaborating through more or less complex ontologies to solve complex problems.

The common denominator of all techniques presented in the Computer Science and Artificial Intelligence literature are simplicity, modularity, portability and what we could call *thin* intelligence (drawing an analogy to the concept of thin client). We have complex problems, we want simple solutions. Simple means that we may dissect our problems into smaller ones, and we can indeed build small processing units capable of little in their individuality, but with greater potentials as a collective.

In this chapter we will only mention methods using indirect communication. Direct communication, might be indeed necessary, but it usually causes higher overheads: definition of ontologies, protocols of communication and layers of abstraction. The main aim here is to show simple solutions, we are concerned with those making use of indirect communication, almost casual, based on, what one might define, random events. Indirect communication is the catalyst between local and global solutions, it is biologically inspired and hopefully mimicking nature means developing robust and effective solutions.

The chapter is organised as follows. Section 2 describes the behaviour of some insect species, with particular stress on the ants. Section 3 presents a survey of theoretical and applied work inspired by the insect mechanisms. Section 4 introduces a novel technique which combines Reinforcement Learning with the chemical-based insect communication. Section 5 describes in details some applications and speculates on how a collective can provide an optimal solution to a complex problem.

2 Insect Behaviour Can Be Inspiring

Insects are capable of complex behaviour through collaboration. For insatnce ants can find the shortest path between their nest and a food source (Beckers *et al.* 1990, Beckers *et al.* 1992, Goss *et al.* 1989) without vision, but through an indirect exchange of information (Hölldobler and Wilson 1995). Ants secrete a pheromone chemical on the trail as they move along in the search for food or resources to construct a nest. The pheromone chemical evaporates with time, nevertheless ants follow a pheromone trail and at branching points prefer to follow the path with the higher concentration. On finding the food source, ants return laden to the nest laying down more pheromone along the way thus reinforcing the pheromone trail. Ants that have followed the shortest path are quicker to return to the nest, reinforcing the pheromone concentration on the shorter trail at a faster rate than those ants that followed an alternative longer route. Further ants arriving at the branching point choose to follow the path with the higher concentrations of pheromone thus reinforcing even further the pheromone concentration and eventually most ants follow the shortest path. The amount of pheromone secreted is a function of an angle between the path and a line joining the food and nest locations (Cammaerts-Tricot 1974) on the return journey. So far two properties of pheromone secretion were mentioned: aggregation and evaporation. The concentration adds when ants secrete pheromone at the same location, and over time the concentration gradually reduces by evaporation. A third property is diffusion. The pheromone at a location diffuses into neighbouring locations. Bees, wasps and termites also show a strong propensity towards a organised collective behaviour. Like any society in this world a colony is made of categories of individuals, each one having a different purpose. Each catagory contributes to the common wealth of the colony. Mathematics and software solutions have indeed been used in similar fashion. Optimisation techniques globally minimise or maximise a functional by applying local computations. Software techniques offer pervasive and virtually unlimited resources by creating an invisible layer that covers, for instance a multi-processor system. World Wide Web search engines solve information mining problems by extracting information with an ensemble of modular processing units; all crawling the web to serve the global purpose of reporting back to the user meaningful information user. In all given examples small processing units are employed to solve local problems and cooperate with other units to solve larger, global problems.

3 Can Nature Be Mimicked?

Insect behaviour has been studied by entomologists, for the understanding of insect species. Computer scientists and engineer have used results of entomology research as an inspiration for developing self-organising problem-solving computational systems. The following sections focus on research whose purpose is to develop artificially intelligent problem-solving modular units. The work described in the following is broadly categorised as solutions to classic AI problems and applications in telecommunications and robotics.

3.1 Applying Real Ant Behaviour to Computational Systems

Earlier in Section 2 the emerging behaviour of an ant colony was mentioned. The ant implicit communication takes place by modifying the environment to *encourage* other ants to choose to perform one action over another. An example of emerging functionality occurs when a single ant or a small group of ants can not move a prey. Through indirect communication ants *recruit* other ants from the colony. The newly recruited ants join the original *team* and together they share total load, each one of them responsible for a smaller portion of it. Ants communicate by secreting a trail of pheromone. They communicate modifying the environment locally. Other ants in the vicinity will be influenced by the action of the ant that passed through that same part of the environment. The ants that arrive at a later time might choose to make use of the information left by others. The nature of the information thus transmitted is very localised and is eventually lost (through evaporation) over time with the result that *poor* solutions can be forgotten. The solution to a problem is possible or faster to find or more optimal only through cooperation and communication amongst a large group of individuals. Each individual, or processing unit takes care of its own current locality, leaves behind a message, which is laid for others to read and decode. Effectively a probabilistic concept of frequency is here implemented: the more an information bit is confirmed and accepted the more likely is to be correct. Only information that is deemed correct over long periods of times contributes to the final solution. The following paragraphs detail the necessary *ingredients* for the desing of an biologically inspired computational behaviour.

Pheromone Secretion

The search for a solution requires a large number of iterations. The pheromone update may be made at the end of each iteration once a solution is found. For example, on the outward journey, the artificial ants do not secrete pheromone, only on the return journey and the quantity is a function of the *quality* of the solution (Dorigo et al. 1999). If the measure of the quality of the solution is distance as in the classic Travelling Salesman Problem, the nodes of the shortest route will receive a higher dose of pheromone. Some implementations of the ant algorithms, allow the artificial ants to secrete pheromone at each step during the search for the solution. The choice is very much problem dependent. It has been shown (Dorigo et al. 1999) that the timing of pheromone secretion affects quality of solution for a given problem, one method producing better solutions than the other.

Pheromone Evaporation

This phenomenon occurs naturally though at a slow rate. It is necessary for ants, real or artificial, not to be lead astray toward an incorrect or sub-optimal solution. The effect of evaporation is to minimise *poor* localised decisions that would affect the overall solution. Work carried out by (Monekosso and Remagnino 2004) demonstrates the influence on speed to solution with different evaporation rates.

Real ants do not have an explicit memory of the environment, past visits. Unlike real ants, artificial ants can have memory. In fact, some ant algorithms require to keep track of past states (Dorigo and Gambardella 1997). In a system that updates pheromone at the end of each iteration, it is necessary to keep track of states for the purpose of determining the size of the pheromone secretion during the update phase for each arc (between nodes/states).

A Suitable State-Space

When possible it is advantageous to discretise the problem state-space. Even in large state spaces one can envisage of the creation of *locales* representing a locality split into a number of states. In discretised state spaces artificial units can secrete pheromone, with diffusion into the local neighbourhood. The solution can then be incrementally built by a colony of artificial ants moving in the state space, initially in a random manner and as time elapses, in a more orderly fashion. A variety of policies can be devised as a function of the concentration in the current state and that in surrounding states. The assumption here is that the artificial ant, similarly to its real counterpart, can only move between adjacent states (for instance cells in an N-dimensional discrete space). This is reminiscent of number of computational techniques, including reinforcement learning (RL) (Sutton and Barto 1998). A number of techniques were discussed in (Gambardella and Dorigo 1995), where a connection between the ant optimisation algorithm and reinforcement learning (RL) is suggested. For instance that the Ant-Q, is related to Q-learning (Dorigo *et al.* 1996, Dorigo and Gambardella 1996). The general ant optimisation algorithm is a special case of the Ant-Q family. In (Monekosso and Remagnino 2001), Q-learning and synthetic pheromone are combined. The standard Q-learning update equation is modified to include a term that takes into account pheromone concentration in cells. This additional term reflect the belief an artificial ant has in the information received. As with all ant algorithms, there is a stochastic element in the action selection process, to prevent the ants following a *sub-optimal* path, where the ant remains trapped in a local minimum. The work of (Kawamura *et al.* 2000) takes this approach one step further, implementing a system where multiple ant colonies cooperate to solve an optimisation problem. Inter-colony interactions take place by exchanging information on pheromone. The pheromone is colony-specific. Its is also either positive or negative, the positive pheromone allows one colony to be influenced by another while negative pheromone disallows the interaction.

3.2 Solving Classic Optimisation Problems

By classic problems, we mean problems that are typically used as benchmarks for evaluating new techniques. These problems may not necessarily be *real-life* problems but reflect the complexity of a real-life problem. Some problems solved may be classified as both classic problems and applications. Some of the mechanisms adopted by foraging ants have been applied to classical NP-hard combinatorial optimisation problems with success. These problems include *the Travelling Salesman Problem - TSP, scheduling - the job-shop scheduling, the Quadratic Assignment Problem - QAP, the Vehicle Routing Problem - VRP,* and *the Network Routing Problem.* The **travelling salesman problem**, TSP, is a NP-Hard combinatorial problem. The problem is to find the shortest route through a number of cities, visiting each city once only and ending at the starting city. Even with a relatively small number of cities, e.g. 15+, the problem is computationally intractable due to the large number of possible routes. Using an ant algorithm, it has been shown that optimal or near-optimal solutions can be found after several iterations during which the ants are allowed to travel from city to city until a complete tour is achieved and the ants return to the starting point (Dorigo *et al.* 1996, Dorigo and Gambardella 1997, Gambardella and Dorigo 1995, Stutzle and Dorigo 1999). On the return journey, the ants secrete pheromone, the amount being a function of the route length. The shorter the path, the higher the concentration of pheromone. Subsequent trips made by the ants will make use of the pheromone in the same manner real ants do. After many iterations, the ants will eventually converge to an optimal solution.

A problem related to the TSP problem is the **Vehicle Routing problem** (VRP). VRP reduces to a TSP for each assigned customer to vehicles (Bullnheimer *et al.* 1997, Bullnheimer *et al.* 1999). An instance of the **Job-Shop Scheduling** problem is found in manufacturing. Briefly described, consider a number, M, of machines and a number, J, of jobs. Each job is an ordered sequence of operations and each operation makes use of a machine for a fixed time. The problem is to assign the operations (for each job) such as to minimise the completion time of all jobs, ensuring that no machine is ever assigned more than one job at a time. In (Colorni *et al.* 1993), the ant algorithm is applied to the job-shop scheduling problem and to bus driver scheduling (Forsyth and Wren 1997). The **Quadratic Assignment Problem** is another optimisation problem that has been solved using an ant algorithm. The generalised QAP problem can be stated as "assigning a set of facilities to a set of locations with given distances between the locations and given flows between the facilities" (Stutzle and Dorigo 1999). One instance of the QAP can be summarised as follows: the manufacturing of goods must take place at different factory sites. The problem is to find the best assignment of goods to factories whilst minimising the distance the goods must travel. A solution to this problem is presented in (Stutzle and Dorigo 1999, Gambardella *et al.* 1998). The last application mentioned in this section is the ant algorithm applied to the *Long Term Car Pooling problem* (Maniezzo *et al.* 2001).

3.3 Telecommunications Applications

The objective of routing in a telecommunication network is to forward packets from source to destination via a number of nodes. There will be more than one route that the packet may take to arrive at the destination. Routing is done by maintaining a table for each node, indicating which path incoming data must take to arrive at its destination (next node). A typical example of network is the world wide web. The algorithm is required to find the best path for messages, avoiding congested areas, maximising throughput (and minimising packet delay) and minimising costs. The ant algorithm applied to network routing is described in the AntNet, AntNet-CL and AntNet-CO, telecommunication routing applications ((Di Caro and Dorigo 1997, Di Caro and Dorigo 1998)).

3.4 Robot Navigation Applications

Work combining ant foraging mechanisms and reinforcement learning is described in (Leerink et al. 1995). A simulated robot navigates a maze using a technique that mimics the mechanisms of ant behaviour. Three mechanisms [2] found in ant trail formation were used as exploration strategy in the robot navigation task. The maze is a grid of cells, corresponding to states in which pheromone aggregates and evaporates over time. The robots receive a positive reward for reaching the goal state and a negative reward for colliding with an obstacle. The objective for the robot is to maximise the rewards received. The action selection mechanism (which cell to visit next) makes use of the pheromone information.

The ant-like self-organising behaviour has also been used to co-ordinate a multi-robot system in (Vaughan et al. 2000, Vaughan et al. 2002) where the robots transport objects between different locations. In this instance, rather than physically laying a trail of synthetic pheromones, the robots communicate path information via shared memory. The (synthetic) pheromone is replaced by way-point coordinates which the robots share information about.

3.5 Other Robotic Applications

In (Parunak and Brueckner 2000, Parunak et al. 2001, Parunak et al. 2002), the pheromone trails are used to construct potential fields. Unmanned vehicles use these to navigate, guided by the potential

---

[2]The mechanisms are: bidirectional trail laying, U-turns, and angle sustained by the path and nest-food axis (Leerink et al. 1995)

gradients. In this implementation of the ant system, there are two or more *flavours* of pheromone. Each pheromone type is unique in that it is associated with a particular feature of the environment and may have its own evaporation and/or diffusion rates resulting in different dynamics. Bee behaviour has also been used for homing applications in robotics. For instance (Rizzi 1998) uses image information to estimate the distance between the current robot position and the homing position. Their technique borrows from the world of flying insects such as bees and wasps, able to return to specific locations for foraging or homing by taking snapshots of the surroundings and matching the current view with the target location. The work of (Srinivasan 1996) has also made strong link with the insect world, in particular the world of bees.

3.6 Image Processing Applications

Insect behaviour has also been applied image processing. Basic image processing can in fact be reproduced by employing simple units, able to analyse the chracteristics of an image, and then leave a trail for others to use or ignore. This communication without interaction has been employed in image processing for edge detection (Ramos *et al.* 2000) and region segmentation (Bourjot *et al.* 2002).

4 Combining Reinforcement Learning and Synthetic Pheromones

In this Section ant algorithms that combine synthetic pheromones with a machine learning technique, namely Reinforcement Learning (RL), will be described as an example. First RL is explained.

4.1 Reinforcement Learning

Reinforcement Learning (RL) (Bertsekas 1996, Sutton and Barto 1998) is a machine learning technique that allows an agent to learn by trial and error which action to perform by interacting with the environment. Models of the agent or environment are not required. At each discrete time step, the agent selects an action given the current state and executes the action, causing the environment to move to the next state. The agent receives a reward that reflects the *value* of the action taken. The objective of the agent is to maximise the sum of rewards received when starting from an initial state and ending in a goal state. One form of RL is Q-Learning (Watkins 1989). The objective of Q-learning is to generate Q-values (quality values) for each state-action pair. At each time step, the agent observes the state s_t, and takes action a. It then receives a reward r dependent on the new state s_{t+1}. The reward may be discounted into the future, meaning that rewards received n time steps into the future are worth less by a factor γ^n than rewards received in the present. Thus the cumulative discounted reward is given by (1)

$$R = r_t + \gamma r_{t+1} + \gamma^2 r_{t+2} + \cdots + \gamma^n r_{t+n} \qquad (1)$$

where $0 \leq \gamma < 1$. The Q-value is updated at each step using the update equation (2) for a non-deterministic Markov Decision Process (MDP)

$$\hat{Q}_n(s_t, a) \longleftarrow (1 - \alpha_n)\hat{Q}_{n-1}(s_t, a) + \\ \alpha_n(r_t + \gamma \cdot max_{a'}\hat{Q}_{n-1}(s_{t+1}, a')) \qquad (2)$$

where $\alpha_n = \frac{1}{1+visits_n(s_t,a)}$. Q-learning can be implemented using a look-up table to store the values of Q for a relatively small state space. Neural networks may be used for the Q-function approximation.

4.2 Synthetic Pheromones and Q-Learning

Synthetic Pheromone and Q-Learning are combined to produce ant inspired algorithms. Unlike Q-Learning and reinforcement learning in general that are based on a single agent, the ant algorithms for

A Collective Can Do Better

their success rely on the emergence of a problem solving behaviour by combining the simple behaviours of a relatively large number of agents/ants. As with RL, the ants learn by trial and error, searching for the optimal (or near-optimal) solution. The characteristics that make a system an ant system are (Dorigo and Gambardella 1996):

1. As with Q-Learning, an action selection rule is required.
2. Whereas the Q-Learning uses a (pseudo) random action selection rule, in the ant inspired algorithm, the random action selection is biased according to heuristics that are problem dependent. For example (Dorigo and Gambardella 1996) use the inverse of the distance for an ATSP problem.
3. A reward (reinforcement) is given to the ants. There are a number of ways to assign rewards, mainly problem dependent. Rewards may be given at each step or at completion of a cycle when a solution is found.

A Q-table is maintained and the values updated according to an update rule cf. the Q-Learning update rule in Equation 2. The algorithm for the ant inspired Q-learning follows that of standard Q-Learning (Sutton and Barto 1998). The main difference being there are several agents and the update rule must be applied to all. Works combining reinforcement learning and synthetic pheromone are (Dorigo and Gambardella 1996, Leerink *et al.* 1995, Vaughan *et al.* 2000, Monekosso and Remagnino 2001). The latter is the authors' work and will be described in greater detail, for further information on Ant-Q please refer to the relevant publications.

Ant-Q The ANT-Q algorithm was developed by (Gambardella and Dorigo 1995). At every step, each ant updates the AQ-values in the AQ-table according to the update rule. This means that each ant contributes to learning a global policy.

Evaluation of ANT-Q The performance measure once a solution is achieved is based on that of the best ant solution for the ANT-Q algorithm. The added value of ant cooperation and the effect of the parameters are analysed for the ANT-Q in (Dorigo and Gambardella 1996). They report that *turning-off* ant cooperation produces poorer results on a given problem. Furthermore by varying the number of ants used to solve a ATSP problem, they showed that performance improved with increasing number of ants whilst maintaining a fixed number of iterations. The performance turned out to be less sensitive to the number of ants when the number of tours was kept fixed however the problem still required a large number of ants to achieve the optimal solution. The variation of the learning parameters indicate that a certain amount of exploration for action selection is necessary to achieve good performance.

Phe-Q The Phe-Q algorithm (Monekosso and Remagnino 2001) combines Q-Learning (Watkins 1989) and synthetic pheromone, by introducing a belief factor into the update equation (see Equation 2). The belief factor is a function of the synthetic pheromone concentration on the trail and reflects the extent to which an ant is influenced by the information communicated by other ants belonging to the same colony. The usefulness of the belief factor is that it allows an ant to selectively make use of the implicit communication from other ants where the information may not be reliable (incomplete and uncertain information are critical issues in the design of real world systems) due to changes in the environment. The Phe-Q algorithm was tested on a obstacle course with food located at one or more cells in the grid. The algorithm was tested on several courses of varying complexity in terms of size and number and placement of obstacles. The goal for the ants was to locate the food and return to the nest. The synthetic pheromone $\Phi(s)$ is a scalar value (where s is a state or cell in a grid) that comprises three components: aggregation, evaporation and diffusion. The cell pheromone concentration $\Phi(s)$ is updated at each step during the search and after a solution is found when the ants are returning to the starting point. Experience showed that better performance was achieved when the amount of pheromone secreted on the return path was higher than that secreted during the search.

As said earlier, the belief factor (B) governs the extent to which an ant believes in the pheromone that it detects. During early training episodes, the ant will believe to a lesser degree in the pheromone map because all ants are biased towards exploration. The belief factor is given by (3)

$$B(s,a) = \frac{\sum_{s \in N_a} \Phi(s)}{\sum_{\sigma \in N_a} \Phi(\sigma)} \qquad (3)$$

where $\Phi(s)$ is the pheromone concentration at a cell, s, in the environment and N_a is the set of neighbouring states for a chosen action a. The Q-Learning update equation modified with synthetic pheromone is given by (4)

$$\begin{aligned}\hat{Q}_n(s_t, a) \longleftarrow &\ (1 - \alpha_n)\hat{Q}_{n-1}(s_t, a) + \\ &\ \alpha_n(r_t + \gamma\prime \cdot max_{a'}(\hat{Q}_{n-1}(s_{t+1}, a') + \\ &\ \xi B(s_{t+1}, a'))\end{aligned} \qquad (4)$$

where the parameter, ξ, is a sigmoid function of time (*epochs* ≥ 0). The value of ξ increases as the number of ants successfully accomplish the task i.e. locate a food source and return to the nest. It can be shown that the pheromone-Q update equation converges for a non-deterministic MDP. [3].

In addition to the Q-Learning parameters, there are a number of other parameters must also be fine tuned. In (Monekosso and Remagnino 2004) the procedure for fine tuning all the parameters is described. The parameters that influence the Phe-Q learning are the number of ants, the secretion rate, diffusion rate, evaporation rate of the pheromone and finally the pheromone saturation level. The coefficients of the sigmoid (function of time) that modulate the belief function were found experimentally to influence the convergence

---

[3]The proof can be found in (Monekosso and Remagnino 2004)

rate. The fine tuning of the parameters for optimum speed of convergence can be cast as an optimisation problem. The function to minimise is the area under the convergence curve (the root mean square of the error between Q-values obtained during successive epochs). This area is a function of the following parameters: the number of agents, the secretion rate, the evaporation rate, the diffusion rate, and saturation level. The search for a global minimum in the data was achieved using a gradient projection method (Bertsekas 1996). A b-spline model produced the function value at arbitrary points together with the first derivative. Several local minima were found by perturbing the starting point for the minimum search. The global minimum was taken as the minimum of all local minima. This optimisation technique was selected based on the fact that the closed form function was not available.

Evaluation of Phe-Q The cooperative behaviour was verified by *turning-off* pheromone communication. In all maze layout, the cooperative ants outperform the non-cooperative ants. The measure of performance being the number of iterations required to locate the food. It was seen that the information exchange allows the agents to learn faster when groups of communicating and non-communicating agents were compared. However there is a price to pay for this information sharing. Too much information i.e. too high a pheromone secretion rate or too low a pheromone evaporation rate resulted in (not unexpectedly) degraded results especially in the earlier learning stages where ants are misled by other exploring ants. If the communication is sufficiently high (high pheromone secretion or high number of ants) then the performance degrades to that of non-communicating ants due to the *ant-misleading-ant* effect. The optimal pheromone parameter set selection must be such as to minimise the misleading effect whilst maximising learning. The effect of misleading information, *interference*, can be minimised by careful choice of the evaporation rate and belief modulating function (for a given secretion rate), particularly in the exploratory phases. Having found the optimum parameter set as described above, it is seen that performance is statistically similar for any ant group size andenvironment complexity. Comparing communicating and non-communicating ants in solving problems of increasing complexity, the performance gap is statistically similar however at the highest complexity, the gap decreases marginally. Using the optimum parameter values, the communicating ants proved to be less sensitive to incorrect information than initially thought.

The issue of adaptability was also investigated. The real-world is dynamic, constantly changing requiring problem solvers that can adapt to change (or recover) with minimal delay when problem solving is taking place in real-time. Because the ant behaviours are relatively simple, these behaviours adapt better to change. This applies to both communicating and non-communicating ants. The communicating ants have an additional advantage in that each individual response to environmental change (such as moving target) is communicated to other ants in the group, thus speeding up the adaptation (re-learning) process. There are two aspects to the adaptation: first the duration of the disturbance and second, the intensity of the disturbance. In both these aspects the communicating ants perform better. The shorter recovery time is a considerable advantage in an on-line system. The communicating ants recover on average 50 episodes earlier (over 100 runs) than the non-communicating ants. The results show that for a given obstacle course, the required number of ants to optimise learning is low, this means that the computational load can be kept relatively low. It must be said though that the computational requirements (processing power) of the non-communicating ants are lighter (no message passing) than that for communicating ants. This is to be expected but this does not offset the learning improvements because of the nature of the implicit communication.

5 Cooperative Robotic Transport

This last Section briefly touches on the field of cooperative robotic transport. This field is still in its infancy, the problems tackled (e.g. box pushing (Kube and Bonabeau 2000), object transport (Vaughan *et al.* 2002)) are simple compared to optimisation problems. This application uses, in common with those described above, stigmergy. It is inspired by the cooperative transport mechanisms of various ant species. Ants have been shown to cooperate and carry a load that a single ant cannot carry. Robotics practitioners, inspired by ants, have implemented cooperative transport (Kube and Bonabeau 1998).

An important problem that robotics practitioners face is stagnation (Bonabeau *et al.* 1999). This occurs when the ants can no longer move the object. The cause may be external, e.g. an obstacle or the ants are positioned such that the resultant force applied is zero. Kube et al in (Kube and Zhang 1995) adopted the approaches used by real ants to overcome this problem, namely realignment and repositioning.

6 Conclusions

Although distributed problem solving has long been employed, the ant system takes that approach one step further adapting mechanisms found in the natural world. The key feature of these natural systems is the stigmergy phenomenon whereby through implicit communication and cooperation, relatively simple individuals can solve complex and large problems. The size and complexity of the problem being such that an individual alone cannot solve it. Alternatively, the problem would be solved more efficiently with several problem solvers combining their effort.

A common feature of all artificial ant systems is the stigmergy resulting from the *work* of the individuals in the colony. It is found that the overall performance is dependent on a number of interacting agents. For all problems used as test beds, the effect of *turning-off* the communication is poorer performance. There is however the risk of interference, one misguided ant misleading others in the colony. This phenomenon can be dealt with using evaporation. All the above mentioned systems rely on pheromone evaporation to enable poor solutions to be eventually *forgotten*. The effective rates of evaporation are problem dependent as are the method and timing of pheromone secretion.

The purpose of these systems is to solve efficiently complex problems taking inspiration from nature. However it is not necessary to follow strictly the real (natural) ant behaviour. These behaviours can be augmented with problem dependent heuristics which improve the quality of the solutions found.

Ant systems have proved successful in optimisation and similar problems. The application to robotics, in particular mobile robotics, is in its infancy. The robotics problems tackled are relatively simple and have been solved more efficiently by using more traditional techniques. The benefits of the ant approach will be most evident with the advent of miniaturised robots on the nano-scale.

References

Anderson, C., Blackwell, P.G., and Cannings, C. (1997), "Simulating ants that forage by expectation", *Proceedings of the 4Th Conference on Artificial Life*, pp. 531-538.

Beckers, R., Deneubourg, J. L., Goss, S., and Pasteels, J. M. (1990), "Collective decision making through food recruitment", *Insect Sociaux*, vol. 37, pp. 258-267.

Beckers, R., Deneubourg, J.L., and Goss, S. (1992), "Trails and Uturns in the selection of the shortest path by the ant Lasius niger", *Journal of Theoretical Biology*, vol. 159, pp. 397-415.

Bertsekas, D.P., and Tsitsiklis, J.N. (1996), "Neuro Dynamic Programming", Athena Scientific.

Bonabeau, E., Dorigo, M., and Theraulaz, G. (1999), "Swarm intelligence, From Natural to Artificial Systems", Oxford University Press.

Bourjot, C., Chervier, V., and Thomas, V. (2002), "How social spiders inspired an approach To region detection", *Proceedings of the first international joint conference on Autonomous agents and multiagent systems*, Part 1, pp. 426-433.

Bullnheimer, B., Hartl, R.F., and Stauss, C. (1997), "An Improved Ant System Algorithm for the Vehicle Routing Problem", *6th Viennese workshop on Optimal Control, Dynamic Games, Nonlinear Dynamics and Adaptive Systems*, vol. 89, pp. 319-328.

Bullnheimer, B., Hartl, R.F., and Stauss, C. (1999), "Applying the Ant System to the Vehicle Routing Problem", *Meta Heuristics: Advances and Trends in Local Search Paradigms for Optimization*, Kluwer.

Cammaerts-Tricot, M.C. (1974), "Piste et Pheromone attraction chez la fourmi myrmica ruba", *Journal of Computational Physiology*, vol. 88, pp. 373-382.

Colorni, A., Dorigo, M., and Theraulaz, G. (1991), "Distributed optimzation by ant colonies", *Proceedings of First European Conference on Artificial Life*, pp. 134-142.

Colorni, A., Dorigo, M., and Maniezzo, V. (1993), "Ant system for job shop scheduling", *Belgian Journal of Operational Research, Statistics and Computer Science*, vol. 34, 39-53.

Deneubourg, J.L., and Goss, S. (1989), "Collective patterns and decision making", *Ethol. Ecol. and Evol.*, vol. 1, pp. 295-311.

Deneubourg, J.L., Beckers, R., and Goss, S. (1992), "Trails and U-turns in the selection of a path by the ant Lasius niger", *Journal of Theoretical Biology*, vol. 159, pp. 397-415.

Di Caro, G., Dorigo, M. (1997), "Ant-Net: a mobile agents approach to adaptive routing", *Technical Report IRIDIA 97 12, Universite Libre de Bruxelles, Belgium*.

Di Caro, G., and Dorigo, M. (1998), "Two Ant Colony Algorithms for Best-Effort Routing in Datagram Networks", *Proceedings of International Conference on Parallel and Distributed Computing and Systems*, pp. 541-546.

Dorigo, M., Maniezzo, V., and Colorni, A. (1996), "The Ant System: Optimization by a colony of cooperatin agents", *IEEE Transactions on Systems, Man, and Cybernetics*, vol. 26, pp. 1-13.

Dorigo, M., and Gambardella, L.M. (1996), "A Study of Some Properties of ANT-Q", *Proceedings of Fourth Conference on Parallel Problem Solving From Nature*, pp. 656-665.

Dorigo, M., and Gambardella, L.M. (1997), "Ant Colony System: A cooperative learning approach to the travelling salesman problem", *IEEE Tranactions on Evolutionary Computing*, vol. 1, pp. 53-66.

Dorigo, M., Di Caro, G., and Gambardella, L.M. (1999), "Ant Algorithms for Discrete Optimisation", *Artificial Life*, vol. 3, pp. 137-172.

Forsyth, P., and Wren, A. (1997), "An Ant System for Bus Driver Scheduling", *7th International Workshop on Computer Aided Scheduling of Public Transport*, pp. 252-260.

Gambardella, L.M., and Dorigo, M. (1995), "AntQ: A Reinforcement Learning approach to the traveling salesman problem", *Proceedings of the 12Th International Conference on Machine Learning*, pp. 252-260.

Gambardella, L.M., Taillard, E.D., and Dorigo, M. (1999), "Ant colonies for the Quadratic Assignment Problem", *Journal of Operational Research society*, vol. 50, pp. 167-176.

Giarratano, J.C. (1998), "Expert Systems: Principles and Programming", Brooks Cole.

Goss, S., Aron, S., Deneubourg, J.L., and Pasteels, J.M. (1989), "Self organized shorcuts in the Argentine ants", *Naturwissenschaften*, pp. 579-581.

Hölldobler, F., and Wilson, R. (1995), "Journey to the Ants, A story of scientific exploration", Library of Congress".

Jaakkola, T., Jordan, M.I., and Singh, S.P. (1994), "On the convergence of stochastic Iterative Dynamic Programming Algorithms", *Neural Computation*, vol. 6, pp. 1185-1201.

Kawamura, H., Yamamoto, M., uzuki, K., and Ohuchi, A. (2000), "Multiple Ant Colonies Algorithm Based on Colony Level Interactions", *IEIEC Transactions Fundamentals*, vol. E83-A, No.2, pp. 371-379.

Kube, C.R., and Zhang, H. (1995), "Stagnation Recovery Procedures for Collective Behaviors", *Proceedings of IEEE/RSJ/GI International Conference on Intelligent Robots and Systems*, pp. 1883-1890.

Kube, C.R., and Bonabeau, E. (1998), "Cooperative transport by ants and robots", *Robotics and Autonomous Systems*, pp. 85-101.

Kube, C.R., and Bonabeau, E. (2000), "Cooperative transport by ants and robots", *Robotics and Autonomous Systems*, pp. 85-101.

Leerink, L.R., Schultz, S.R., and Jabri, M.A. (1995), "A Reinforcement Learning Exploration Strategy based on Ant Foraging Mechanisms", *Proceedings 6Th Australian Conference on Neural Nets*, pp.217-220.

Maniezzo, V., Carbonaro, A., and Hildmann, H. (2001), "An ANTS Heuristic for the Long Term Car Pooling Problem", *Proceedings of Second International Workshop on Ant Colony Optimisation*.

Mitchell, T.M. (1997), "Machine Learning", *McGraw-Hill*.

Monekosso, N., and Remagnino, P. (2001), "Phe-Q: A pheromone based Q learning", *Proceedings of 14Th Australian Joint Conference on Artificial Intelligence*, pp. 345-355.

Monekosso, N., and Remagnino, P. (2002), "An analysis of the Pheromone Q Learning algorithm", *Proceedings of the 8th Iberoamerican Conference on Artificial Intelligence, IBERAMIA 2002*, pp.224-232.

Monekosso, N., and Remagnino, P. (2004), "The Analysis and Performance Evaluation of the Pheromone Q Learning Algorithm", *to appear in Expert Systems*, vol. 21(2), pp. .

Ollason, J.G. (1980), "Learning to forage optimally?", *Theoretical Population Biology*, vol. 18, pp. 44-56.

Ollason, J.G. (1987), "Learning to forage in a regenerating patchy environment: can it fail to be optimal?", *Theoretical Population Biology*, vol. 31, pp. 13-32.

Van Dyke Parunak, H., and Brueckner, S. (2000), "Ant-Like Missionnaries and Cannibals: Synthetic pheromones for distributed motion control", *Proceedings of Fifth International Conference on Autonomous Agents*, pp. 467-474.

Van Dyke Parunak, H., Brueckner, S., Sauter, J., and Posdamer, J. (2001), "Mechanisms and Military Applications for Synthetic Pheromones", *Proceedings of workshop on Autonomy Oriented Computation at the Fifth International Conference Autonomous Agents*, pp. 58-67.

Van Dyke Parunak, H., Brueckner, S., and Sauter, J.A. (2002), "Digital pheromone mechanisms for coordination of unmanned vehicles", *Proceedings of the first international joint conference on Autonomous agents and multiagent systems*, pp. 449-450.

Ramos, V., Almeida, F., "Artificial Ant Colonies in Digital Image Habitats - A Mass Behaviour Effect Study on Pattern Recognition", *Proceedings of ANTS'2000, Int. Workshop on Ant Algorithms (From Ant Colonies to Artificial Ants)*, pp. 113-116.

Rizzi, A., Biam G., Cassinis R. (1998), "A bee-inspired visual honfing using color images", *Robotics and Autonomous Systems*, 25, pp. 159-164.

Sabouret, N., and Sansonnet, J.P. (2001), "Learning Collective Behaviour from Local Interaction", *Proceedings of CEEMAS 2001*, LNAI-LNCS 2296, Springer Verlag, pp. 273-282.

Sauter, J.A., and Matthews, R., and Van Dyke Parunak, H., and Brueckner, S. (2002), "Evolving adaptive pheromone path planning mechanisms", *Proceedings of the first international joint conference on Autonomous agents and multiagent systems*, pp. 434-440.

Srinivasan, M.V., Zhang, S.W., Lehrer, M., and Collett, T.S. (1996), "Honeybee Navigation en route to the goal: visual flight control and odometry", *The Journal of Experimental Biology*, 199, pp. 237-244.

Stutzle, T., and Dorigo, M. (1999), "ACO Algorithms for the Quadratic Assignment Problem, New Ideas in Optimization", *McGraw Hill Press*.

Stutzle, T., and Dorigo, M. (1999), "ACO Algorithms for the the Traveling Salesman Problem, Evolutionary Algorithms in Engineering and Computer Science", *John Wiley and Sons*.

Sutton, R.S., and Barto, A.G. (1998), "Reinforcement Learning", *MIT Press*.

Vaughan, R. T., Stoy, K., Sukhatme, G.S., and Mataric, M. J. (2000), "Whistling in the dark: Cooperative trail following in uncertain localization space", *Proceedings of 4Th International Conference on Autonomous Agents*, pp. 373-380.

Vaughan, R. T., Stoy, K., Sukhatme, G. and Mataric, M.J. (2002), "LOST: Localization Space Trails for Robot Teams", *IEEE Transactions on Robotics and Automation, special issue on Advances in MultiRobot Systems*, vol. 18, pp. 796-812.

Watkins, C. J. C. H. (1989), "Learning with delayed rewards", *University of Cambridge*.

Chapter 8

Coordinating Multi-Agent Assistants with an Application by Means of Computational Reflection

A. Di Stefano, G. Pappalardo, C. Santoro, and E. Tramontana

Assistant agents are employed to provide users working with an application with various help services aimed at, e.g., organising user data, performing useful actions on the application on users' behalf, and suggesting suitable, independently obtained, information. In order to better structure the implementation of the (frequently heterogeneous) assistance functionalities, it would often be desirable to introduce, in lieu of a monolithic agent, a group of independent ones, each specialised for a simple, well-defined task. Such assistant agents need to interact with each other and with the application they extend, for the sake of exchanging data and coordinating their tasks.

We tackle these issues by devising a software architecture that provides the necessary support for agents to communicate and synchronise with the application in a totally transparent way for the latter. The key architecture component is the *Coordinator*, which has been designed with the objective to be general enough and independent of the particular group of agents. This allows new assistant agents to be plugged in, or existing ones to be removed, even at runtime. Moreover, the Coordinator is easily reusable to assist a large class of applications, essentially by replacing the component responsible for interacting with the application.

Connection with the application is made possible by the adoption of *computational reflection*. This allows operations of a system part to be observed by another part, which is also capable of influencing the former, when needed. In our architecture, the assisted application represents the observed part, while the Coordinator and the assistant agents are the observing part. In such a way, the Coordinator has been designed and structured in strict separation from the specific assisted application, but can nevertheless be seamlessly connected with it at runtime.

As a case study we consider the design of multi-agent assistance aimed at facilitating e-commerce activities during web browsing. The Coordinator implements the interface with the extended application, namely a web browser, and manages interaction among it and a set of agents, each dedicated to a specific assistance task. Such tasks include: understanding user preferences from visited web pages, extracting data from web pages to collect features of interesting goods, creating a client-side virtual cart that stores potential user's purchases.

1 Introduction

Customisation is an important aspect of application design, development and maintenance. Unfortunately, any application consists of a limited amount of lines of code that cannot take into account all possible user requirements. In order to painlessly customise and extend applications, *software agents* have been used as *assistants*. These assistant agents help users as these work with the application, and are able to take decisions by observing users' activities and inferring their needs. In adding new services to existing applications, assistants react to events generated by users and their environment, executing their own thread of control.

In this context, many authors propose assistants which help users during web browsing (Lieberman 1995, Di Stefano and Santoro

2000), web mining (Kitamura et al. 2001), chatting (Lieberman et al. 1999), etc. In addition, it is worth recalling one of the most widespread assistants, the Microsoft Office assistant (MSAgent). In the cited approaches, assistance tasks are usually accomplished by a single assistant agent. In some cases, an application to be assisted by an agents embeds in its code statements meant to implement, or invoke, agents activities. In other cases, connection between application and assistants is obtained by exploiting operating system facilities, like DCOM (Eddon and Eddon 1998) connections (as for the MSAgent), or scripting services. Such solutions are rather application- or environment-dependent and make assistants tightly connected with the application, hence considerably more difficult to develop and reuse. Unfortunately, no general technique has been available so far to interface applications and assistants.

The proposed approach overcomes such limitations by allowing an application to be extended using *a set* of software agents, each responsible for a specific assistance task. Moreover, our assistant agents can be developed independently of the application and easily connected with it, even at runtime. In order to make such a connection possible, we exploit *computational reflection*, a technology whereby a software component can *intercept* events taking place for an application, and exploit this chance to *extend* the application with further functionalities (Maes 1987). We use reflection to connect applications with a component, the *Coordinator*, that regulates the work of assistant agents. Stipulating that the application needing extension has been developed in an object-oriented programming language, we adopt the *metaobject* model (Ferber 1989) to capture control from an application object whenever an operation (e.g. a method call) is performed on it, and bring control within the associated *metaobject*, which can choose to modify the behaviour of the application object.

More specifically, in our architecture the *Coordinator* is a special agent whose tasks are: (i) to handle interaction between the application and a set of assistant agents, each one implementing a specific assistance activity, and (ii) to serve as a coordination means for the activities of assistant agents that need synchronisation or data sharing. To accomplish the former task, the *Coordinator* embeds the metaobjects needed to intercept control from application objects, trigger assistant activities, and use their outcomes to change or enrich the behaviour of the application.

As we shall argue in Section 3, the architecture we propose affords a high degree of *flexibility* and *modularity* in the design and implementation of assistants: each assistant agent is treated as a "plug-in" that can be added or removed (even at runtime) according as the service it offers is needed, without affecting the functioning of the entire system.

We have experimented our approach to coordinating several assistants and applications, and found it advantageous in various contexts (Di Stefano et al. 2002b, Di Stefano et al. 2002c). In the sequel, its effectiveness will be demonstrated by presenting a case-study dealing with multi-agent assistance for e-commerce. For the purpose of extending a web browser, a set of assistants will be employed to *profile users*, thus understanding their preferences; *extract data* from visited web pages that contain information on goods of interest for the user; and *manage a virtual cart*, which stores at the client side information on goods, performs comparisons, provides price trends, etc.

The outline of this chapter is as follows. Section 2 motivates the recourse to multi-agent assistance by arguing its usefulness and discussing an example. Section 3 describes a reflective software architecture for coordinating multiple web assistants. Section 4 presents a set of assistants cooperating within the proposed architecture, in order to help users performing e-commerce related activities. Section 5 expounds some concluding remarks.

2 The Motivation for a Multi-Assistant Architecture

As stated in Section 1, the literature presents many research works (see e.g. (Di Stefano and Santoro 2000, Kitamura et al. 2001, Joachims et al. 1997, Bollacker et al. 1998)) describing the enhancement of an application with a single *assistant agent*, explicitly designed for a specific assistance task. E.g., for a web browser, examples include agents performing page prefetching (on the basis of user preferences) (Joachims et al. 1997, Di Stefano and Santoro 2000), crawlers searching pages on specific topics (Kitamura et al. 2001), etc.

However, in practice, browsers and other complex applications demand multiple, and to some extent heterogeneous, assistance functionalities. In order to better structure the development of such complex assistance requirements, it seems reasonable to envisage a design consisting in a group of independent agents, each specialised for a well-defined task, but potentially interested in interacting and coordinating with the others. For this purpose, the supporting architecture should provide assistant agents with convenient ways to interoperate with the application and with each other, for the sake of exchanging data, and coordinating their tasks.

In order to achieve an adequate degree of *flexibility* and *modularity* in the design, implementation and connection of different assistants, it seems best to adopt a *component-based* approach, where each assistant agent can be easily added or removed[1], as dictated by user preferences or requirements, without affecting the rest of the system or requiring a re-engineering process.

The noted design choices are also advantageous in that they favour

---
[1] Of course, in principle, unplugging an assistant might be unfeasible or harmful if its activity is relied upon by another one.

separation of concerns among agents and between them and the application.

The adoption of multiple assistant agents is also in good agreement with current practices in the design of general agent-based systems (Wooldridge 1999), where an emerging trend is to split a global goal into several sub-goals and then use a set of co-operating agents (thus constructing a so-called "multi-agent system"), each entrusted with one of the identified sub-goals.

As a solution to the issues discussed, an architecture is needed to support seamless integration of assistant agents with the application, also at runtime, and an effective interaction among agents when needed. As explained in depth in Section 3, in the proposed architecture, agents communicate by means of an Agent Communication Language (ACL) (Labrou et al. 1999, FIPA (Foundation for Intelligent Physical Agents) _a) and interact with the application only through a special agent called *Coordinator*. The design of the *Coordinator* has been devised to offer a uniform scheme to connect assistant agents with the application. In particular, the interface provided by the *Coordinator* agent is sufficiently general to fit the needs of most assistant agents.

2.1 Example: Extending a Web Browser with Assistant Agents

As an example of an application that can be extended with suitable assistant agents, we consider a web browser, which, as well-known, allows users to surf the web to find interesting data. A web browser (even a very basic one) should provide functionalities for rendering web pages; establishing connections with remote web sites and downloading text, images, etc.; keeping track of visited web pages. A GUI-based browser operates through a series of events initiated by either itself or the user, and involving both (e.g. dialog windows opening, mouse actions, etc.).

A set of assistant agents intended to enrich a browser might provide, e.g., the following additional functionalities and/or services.

- Guessing or recording the user preferences, while he surfs, so as to collect a weighted list of topics that may be of interest for him.
- Customising the behaviour of the browser by giving it the ability to highlight page sections containing interesting topics.
- Looking for web pages containing pieces of information related to the topics of interest for the user.
- Ranking, according to previously recorded user interests, the results of searches received from web engines.
- Identifying in web pages data representing goods of interest for the user, and extracting them in a structured form.
- Presenting tables comparing homogeneous data extracted from different web pages (cf. previous item).
- Notifying the user of changes occurred in selected web pages, possibly extracting from them structured data of interest. This would implement a sort of client-side push mode.
- Saving a mirror copy of text typed by the user in HTML forms, in order to retain a "blueprint" and also back up information that a malfunctioning could abruptly wipe out or discard (this could be especially useful, say, for web mail applications, where outgoing mails are composed in HTML forms).

Clearly, the above items can be considered well defined tasks that may be assigned to distinct assistant agents.

In order to support the work of such agents, the appropriate events involving the web browser need to be notified to agents. These events include e.g.: (a) downloading new web pages; (b) rendering a web page; (c) clicking on a web page. The first event allows a web page to be analysed and re-engineered, so as e.g. to strip them of images or other parts or change the ranking of items. The second event triggers, on the web page shown, modifications decided by assistants (e.g. keyword highlighting), exactly when operations performed by the user demand the page to be (re-)rendered (not before). This solves the problem of synchronising rendering changes decided by an assistant with user activities. The third event, on the basis of the string selected and clicked upon by the pointing device, allows the appropriate assistant to update the list of topics of interest or perform web searches of the selection.

Most of the example assistants described also need mechanisms to influence the operations and behaviour of the assisted browser. E.g., they might want to alter the HTML code rendered by the browser, interfere with rendering parameters, pop up message windows, etc.

3 The Multi-Agent Reflective Architecture

Assistants aim at providing applications with additional functionalities, in order to help users while they work. Enriching a software system with further functionalities raises the issue of integrating separated parts to make them work together. For the sake of integration, we propose the adoption of *computational reflection* in the form of the so-called *metaobject model*. This allows our architecture to be split into a *baselevel*, consisting simply in the application, and a *metalevel*, consisting of several specialised assistant agents, and an additional agent, the *Coordinator*. This is dedicated to interfacing the assistants with the application, keeping them separated until runtime, when a connection is realised, via the *Coordinator*, by means of reflective mechanisms. This approach offers a good degree of modularisation and concern separation in the development of the whole system, since the original application source code does not require any modification, and assistant agents can be plugged in and out dynamically, as desired.

In contemporary software development scenarios, complex applications, such as those our approach targets, are implemented in object-oriented languages. This ideally matches the trend for object-oriented agent platforms, and the availability of excellent object-oriented reflective environments (witness e.g. tools like those proposed in (Chiba 1995, Gowing and Cahill 1996, Chiba 2000), and the reflective features already integrated in Java (Bloch 2001) and Smalltalk (Liu 2000)). As a consequence, the methodology we propose, and the examples and case study we present, are thoroughly set in an object-oriented framework.

3.1 Computational Reflection

A *reflective* software system contains structures, representing some of its own aspects, which allow it to observe, and operate on, itself (Maes 1987). Such a system typically comes in the form of a two-level system, whereby a *baselevel* implements some functionalities, and a *metalevel* observes and acts on the baselevel. A widespread reflective model for object-oriented environments is the *metaobject model*, which associates each baselevel object for which this is deemed useful, with a corresponding metalevel object called a *metaobject*. As Figure 1 shows, a metaobject *intercepts* control from its associated baselevel object whenever the latter experiences a method invocation (see (1) of Figure 1) or a state change (Maes 1987, Ferber 1989). Once control is within the metaobject (2), this can modify the baselevel object behaviour by e.g. calling baselevel methods, changing parameters, etc. Finally, the metaobject usually gives control back to the associated object (3), which will carry on with the trapped activity.

Software evolution is a natural application of reflection, which can be exploited to transparently enhance systems with features including, besides assistance, synchronisation (Tramontana 2000*a*), adaptation to changing environment conditions (Tramontana 2000*b*), fine grained allocation of objects in a distributed environment (Di Stefano et al. 2002*d*), etc.

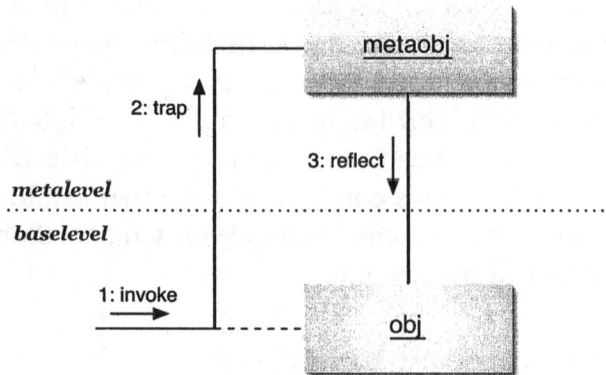

Figure 1. Metaobject model

The metaobject model can be made available for standard object-oriented programming languages by add-ons such as OpenC++ (Chiba 1995), Javassist (Chiba 2000), etc. The former is a reflective version of C++, which relies on inserting into C++ code keywords specifying how to connect baselevel application objects with metaobjects; the resulting code is then processed by a front-end to produce pure C++ code. Javassist is instead a Java library that allows baselevel bytecode to be injected with statements that notify the appropriate metaobject of any events regarding the associated baselevel object (see Section 3.1.1 for more details).

3.1.1 Using Javassist

At this stage, for a better understanding of the code presented in the following, it is useful to briefly sketch how Javassist allows an existing application to be made reflective. Further information can be obtained from the Javassist documentation (Chiba 2002).

Assuming that the application's `main()` method is implemented in a Java class `App`, all that is required is to write for it a wrapper class `WrApp` that:

1. instantiates the Javassist class loader as, say, object `cl`,

```
import javassist.reflect.ClassMetaobject;
import javassist.reflect.Loader;
public class WrApp
{
  public static void main(String[] args)
    throws Throwable
  {
    Loader cl = new Loader();
    cl.makeReflective("Foo", "MetaFoo",
              "javassist.reflect.ClassMetaobject");
    cl.run("App", args);
  }
}
```

Figure 2. Class WrApp associating baselevel and metalevel classes

2. calls the cl.makeReflective() method as many times as necessary, to connect each baselevel class to be made reflective, with the relevant metalevel class, and
3. calls the cl.run() method with argument App in order to launch the application[2].

These steps are concretely illustrated by the code in Figure 2, which connects class Foo to metalevel class MetaFoo, and then calls method run(), which will start up the application, i.e. the App class, by invoking its original main() method.

Let us now show the specific mechanism whereby Javassist allows baselevel method invocations to be trapped by an interested metaobject. The latter is an instance of a metalevel class, which must be a subclass of Javassist's Metaobject class and override its trapMethodcall() method. This method will be automatically

---

[2]The approach discussed performs runtime, on-the-fly baselevel-metalevel connection. Another solution offered by Javassist is to employ a command line tool that modifies the bytecode of the appropriate baselevel class files by persistently incorporating reflective behaviour into them.

invoked first, whenever any method of an associated baselevel object is called. This is demonstrated by the code fragment in Figure 3

```
public class MyMetaobject extends Metaobject
{
  ...

  public Object trapMethodcall(int identifier,
                               Object[] args)
    throws Throwable
  // whenever a method is invoked on an instance of the baselevel class
  // associated with this one, this method is invoked first
  {
    // perform desired metalevel operations
    ...
    // now call the trapped baselevel method, through Metaobject's
    // (the parent class') trapMethodCall() method
    return super.trapMethodcall(identifier, args);
  }
  ...
}
```

Figure 3. Method trapping in Javassist

The `identifier` argument allows the code of `trapMethodcall()` to recognise the baselevel method intercepted; the original arguments passed to the intercepted method can be accessed through the `args[]` array parameter, and, once the desired metalevel activity has been performed, the baselevel method can be made to return by invoking `Metaobject`'s `trapMethodcall()`; finally, the baselevel object that the intercepted method call was directed to can be identified by invoking Javassist's `getObject()` (not shown in Figure 3).

In this respect, it should be noted that at runtime, whenever a new instance object of a baselevel class like `Foo` is requested, `Foo`'s reflectively modified bytecode will trap to Javassist runtime support, which will in turn instantiate an object of the associated class `MetaFoo`. In this way, as the metaobject model requires, each reflective baselevel object is accompanied by an associated metalevel object right from the start of its lifetime. The metalevel programmer need not take any special provision to recognise baselevel instantiation: the constructor `MetaFoo()` will be automatically trapped to, when the constructor `Foo()` is invoked at instantiation time, and before. `Foo()` is executed. An example of metalevel constructor handling is provided by the code in Figure 8.

3.2 The Architecture

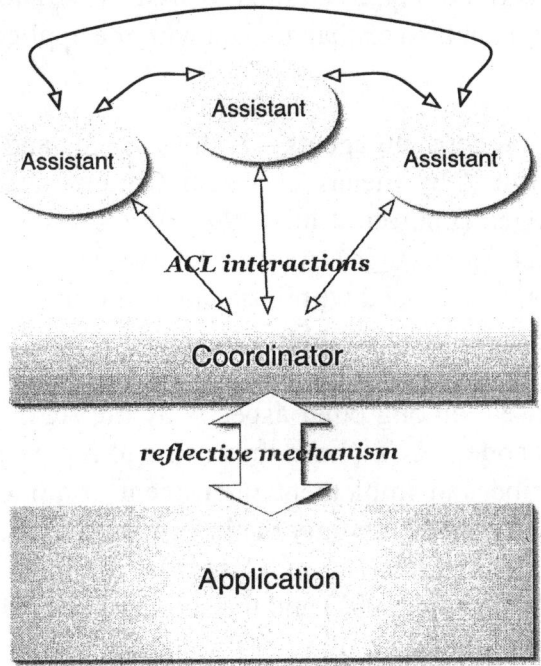

Figure 4. A Reflective software architecture to coordinate several assistants

The proposed reflective architecture consists of the application to be extended at the baselevel, and a set of agents at the metalevel. The metalevel agents are a *Coordinator* and several assistants (Figure 4), each one built as an autonomous component dedicated to a single task. The architecture has been designed in such a way to be general-purpose; in particular, in the case study presented in Section 4, the baselevel application considered is a web browser.

The core architecture component is the *Coordinator* agent, whose tasks are:

1. to inform assistants of the interesting events involving the application
2. to trigger suitable actions on the application, based on the results produced and the requests issued by assistants, and
3. to allow assistants to exchange data with the application and other assistants.

Communication between agents—both assistants and the *Coordinator*—is performed by means of Agent Communication Language (ACL) messages (Labrou et al. 1999, FIPA (Foundation for Intelligent Physical Agents) _a). As well-known, nowadays multi-agent applications are developed within specialised frameworks, which are rapidly converging towards reference standards such as FIPA (FIPA (Foundation for Intelligent Physical Agents) _d). Such frameworks are characterised, among other aspects, by the ACL in which agent interaction is coded. ACLs have thus come to represent the standard way to describe and implement inter-agent communication, since they manage to convey not only the information exchanged, but also the semantics of the interaction. Currently, the most popular ACLs are KQML (Finin et al. 1997) and the standardised FIPA-ACL (FIPA (Foundation for Intelligent Physical Agents) _a).

The proposed reflective software architecture has been designed to allow for plugging-in various assistants, even at run-time, in accordance with user requirements. This has been obtained through various means, the foremost being the introduction of the *Coordinator* component, as we now discuss. Firstly, the reflection technology employed lets the application stay unaware of assistants changing its behaviour. Moreover, in designing the *Coordinator*, we have only provided means for assistants to be notified of application events, and to affect and supply data to the application, but have avoided wiring in any knowledge about their number or specific assistance task. This allows assistants to be of any type and be created incrementally and deployed dynamically.

The *Coordinator* is also of benefit for assistants (and their designers), by decoupling them from the specific application to be assisted. There is only one constraint to be satisfied by assistants, so that they can interact with each other and the application through the *Coordinator*: they should be able to exchange—and exploit for their purposes—those events, data and actions on the application that the *Coordinator* makes available through appropriate *protocols*. We feel that the interface such protocols define for assistant designers is more general and easy-to-use than what would result, if they had to interface directly with the variety of applications for which assistance may be desired. In some sense, a *Coordinator*-based interface could even be chosen as a standard by an organisation, or developers team, needing to support several applications by multiple agents, some of which possibly targeted to more than one application.

3.3 Coordinator Agent

As noted, the *Coordinator* should be able to capture some events involving the application, let the agents be informed of these, and perform on the application, on behalf of the agents, the operations these request. To support these requirements, the *Coordinator* has been designed to consist of three metalevel classes a `Switcher`, a `Merger` and a `Blackboard` (see Figure 5).

Metaobject class `Switcher` is responsible for detecting interesting application events. For this purpose, each application class whose

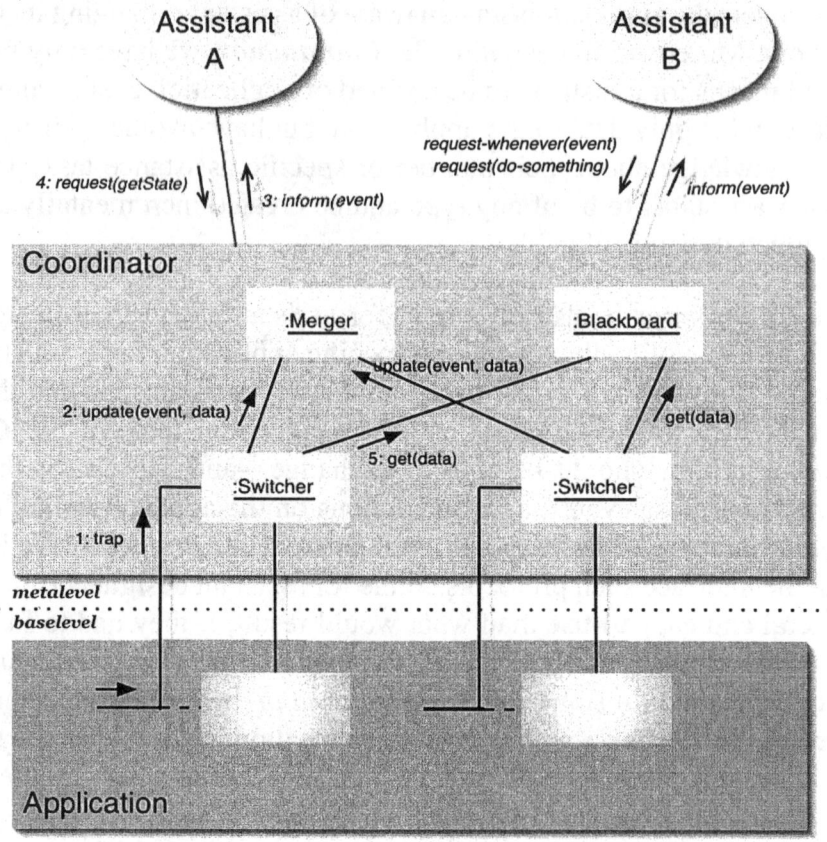

Figure 5. Coordinator functionalities and interactions with assistant agents

events are to be monitored gets associated with Switcher. Thus, at run time, method invocations on a monitored application class instance are trapped by the associated Switcher instance (see (1), Figure 5), and notified to the Merger component (which in turn will contact assistants—see below). E.g., as previously mentioned, for a web browser application, the events that might need to be captured by Switchers include: downloading a new web page, rendering a web page, inserting a word in a web form. Some trapped events might require, besides notifying assistants, further intervention that may well be carried out by metaobject Switcher, in a typical reflective style. E.g., when a Switcher intercepts a web page rendering method, it could tamper with its arguments to introduce effects that an assistant had requested. Finally, Switcher instances have also the task of carrying out operations on the application (by acting on the associated application objects) in accordance with the requests performed by assistants. E.g., these could request a browser to store a bookmark for an interesting web page, open a new window to display an outline of a web page, or fill out an HTML form in a context-sensitive way for the benefit of the user.

A single instance of class Merger is employed, to collect, from all Switcher instances, the captured application events with the related data, and to notify the assistant agents that had shown interest in those events. In order to decouple users' operations from agents' as much as possible, Merger is designed to play the role of a *concrete subject* in the *Observer* Design Pattern (Gamma et al. 1994). In this way, assistant agents and the application can act asynchronously, so as to allow users to work independently of agents' operations, including those that agents request, through the Merger, to be carried out on the application. Thus, when an agent is notified of an event (ultimately by means of ACL message passing), its work is performed in a separate thread of control; this allows the agent to operate autonomously and the application to carry on without being forced to wait for agents results.

Metaobject Blackboard allows agents to communicate with the application and among them. It acts as a repository of data for assistant agents that want to communicate their results to the application, and also as a common *knowledge base* which assistant agents use to share information. Since assistants are meant to work autonomously, their results are made available for the application, asynchronously, by updating this common repository (potential concurrent accesses by multiple agents are easily regulated thanks to Java synchronisation capabilities). Depending on their requirements, assistants can tag data inserted in the Blackboard as *persistent*. In this case, the Blackboard will store those data into stable storage, in order to have them available at the next program startup. This can be useful for data like user profiles, which, once built, must be ready for use each time the application and the assistants are started up.

It is worth emphasising again, at this stage, that, as assistant agents produce their outcomes (data, or requests of actions on the application), these are stored into the Blackboard and will get to influence the application behaviour later on, through, the relevant Switcher metaobject. For example, an assistance activity aimed at highlighting some keywords when a web page is rendered would involve a Switcher which (i) captures the rendering event, (ii) checks the Blackboard for keywords to highlight (perhaps previously stored by an assistant), and (iii) accordingly influences the rendering activity. Finally, it is worth noting that Switchers should periodically check the Blackboard for requests from the assistants to act on the application.

In the light of the characteristics and roles of its constituent parts, the *Coordinator* can be viewed as a two-layer system, which seamlessly interfaces assistants with the application. The lower layer, which is application-dependent, comprises the Switchers, which are tightly coupled with the application, being directly connected with application classes. The upper layer consists of Merger and Blackboard, which offer a standardised interface to assistants (based on ACL speech acts, see Section 3.4); this layer is application-independent in the sense that its design need only decide which application operations are related with assistants' activity, and how they are to be encoded at the interface with the assistants, but need not know how they affect application behaviour.

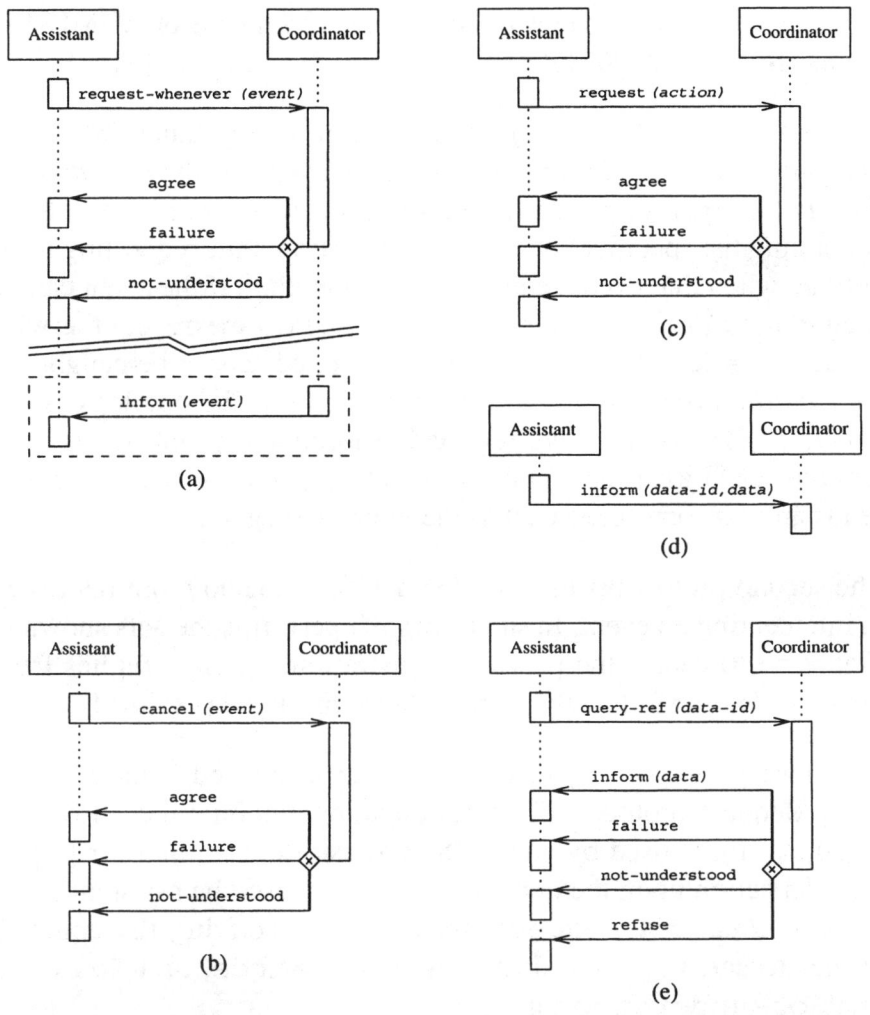

Figure 6. Conversational protocol for the interaction with Coordinator

3.4 Coordinator-Assistants Interactions

As stated previously, the *Coordinator* is intended to connect the assistant agents with the application. In order to regulate this interaction, a set of conversation protocols, based on ACL speech acts, are implemented; they are depicted in Figure 6 by means of AUML diagrams (Bauer et al. 2001).

The first protocol (Figure 6a) is started when an assistant is interested in a particular event. In this case, the assistant sends the *Coordinator* (i.e. its Merger component) a request-whenever speech act containing the specification of the event to trap, thus requesting to be notified when that event occurs in the application. If the event can be caught, the *Coordinator* replies with an agree message, otherwise a failure (e.g. the event is unknown to, or cannot be caught by, the *Coordinator*) or a not-understood (e.g. the ontology is unknown to *Coordinator*) are reported. Subsequently, during user operations, each time the *Coordinator* intercepts a registered event, it sends an inform speech act to the interested agent.

The second protocol is used when an assistant is no more interested in intercepting an event. In such a case, it performs the acts shown in Figure 6b to cancel the previous registration. Possible replies from the *Coordinator* follow the same pattern discussed previously.

The third protocol (Figure 6c), which is a reduced version of the FIPA Request protocol (FIPA (Foundation for Intelligent Physical Agents) _c), is used by an assistant to ask the *Coordinator* to perform an action upon the application. In this case, the assistant sends the *Coordinator* a request speech act, specifying the action it wishes to carry out, and then waits for an agree, or a failure or a not-understood messages.

The last two protocols (Figure 6d, e) refer to the shared knowledge-base functionality provided by the Blackboard component of the *Coordinator*. The protocol in Figure 6d allows assistant agents to store useful data into the Blackboard, whereas that in Figure 6e (based on the FIPA Query protocol (FIPA (Foundation for Intelligent Physical Agents) _b)) is used to retrieve previously stored ones.

3.5 A Concrete Example: an Assistant that Highlights Keywords for a Web Browser

In order to better understand the mechanisms operating within our architecture, in this Section we provide a simple example showing how a web browser can be extended by adding user assistance functionalities to it.

For the sake of simplicity, we consider here only an assistant agent charged with the task of finding and highlighting keywords each time a new web page is loaded and displayed by the browser. The assistant builds a ranked list of keywords by means of a term-frequency algorithm, which updates the list each time a new page is loaded. The top elements of the list constructed are used to determine the words to highlight.

A prototype of this assistant has been implemented, and later evolved (after a thorough functional re-design) into the User Profiler Assistant of Section 4.1. For its development, as well as for all the assistants presented in Section 4, we have employed the Jade (Bellifemine et al. 2001) agent platform. Jade has been chosen because it is FIPA-compliant and, most importantly, Java-based, which permits an effortless integration with the reflective support provided by the Javassist Java library.

The web browser we have enhanced by this example assistant is Jazilla (SourceForge 2002), the Java-based Mozilla clone. Reflection is introduced into Jazilla by using the Javassist package, which provides suitable mechanisms to add to existing Java classes patches that support the implementation of a metaobject model (cf. Section 3.1). For our purposes, Javassist provides a class loader capable of modifying Java bytecode on-the-fly.

The code shown in Figure 7 connects Jazilla's class `org.jazilla.javafe.renderer.athens.JAZDocument` to our met-

```java
import javassist.reflect.ClassMetaobject;
import javassist.reflect.Loader;
public class Main
{
  public static void main(String[] args)
    throws Throwable
  {
    Loader cl = new Loader();
    cl.makeReflective(
      "org.jazilla.javafe.renderer.athens.JAZDocument",
      "Switcher",
      "javassist.reflect.ClassMetaobject");
    cl.run("org.jazilla.Jazilla", args);
  }
}
```

Figure 7. Connecting the Switcher metaobject class with a Jazilla class

alevel class `Switcher`, and then calls method `run()`, which will start up Jazilla, i.e. the `org.jazilla.Jazilla` class, by invoking its original `main()` method. (Relevant Javassist usage information is given in Section 3.1.1.)

As discussed in (Di Stefano et al. 2002a), for a programmer wishing to connect an assistant to an existing application, the first step to tackle is *identifying* those user-initiated application *events* that warrant assistance activity. In this case, the relevant events are: *downloading* a new web page and *rendering* a web page. From an exploration of Jazilla source, documentation and behaviour, we found that class `JAZDocument` represents a parser for HTML documents, and is capable of displaying them by its `renderHTML()` method, whose single parameter is the URL of the page to be displayed. We also found out that the second parameter of the `JAZDocument` constructor is an object representing the browser frame—an instance of class `BrowserFrame`—where the page is displayed; the latter class also provides some methods to access the whole page content.

Given this background, in order to achieve our goal, the base-level class JAZDocument is associated with a metalevel class Switcher, which is programmed (see Figure 8) to capture invocations of the constructor and the renderHTML() methods on base-level JAZDocument objects. Capturing constructor invocations is needed to obtain a reference to the BrowserFrame involved. Capturing renderHTML() allows the metalevel to know when a page is being rendered and (with some extra tests) downloaded. This information is used both to forward the page content to the assistant, which can analyse it to update the keyword list, and to modify the HTML page source in order to inject keyword highlighting tags.

The next design step, after identifying and intercepting interesting application events and possibly modifying application behaviour as seems fit, is to connect the *Coordinator* with the appropriate assistant in order to support the activity detected. As reported in the previous subsection, this is accomplished by means of conversational protocols based on ACL speech acts. The details of interactions between the assistant and the *Coordinator*, and between the latter and the application, are sketched below for some typical scenarios.

1. *Downloading a new web page.* The assistant needs to obtain, from the *Coordinator*, a notification each time a new page is requested by the user and loaded by the browser. To this aim, at startup, the assistant contacts the *Coordinator* (i.e. the Merger) using a request-whenever speech act which expresses the assistant's interest in receiving this notification. E.g.:

```
(request-whenever
  :ontology web-assistance
  :content (new-page-loaded :action inform
           :parameters (set (url pagesource)) ) )
```

This speech act also instructs the Merger component of the *Coordinator* to send the assistant, besides a notification, the page content and URL each time a new page arrives.

In the example code in Figure 8, a metaobject `Switcher` uses method `trapMethodCall()`) to capture invocations of base-level Jazilla's `renderHTML()` method. In fact, these invocations do not mean the page about to be rendered is a new one, but the `Switcher` can easily ascertain this by its reflective inspection capabilities (details are not shown in Figure 8).

`Switcher` then forwards the URL and the content of the loaded page to the `Merger` by calling its (static) `update()` method. `Merger`, in turn, sends the assistant an `inform` speech act such as the following, containing the downloaded HTML source page:

```
(inform
  :ontology web-assistance
  :content
    (page :url "http://www.unict.it/staff.html"
          :pagesource "<HTML><HEAD>.....</HTML>" ))
```

2. *Updating the keyword list.* The assistant analyses the HTML pages received, looking for keywords and organising them into a ranked list. It will then execute a `request` speech act that supplies the *Coordinator* with a list of tuples, in the form (*keyword,colour*), where the colour indicates importance; e.g.:

```
(request
  :ontology web-assistance
  :content (highlight-keywords
            (set (kw "garage" "red")
                 (kw "engine" "green")
                 (kw "tyres"  "yellow") ) )
)
```

As a result, the tuples will be stored into the `Blackboard`, for later use.

3. *Keyword highlighting.* In order to enforce keyword highlighting, metaobject `Switcher`, after trapping a call to Jazilla's `renderHTML()` method, retrieves the keyword-colour list from the `Blackboard` (see Figure 8) and accordingly modifies the page source (method `injectHighlighters()` in Figure 8).

```
import javassist.*;
import javassist.reflect.Metaobject;
import java.lang.reflect.*;
// other imports

public class Switcher extends Metaobject
{
  private BrowserFrame browserFrame;

  public Switcher(Object self, Object[] args)
  // a Javassist metaobject class constructor should be declared thus;
  {
     // whenever the constructor of a baselevel object JazDocument is called,
     // this constructor is first trapped to
     super(self, args);  // call Metaobject's constructor
     // the 2nd argument of the trapped JazDocument() constructor
     // is the browser frame for the JazDocument about to be instantiated
     browserFrame = (BrowserFrame) args[1];
  }

  public Object trapMethodcall(int id, Object[] args)
     throws Throwable
  // each time a method is invoked on a JazDocument object, this
  // method of the associated Switcher is invoked first
  {
     if(getMethodName(id).compareTo("renderHTML") == 0)
     {  // renderHTML() method was called, so ...
        if ( // rendering was actually called for a new page (case 1, page 261)
           ) {
           // get HTML source and page URL
           String pageSource = (String)
              browserFrame.getRenderer().getPageSource();
           URL url = (URL)args[0];
           // notify the merger and pass it URL and source
           Merger.update(Events.NEW_PAGE_LOADED,
                          new Object[2]{pageSource,url});
           ...
```

Figure 8. Metaobject Switcher for browsing assistance

```
        ...
    } // end of download handling

    // get the keyword list from the Blackboard (case 3, page 262)
    Hashtable keyword_list = (Hashtable)
        Blackboard.get(DataIdentifiers.KEYWORD_LIST);
    // modify HTML source to inject keyword highlighting
    injectHighlighters (pageSource, keyword_list);
    }

    // call the baselevel object method
    return super.trapMethodcall (id, args);
}

void injectHighlighters (String pageSource,
                        Hashtable keywords)
{
    ...
}
}
```

Figure 8. Metaobject `Switcher` for browsing assistance

In order to fully appreciate the interoperation mechanisms described, the reader should examine, together with the above ACL speech acts, the code in Figure 8, which provides a significant fragment of metaclass `Switcher`.

4 A Case Study: E-Commerce Assistants for a Web Browser

As discussed above, several assistant agents can cooperate to extend a web browser with new functionalities. We have realised a group of assistant agents to help users performing e-commerce activities. Three assistants have been developed, to perform the following tasks respectively: (a) building a user profile describing preferences and interests; (b) extracting from web pages data on goods that might interest the user, and organising such data in a structured form; (c) providing one or more visual representations that allow the captured data to be compared, analysed and manipulated.

As a benefit of the proposed architecture, these assistants have been developed independently of the web browser extended. An implementation of the *Coordinator* component has also been realised to interface the assistants with the Jazilla web browser.

The set of assistant agents proposed, shown in Figure 9, is described in the rest of this section.

4.1 User Profiler Assistant

The User Profiler Assistant agent is meant to understand users' preferences while they surf the web. To carry out its task, it analyses the web pages visited by users, whenever it is informed by the *Coordinator* that a new page has been downloaded. This analysis is performed *autonomously* and *asynchronously* with respect to the web browser and its user, and aims at characterising web pages in order to determine which topics are of interest for the user and how important each topic is.

The User Profiler Assistant employs a knowledge base that is constructed off-line in a training phase. This knowledge base is used to categorise the web pages visited by the user while s/he surfs and to generate the user profile. Such a profile is determined by the categories of the visited web pages, and is constantly updated, as new web pages are downloaded.

Mase's approach (Mase 1998) has been used to categorise web pages. Accordingly, a knowledge base is built that consists in a set of categories, each characterised by a name and a list of keywords. Each keyword is associated with a score that indicates how important the

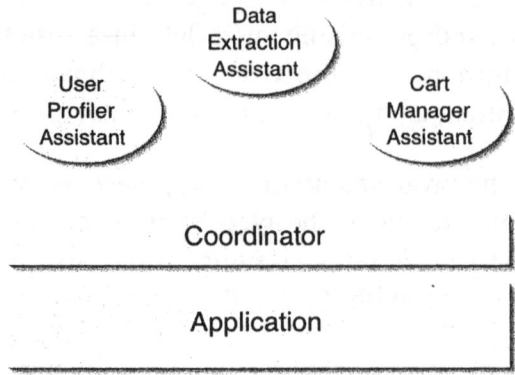

Figure 9. E-commerce assistants for a web browser

related keyword is for identifying its category. The training phase is usually performed off-line, but can also be activated on-line by the user, in order to improve the knowledge base; indeed, the larger the knowledge base available the better this assistant will perform in categorising the web pages. Each time the training phase is activated, this assistant is informed that a new category has to be created and which pages characterise it; alternatively, it accepts suggestions about additional page sets to be used to extend the characterisation of a known category.

During the training phase a large set of web pages, belonging to known categories, must be analysed to extract the most recurrent meaningful words. As described in (Mase 1998), once HTML tags, scripting code and stop words (i.e. articles, prepositions, conjunctions, common words, etc.) have been removed from web pages, the remaining words, which are the significant ones for page characterisation, are "truncated" so as to retain only their root, by using a *stemming* algorithm (Porter 1997). For each word, we then calculate the *frequency* as the ratio between the number of occurrences and the total amount of significant words in the page. Since not all the pages have the same length, this ratio allows the importance of a word to be evaluated independently of the web page length. Those words whose frequency exceeds a fixed threshold, which we have set at 5% on the basis of extensive testing, are selected as the keywords characterising the category.

In our experiments we have considered around 4000 web pages, in English, each of which has been classified as belonging to one or more of the following categories: *book, car, computing, photography, sport, travelling, videogame*. E.g. the web pages selected for the category *car* concerned car buying and selling, insurance, and components. By means of the above described training phase, keywords for this category have been found to include: *accessory, alarm, battery, brake, car, engine, fuel, fiat, garage, honda, insurance, lorry, mercedes, renault, rent, rover, sport, suspension, tyre, van, vehicle, wagon, wheel*.

When the user visits a new web page, this is classified according to the assistant's knowledge base. As in the training phase, firstly the web page is cleaned by removing HTML tags, scripting code and stop words, then the stemming algorithm is used and a list of words with their frequency is created. In order to categorise a page, a "similarity value" is calculated between such a list of extracted words and the set of keywords of each known category, following the approach described in (Mase 1998). It is assumed that the web page belongs to the categories whose similarity value lies above a fixed threshold. As a first outcome, this assistant updates a list of *page categories*, i.e. list items consisting each of the following data: a web page URL, the categories found in it and a checksum value for the web page source. Such a list allows this assistant to remember the categorisation of previously downloaded pages. A copy of this list is sent to the `Blackboard`.

The categories to which visited web pages belong are inserted into the list that describes the user profile. Inside the *user profile* list, each category has a specific value that scores the different degree of user interest for it. Such a value is determined as the normalised count of pages belonging to this category that the user has visited so far; moreover, as the browsing activity goes on, an aging mechanism is applied to progressively decrease the score of least recently referenced categories.

The second outcome produced by this assistant is therefore the *user profile* list, containing a certain number of ranked categories. Once updated, this list is passed to the *Coordinator*, which stores it into the `Blackboard` in order to make it available to other assistants and to the browser application.

It is worth noting that the activity of this assistant is not focused specifically on e-commerce. It can be employed to support many of the assistance tasks we described in Section 2.1 and, as it is, can be reused in several contexts. We have experienced that its ability to effectively recognise categories of web pages, and so build proper user profiles, depends on the extent of the training phase. The larger the set of categories and the keywords identifying them, the better the assistant's judgement in classifying new web pages.

4.2 Data Extraction Assistant

The Data Extraction Assistant is entrusted with the tasks of extracting from a web page the data referring to commercial goods, and putting these data into a structured form, so that they can be easily handled. For these purposes, it autonomously performs, in a separate thread, a page analysis, whenever it is informed by the *Coordinator* that a new web page has been downloaded by the browser.

This assistant has an a priori knowledge of some goods that can be considered interesting. This knowledge consists of the set of attributes characterising the goods and some rules that allow the assistant to identify the attribute values, given an unstructured text. Our current implementation includes attribute lists for the following goods: *book, car, computing* and *photography*. E.g., for *car* the attributes considered are: *model, manufacturer, cost, mileage, year, colour, options, phone number, description*. To these, two attributes have been added, namely *website* and *download date*, to keep track of where and when data were found.

The first step performed by this assistant is to decide whether the page is a likely candidate to contain any data about goods. For this purpose, a page is taken into consideration only if it contains a HTML table or enough meaningful keywords (those defining the known attributes).

The second step performed is checking whether the page category, as determined by the User Profiler Assistant, is among those the latter assistant considered interesting for the assisted user. For this check, this assistant needs to retrieve from the Blackboard and compare the user profile and the category of the web page under scrutiny.

As a third step, if the page category is considered of interest, it is used to select, among the goods known to the assistant, the one for which data are most likely to be found in the page. This selection is made by looking for the page category within a small set of synonyms for each known good. The assistant can now apply the rules it is a priori endowed with, to identify within the web page any instances of the selected good, We shall term *article* any such instance. Data related to any article detected in the web page are gathered according to the attribute list for the relevant good. This information is subsequently used by this assistant to update, with a "least recently accessed" policy, a data structure termed the *goods catalog*. This transformation of text into a structure is accomplished by Embley et al's algorithm (Embley et al. 1999). The goods catalog is finally sent to the *Coordinator*, which stores it into the Blackboard so that other assistants can access it when needed even throughout different sessions.

Extracting goods data from several web pages visited by the user provides the benefit of a local repository of organised data. Since such data are stored exclusively at the client side, remote web sites are cut off from their handling, which ensures security and privacy, unlike a server side shopping cart. This local repository allows the user (and other assistants) to easily access goods data, avoiding their fragmentation among several web sites, and independently of the existence of a working network connection. Moreover, the availability of structured data enables the user to further process them, after importing them into the preferred application, or with the help of another assistant, such as the one described in the following.

4.3 Cart Manager Assistant

The Cart Manager Assistant provides a set of facilities intended to show goods data collected by the previous assistant, and to allow the user to search and manipulate these data. It handles a graphical window, which the user can interact with to obtain a different view for data, a selection of them, etc.

Once this assistant is activated by the user, it retrieves the *goods catalog* (cf. Data Extraction Assistant) from the `Blackboard` by asking the *Coordinator*. It can thus build a list of articles, and show a table where selected data on articles are grouped. The table only shows, for each article displayed in a row, the attributes (columns) that are common to all the goods, plus an extra column (the first) showing the good each article is an instance of. For the goods that we have considered (cf. Section 4.2), the common attributes turn out to be: *model, cost, description, website, download date*, and the first table column will contain one of the goods, namely *book, car, computing* and *photography*.

The Cart Manager Assistant provides the user with several interaction facilities accessed by means of a graphical user interface. This allows the user to request alternative views of the data, e.g. a graphical bar chart. In addition, users can select from a choice list one of the goods previously identified, to obtain a detailed view of the relevant articles. This is represented by means of a table, with a found article in each row, whose columns depend on the selected good. E.g. when the good *car* is selected, the following columns will be displayed: *model, manufacturer, cost, mileage, year, colour, options, phone number, description, website, download date*.

Another GUI facility enables users to select a subset of either the data currently displayed or the whole data available. The Cart Manager Assistant enables users to perform a comparison among selected articles. When this facility is activated, a window is shown that graphically compares, for each article, prices captured from different web pages or, depending on the data available, plots the trend of prices over time.

Attribute of goods handled by this assistant can be modified by the user: s/he is allowed to remove articles (table rows), edit attribute values for articles (table cells), add or remove attributes to/from those characterising a good (i.e. add/remove table columns). Extra attributes let users insert personal data such as a budget for a good to be purchased, custom keywords to better identify a good, etc. The extra attributes of goods that this assistant will have thus collected are coherently merged with existing data and persistently stored in a suitable format.

The benefits provided by this client-side cart manager include the assistance it offers the user in exploring and analysing collected data, and the availability within a single virtual cart of data originating from various web sites. Moreover, the manager provides a common view of heterogeneous data and a set of facilities to effectively handle them.

5 Concluding Remarks

The work presented aims at proposing a software architecture capable of supporting seamless integration between an application and a set of specialised assistant agents. Before concluding, it is worthwhile to cite some related works and compare them with ours. We consider the approaches of (Bradshaw 1997, Maes 1997, Lieberman 1995, Di Stefano and Santoro 2000, Lieberman et al. 1999, Kitamura et al. 2001), all concerning agents essentially aimed at assisting users accessing various Internet services. In the discussion below, instead of dealing with the assistance aspects of the cited works, we concentrate on the description of the mechanisms whereby assistant agents are connected with the application. This allows an effective assessment of our contribution with respect to the state-of-the-art on this particular issue.

One of the most renowned assistants, the Microsoft Office Assistant, is an ActiveX object (MSAgent) providing functionalities to handle user interactions in the context of assistance to applications (Microsoft Corporation 2000). MSAgent is activated by means of the DCOM (Eddon and Eddon 1998) protocol, and can also be used within a browser through JavaScript/VBScript code embedded in the current web page (provided the browser is JavaScript/VBScript-capable, like Internet Explorer). This approach, which is adopted in (Kitamura et al. 2001), can however be followed only in a Win32 environment (since the MSAgent is a Win32-ActiveX object). In any case, it can serve only to handle user interactions, while assistance activity must be already present and coded inside the application.

Assistants proposed in (Bradshaw 1997, Lieberman 1995, Lieberman et al. 1999, Maes 1997) are designed for the Apple Macintosh platform and adopt the AppleScript (Goodman 1994) technology to interface the application. As a result, this approach is not only strictly tied to that platform, but also requires the application to be controllable through the AppleScript language.

The solution adopted in (Di Stefano and Santoro 2000), for a system providing web browsing assistance, is instead based on an agent acting as HTTP-proxy and thus intercepting all HTTP traffic (requests and responses) between the browser and the web. This allows for some assistance activities (e.g. user profiling and keyword highlighting), but does not permit some events to be captured, e.g. bookmark adding/removal, which could be useful to categorise a web page thus improving the accuracy of user profiles.

With respect to the cited approaches, the architecture proposed here exhibits many important advantages.

First of all, it exploits a methodology—computational reflection—which is general in a twofold sense: (i) it does not require the application, nor the operating system (or GUI libraries) to provide special interface "hooks"; (ii) it is platform-independent, since it only involves the programming language level, and the main reflective toolkits are based on standard languages, available across platforms. Moreover, reflection in a Java framework does not even require availability of the baselevel source code, but can be directly introduced into the bytecode.

Undoubtedly, reflection may introduce an overhead, and overusing it could adversely impact the extended application performance. However, it is up to the metalevel programmer to make judicious use of reflection through sensible metalevel design principles such as e.g.: (i) making sparing recourse to interception, (ii) refraining from introducing overly complex yet frequently triggered metaobjects, and (iii) entrusting to dedicated separate threads baselevel activities liable to be intercepted and the associated metalevel activities, so as to make them as asynchronous as possible with respect to those critical for user interaction and perceived responsiveness.

Secondly, reflection and our modular architecture allow assistant agents to be added or removed (if unnecessary) even at runtime, according to user requirements, by connecting them on-the-fly to the *Coordinator*. This component allows a complete separation between *interfacing issues*, addressed by itself, and *assistance activities*, which are handled by agents that could even be unaware of the particular application they aid (this could be indifferently, e.g., the Internet Explorer, Netscape, or KDE Konqueror browser).

In this work, the usefulness and applicability of the proposed architecture have been demonstrated mainly for web browsing assistance, but obviously transfer to different contexts, i.e. given a different application, or number and/or functionalities of the assistants.

Finally, our architecture painlessly enables *distribution*, since agents interact by means of ACL message passing, a paradigm and notation for which the location of communicating agents is absolutely transparent. For example, we can enrich the e-commerce assistant bundle presented in Section 4 with a "Goods Finder Assistant", an agent that autonomously crawls the web, searching for pages containing information on goods of interest for the user; this agent could also inform the Data Extraction Assistant and/or the Cart Manager Assistant about interesting new or updated pages. The Goods Finder Assistant could operate in a "proxy server", working autonomously even if the intended user is not browsing the web. Then, when a browsing session eventually starts, any interested e-commerce assistant would contact the Goods Finder Assistants and obtain the results of the latter's work, to use them in pursuit of its own goals.

References

Bauer, B., Müller, J. P. and Odell, J. (2001), "Agent UML: A formalism for specifying multiagent interaction," P. Ciancarini and M. Wooldridge, eds, *Agent-Oriented Software Engineer-*

ing, Springer, pp. 91–103.

Bellifemine, F., Poggi, A. and Rimassa, G. (2001), "Developing multi agent systems with a FIPA compliant agent framework," *Software—Practice and Experience*, vol. 31, pp. 103–128.

Bloch, J. (2001), *Effective Java Programming Language Guide*, Addison-Wesley.

Bollacker, K., Lawrence, S. and Giles, C. L. (1998), "An autonomous web agent for automatic retrieval and identification of interesting publications," K. P. Sycara and M. Wooldridge, eds, *Proceedings of the Second International Conference on Autonomous Agents*, ACM Press, New York, pp. 116–123.

Bradshaw, J., ed. (1997), *Software Agents*, AAAI Press, Cambrigde, Mass.

Chiba, S. (1995), "A metaobject protocol for C++," *Proceedings of the Conference on Object-Oriented Programming Systems, Languages and Applications (OOPSLA'95)*, pp. 285–299.

Chiba, S. (2000), "Load-time structural reflection in Java," *Proceedings of the ECOOP 2000*, vol. 1850 of *Lecture Notes in Computer Science*.

Chiba, S. (2002), *Javassist tutorial*.
*http://www.csg.is.titech.ac.jp/~chiba/javassist/tutorial/tutorial.html

Di Stefano, A., Pappalardo, G., Santoro, C. and Tramontana, E. (2002a), "Extending applications using reflective assistant agents," *Proceedings of the 26th Annual International Computer Software and Applications Conference (Compsac'02)*, Oxford.

Di Stefano, A., Pappalardo, G., Santoro, C. and Tramontana, E. (2002b), "A multi-agent reflective architecture for user assistance and its application to e-commerce," M. Klush, S. Ossowski and O. Shehory, eds, *Proceedings of the 6th Cooperative Information Agents Conference (CIA'02)*, vol. 2446 of *Lecture Notes in Artificial Intelligence*, Springer-Verlag, Madrid, Spain, pp. 90–103.

Di Stefano, A., Pappalardo, G., Santoro, C. and Tramontana, E. (2002c), "A multi-agent reflective architecture for web search assistance," *Proceedings of the 8th Symposium of the Italian Association for Artificial Intelligence (AIIA'02)*, Siena, Italy.

Di Stefano, A., Pappalardo, G. and Tramontana, E. (2002d), "Introducing distribution into applications: a reflective approach for transparency and dynamic fine-grained object allocation," *Proceedings of the Seventh IEEE Symposium on Computers and Communications (ISCC'02)*, Taormina, Italy.

Di Stefano, A. and Santoro, C. (2000), "NetChaser: Agent support for personal mobility," *IEEE Internet Computing*, vol. 4.

Eddon, G. and Eddon, H. (1998), *Inside Distributed COM*, Microsoft Press.

Embley, D. W., Campbell, D. M., Jiang, Y. S., Liddle, S. W., Ng, Y.-K., Quass, D. and Smith, R. D. (1999), "Conceptual model-based data extraction from multiple-record web pages," *Data Knowledge Engineering*, vol. 31, pp. 227–251.

Ferber, J. (1989), "Computational reflection in class based object oriented languages," *Proceedings of the ACM Conference on Object-Oriented Programming Systems, Languages and Applications (OOPSLA'89)*, vol. 24 of *Sigplan Notices*, New York, pp. 317–326.

Finin, T., Labrou, Y. and Mayfield, J. (1997), "KQML as an agent communication language," J. Bradshaw, ed., *Software Agents*, MIT Press, Cambridge.

FIPA (Foundation for Intelligent Physical Agents) (_a), *FIPA-ACL specification*.
*http://www.fipa.org/specs/fipa00061/

FIPA (Foundation for Intelligent Physical Agents) (_b), *FIPA query interaction protocol specification*.
*http://www.fipa.org/specs/fipa00027/

FIPA (Foundation for Intelligent Physical Agents) (_c), *FIPA request interaction protocol specification*.
*http://www.fipa.org/specs/fipa00026/

FIPA (Foundation for Intelligent Physical Agents) (_d), *FIPA specification.*
*http://www.fipa.org

Gamma, E., Helm, R., Johnson, R. and Vlissides, R. (1994), *Design Patterns: Elements of Reusable Object-Oriented Software*, Addison-Wesley, Reading, MA.

Goodman, D. (1994), *Danny Goodman's AppleScript Handbook*, Random House, New York.

Gowing, B. and Cahill, V. (1996), "Meta-object protocols for C++: The iguana approach," *Proceedings of Reflection'96*, San Francisco.

Joachims, T., Freitag, D. and Mitchell, T. M. (1997), "Web watcher: A tour guide for the world wide web," *IJCAI*, vol. 1, pp. 770–777.

Kitamura, Y., Yamada, T., Kokubo, T., Mawarimichi, Y., Yamamotom, T. and Ishida, T. (2001), "Interactive integration of information agents on the web," *Proceedings of CIA 2001*, vol. 2182 of *Lecture Notes in Artificial Intelligence*, Springer.

Labrou, Y., Finin, T. and Peng, Y. (1999), "Agent communication languages: the current landscape," *IEEE Intelligent Systems*.

Lieberman, H. (1995), "Letizia: an agent that assists web browsing," *International Joint Conference on Artificial Intelligence*, Montreal.

Lieberman, H., Maes, P. and Van Dyke, N. (1999), "Butterfly: A conversation-finding agent for internet relay chat," *International Conference on Intelligent User Interfaces*, Los Angeles.

Liu, C. (2000), *Smalltalk, Objects, and Design*, iUniverse.com.

Maes, P. (1987), "Concepts and experiments in computational reflection," *Proceedings of the Conference on Object-Oriented Programming Systems, Languages and Applications (OOPSLA'87)*, vol. 22 (12) of *Sigplan Notices*, Orlando, FA, pp. 147–155.

Maes, P. (1997), "Agents that reduce work and information overload," J. Bradshaw, ed., *Software Agents*, AAAI Press/The MIT Press.

Mase, H. (1998), *Experiments on automatic web page categorization for IR system*, Technical report, Stanford University.

Microsoft Corporation (2000), *Microsoft Developer Network Library*.

Porter, M. F. (1997), "An algorithm for suffix stripping program," *Readings in Information Retrieval*, Morgan Kaufmann, San Francisco.

SourceForge (2002), *Jazilla home page*.
*http://jazilla.sourceforge.net

Tramontana, E. (2000a), "Managing evolution using cooperative designs and a reflective architecture," W. Cazzola, R. J. Stroud and F. Tisato, eds, *Reflection and Software Engineering*, vol. 1826 of *Lecture Notes in Computer Science*, Springer-Verlag.

Tramontana, E. (2000b), "Reflective architecture for changing objects," *Proceeding of the ECOOP Workshop on Reflection and Metalevel Architectures (RMA'00)*, Nice, France.

Wooldridge, M. J. (1999), *Multiagent Systems*, MIT Press.

Chapter 9

Learning by Exchanging Advice

Eugénio Oliveira and Luís Nunes

The emergence of Multiagent systems brought new challenges to the field of Machine Learning, as it did to many others. One of the main challenges is to take advantage of the information available when several agents, possibly using different learning techniques, are dealing with similar problems, either in the same location (i.e. acting as a team) or in different ones. This work aims at studying the possible advantages and pitfalls of exchanging information during the learning process, leading to better adaptation. We will discuss the subject of when, how and to whom ask for advice, and present the results obtained in two experimental scenarios: the Pursuit (Predator-Prey) Domain and a Traffic Control simulation. Results show that exchange of information can improve the average performance of learning agents enabling them to escape from local maxima in some cases, although it may reduce the exploration of the space, preventing successful agents from finding better local maxima of the quality function.

1 Introduction

Decentralization and distribution of processes became an issue in *Artificial Intelligence*, as well as in several other areas of Computer Science, in the past decades. In response to the new challenges a new type of software entities was created and later labelled as *agents*. These new entities are usually defined as being more autonomous, distributed and intelligent than previous software tools.

Some problems require that agents adapt to new circumstances or behave *"intelligently"*. Intelligence, as we perceive it, is strongly related to the capability of learning from previous experience and using stored knowledge to improve future behavior. Intelligent (or adaptive) behavior is becoming a competitive factor in today's software. One of the main challenges is to expand the Machine Learning (ML) paradigms from, the old single-agent perspective, to this new world. Currently, software agents inhabit dynamic environments. Often, they must provide answers even when they have only a partial and noisy view of the problem. The extension of learning to these new environments must overcome the difficulties of this new paradigm, but should also take advantage of its benefits. The fact that a multitude of agents populates the software environments, and in some cases they are learning to solve similar problems, leads to the current research issue: *"(How) can agents benefit from the exchange of information during the learning process?"*

In the following sections we will present and discuss a set of techniques for selection, exchange and incorporation of information from multiple sources. These techniques may provide a way to help a learning agent achieve its goal more efficiently than if it was learning only from the information generated by the environment. The main focus will be on *advice-exchange*, a technique introduced in Nunes and Oliveira (2002b) to exchange information within heterogeneous groups of learning agents that are solving similar problems. The heterogeneity constraint is one to which very little attention has been given until now. Results obtained in this research direction may point the way to more powerful learning paradigms.

The next section contains a brief review of related work. In section 3, the techniques that compose *advice-exchange* are explained and in the following section (4), the experimental setup is described. Section 5 presents the results and its discussion and finally, in section 6, we have the conclusions and a brief word on the future work.

2 Communicating to Improve Learning: Historical Notes and Review

The work presented below touches several points such as: cooperation between learning agents; mixed use of different learning paradigms at several levels; learning trust relationships between agents. Several of these issues have been approached in the past by other authors. In the following sections the reader can find a summary of the main contributions in each of these subjects.

2.1 Early Work on Exchange of Information During Learning

The work on information exchange between QL-agents (agents that use *Q-Learning* (Watkins and Dayan, 1992) as a basis for their learning skills), started in the early nineties. (Whitehead, 1991) created a cooperative learning architecture labelled *Learning By Watching*. In this architecture the agent learns by watching its peers' behavior (which is equivalent to sharing series of state, action, quality triplets). The work presented in (Clouse and Utgoff, 1991) is reviewed and expanded in Clouse's Ph.D. thesis (Clouse, 1997). This important contribution reports the results of a strategy labeled *Ask for Help*, in which QL-agents learn by asking other agents of the same type for suggestions and perform the suggested actions. The work presented by Lin (1992) uses an expert trainer to teach lessons to a QL-agent that is starting its own training. Tan (1993) reports the results of sharing several types of information in the predator-prey problem. In these experiments QL-agents shared policies (internal solution parameters), episodes (series of state, action, quality triplets), and sensation (observed states). All these experiments showed improvements in the average performance of agents that exchange information.

2.2 Recent Related Work

The work on exchange of information between QL-agents continued in several fronts. The term *joint learning* was used by Berenji and Vengerov (2000) when referring to agents that update concurrently the same quality values. These authors presented a study of a "fuzzy" variant of QL-agents that cooperate during learning by updating a common table of Q-values. Several researchers studied applications of *Reinforcement Learning* (RL) (Sutton and Barto, 1987) variants to stochastic-games. The agents used in these domains are often referred to as *Joint-Action Learners* (JAL). JAL are QL-agents that learn, each on its own, the quality of joint actions. A joint action is composed of its own action plus the actions chosen by all the peers at a given time. This approach presupposes full observability of the actions done by all agents. The first references to this concept are in the work of Littman (1994) where agents of this type are used to solve (i.e. attain optimal equilibrium in) zero-sum games. This approach was labeled *minimax-Q*. Litmann (2001) presents a summary and comparison of the convergence properties of several types of JAL.

The research on trust relationships between agents was mainly developed by Sen's research group (Sen, 1996; Biswas et al., 2000; Banerjee et al., 2000). These, as well as other related papers, focus on several aspects of learning to trust/distrust other agents.

Many researchers have focused on the use of human advice, or pre-programmed teachers, to help QL-agents. These approaches range from using high-level languages to encode the advice, as in (Maclin and Shavlik, 1994), to direct observation of a human solving a problem and replication of this behavior (Nicolescu and Matarić, 2001).

One of the most interesting works in the subject is (Price, 2003), in which QL-agents learn by *implicit imitation* of pre-trained expert-agents. This work has some very interesting characteristics: the student agent has no knowledge of the actions done by the expert, it can only observe its state transitions; there is no explicit communication between the expert and the student; the goals and actions of both agents can be different.

3 Advice Exchange

The problems considered in our research have several characteristics, they are partially-observable, non-static and distributed. A problem is called *partially-observable* if any given agent can observe only a part of the state, or have only a summarized view of the variables that may be important to the evolution of the environment's state. By *non-static* it is meant that, from an agent's point of view, the same action in the same state can have different outcomes at different times. This is a consequence of the environment's stochastic nature and the interaction between different agents. Distributedness is considered at two different levels: first, each problem is solved by a team of agents; second, several teams are working in similar problems in different locations. Members of the same team can interact either by communicating, or by the consequences of their actions in a local environment (we will refer to these as *partners*). Members of different teams only interact through explicit communication. The agent's objective is to maximize the average reward obtained in a given period. These periods will be called *epochs*. In this work we try to make as few assumptions as possible regarding the learning algorithms used by the agents. This will allow the use of heterogeneous groups, in which each team of agents uses different learning algorithms.

3.1 Exchanging Information During Learning

As mentioned above, the main question addressed here is: "(How) can communication between agents improve learning performance?" This question can be divided into several others:

- What information to exchange?
- How to integrate this information with the usual learning process?
- When should an agent request/accept information?
- How should an agent decide where to get/send information?

In the following subsections we will address each of these questions and propose an approach to these problems.

3.1.1 What Type of Information?

The most obvious way to exchange the learned knowledge is to send a complete description of the solution to another agent (i.e. a complete set of parameters for its learning structure), but this has two major drawbacks: first, all agents would either have to be of the same type, or be able to use different learning structures, thus heterogeneity would be lost; second, the solution was constructed to fit the dynamics of the local situation of the advisor, so, even the smallest differences in the dynamics of the problem could render the solution useless and destroy what was learned by the advisee.

Advice-exchange Nunes and Oliveira (2002b) uses the information available in the environment, such as: states, actions and rewards. Using only these types of information, and statistics based on them, the agents can communicate: states that they experience more often; the quality of the actions they performed at a given state; the action they would choose for a state; sequences of actions that produced good results; etc. Different combinations of these types of information can provide a description of the environment, of the agents' policies and the characteristics of the quality function that agents are trying to maximize. The most common type of communication is for an agent to send another its current state and receive as a reply the action the other agent advises for that particular situation. The answer to a query is computed using the parameters that have obtained a good score in a previous epoch.

3.1.2 How to Integrate This Information with the Usual Learning Process?

The way to use exchanged data depends highly on the type information that is exchanged. It is difficult to envision a process that would be applicable to any type of information, learning algorithm and problem. Our approach simply points a general way and exemplifies its application. The extension to other learning algorithms would necessarily require adaptations.

The integration of advice with the knowledge gathered from acting in the environment is done in a different way for each type of agent. Nevertheless, it uses either imitation or a form of *supervised learning*. Most learning structures are able to integrate information given in this form and learning from supervision information is typically faster than from reinforcement.

When the advised action is different from the one the agent would have chosen it can either imitate the incoming action and learn from the result, or integrate the knowledge with its own hypothesis and then select an action. When an agent is rewarded this information can also be used in different ways, depending on how the action selection was performed. In some cases, just one presentation of the advice is not sufficient to achieve a reasonable effect and it is necessary to replay it. *Advice replay* consists in storing and replaying advice at specific times to improve the effectiveness of the procedure.

A more detailed explanation of how integration and imitation are merged with each of the learning algorithms can be found in section 4.3.

3.1.3 When Should an Agent Request/Accept Information?

The answer to this question may depend on several factors: a) the comparison between the advisor and advisee's performances; b) the comparison of the experience each has in dealing with the current situation; c) how well defined is the response. The problem with a) and b) is that the environment is noisy and partially-observable. This makes it difficult to have a clear evaluation of performance and state. Consider, for instance, the case of controlling a traffic light at a crossing. The traffic volume is different depending on the time of day. At times there are no cars in the incoming lanes for a period of time, and on other occasions the traffic volume is over the limit of saturation. How can we compare the agent's policy in those two periods? What if all surrounding agents are, by a fault in their own policy, diverting all the traffic to one intersection. The agent at that intersection cannot cope with so much traffic no matter what policy it chooses. Should it be penalized by that?

Some of these matters can be dealt with by a careful choice of the state representation, quality-function and times of evaluation. Nevertheless, it is necessary to introduce mechanisms that provide several perspectives of the performance to overcome the noise generated by the dynamics of the environment. Using different statistics of performance, measured over different periods of time, compensates for some of these effects, specially in the above mentioned case a). Each agent calculates the following measures of performance:

- Average reward per epoch (R_n);
- Infinite discounted reward;
- Recent best reward;
- Short, mid and long-term average reward;
- Average reward evolution;
- Self-confidence;

The infinite discounted reward (idr_n) in epoch n is defined as:

$$idr_n = \alpha \, idr_{n-1} + (1 - \alpha) R_n, \qquad (1)$$

where $\alpha \in [0, 1[$ determines the balance between current and previous experience. Recent best reward (b_n) is:

$$b_n = \max(\beta b_{n-1}, R_n), \qquad (2)$$

where $\beta \in [0.9, 1[$ is a decay parameter. The short, mid and long-term average rewards are based in R_n and calculated for fixed periods. Their calculation is, as follows:

$$RT_{n0+p} = \sum_{n=n0}^{n0+p} R_n/p, \qquad (3)$$

where p represents the period which is different for short, mid and long-term measures. This allows the calculation of the average reward evolution e_n in a given epoch n, as:

$$e_n = \sum_{k=0}^{k=K} (RT_{n-(k+1)p} - RT_{n-kp}), \qquad (4)$$

where K is the number of periods to use in the calculation (in these experiments $K = 3$).

An important measure in this case is the comparison of an agent's performance with that of others. This can help the agent to decide whether its own actions are becoming less adequate, or if the problem is in a more difficult state due to some external reason. To help in this respect we have introduced a parameter labelled *self-confidence*. Self-confidence is increased when the performance of a given agent is good in comparison to that of its peers. Self-confidence is decreased when the performance is poor in comparison to others'. This increase or decrease is done by multiplying the parameter by a pre-specified value that is either slightly above, or below 1.0.

In what concerns the comparison of experience it is necessary to know if the advisor has experienced a situation similar to the one it is evaluating. Some algorithms keep records of the states they evaluate while others do not. To normalize this situation all agents keep records of their experiences in certain epochs. The recorded epochs are: current, last and best. The epoch in which an agent saved the parameters it uses to give advice is also recorded. With this knowledge, an advisor can say how similar was the case in which it has used the same option and what was its reward. This will help the advisee to decide whether the situation is similar enough and, after accepting the advice, what is the difference between the announced reward and the one actually received.

To know how well defined an action is corresponds to calculating the (un)certainty in the choice of an action. This uncertainty measure can be represented differently depending on the learning algorithm used. When the agent produces a quality estimate for each of the available actions, if the variance of these qualities is low it may be interpreted as a high uncertainty coefficient, as was done in Clouse (1997).

Another important factor in the decision of when to request/accept information, concerns the agent's learning stage. At different stages, agents need different types and volumes of information. It is known that humans go through several stages when working in a team and learning from others. First they explore solutions individually until someone finds a way to solve the problem. Then others observe, in detail, how the problem was solved and try to mimic the solving behavior. As this phase progresses the "students" gain a growing degree of autonomy. They tend, first, to ask for a complete demonstration and afterwards to ask only for small pieces of information to clear the doubts concerning a particular aspect of the solution. At this point, if there are several experts with different solutions, they may try to compose a new solution from parts of different approaches. After they have mastered the current solution the members of the group start acting in a more autonomous way and try, each individually, to improve on the previous solution, or find a better one. When one succeeds to find a solution that is proven better than the previous ones, the learning process of its peers may go back to one of the previous phases and some of the members of the group will start, once again, to learn this new solution by imitation of the new expert's behavior.

This process was transposed to the domain of software learning agents, by the introduction of four different learning stages:

1. Exploration: Agents do not exchange information with their peers and learn to choose their actions using only the reward information provided by the environment. This stage ends when the learning performance of the agent stabilizes.
2. Novice: At this stage an agent will ask for advice frequently (in this case for all the actions of an epoch) and keeps the same advisor for long periods. Agents in this stage have a considerably lower performance than the best of their peer's. This phase is ended when the performance stabilizes or when the advisee reaches a level of performance that is very similar to that of its advisor.
3. Stable: When an agent reaches a performance level that is close enough to that of its advisor it will only request advice when the choice of the next action is unclear, or when another peer reports a higher performance in acting from a given state. A Stable agent may choose different advisors for each new situation.
4. Expert: Expert agents are those with the best performances for a particular role. They may request advice when it appears to be much better than its own options, but most of the time they are engaged in individual exploration.

Agents evaluate the conditions to switch from one stage to the other at the end of each training epoch. An agent can move one level down in each epoch or as many levels up as it can by meeting the appropriate conditions. The conditions concern the comparison of current or best performances with the best that was achieved by other agents, for example: an agent will transit from Novice to a Stable stage if $idr_{i,n} > \lambda\, idr_{k,n}$, where, $idr_{i,n}$, is the infinite discounted reward for agent i at epoch n, k is the index of the agent with best performance and $\lambda \in [0, 1]$ a discount parameter. To change learning stages an agent is also required to have a certain self-confidence level. This condition prevents agents from changing stages due to a result acquired in exceptional conditions and provides a more stable evaluation. Different learning stages may also imply different learning parameterizations. For example, when an agent enters a Novice stage its reinforcement learning rates are lowered and the supervised ones are increased, so that it learns more from advice than from the environment feedback. The definition of learning stages to set learning rates and other parameters is reported in Dorigo and Colombetti (1994). The transitions from the Exploration and Novice stages also include another type of evaluation. An agent will leave these stages when the, short, mid and long-term average reward evolutions are lower than a certain threshold. This will guaranty that the learning process is stabilized when the agent is promoted.

In the experiments reported below agents have used two different ways to decide whether or not to get advice. The first, used in the Predator Prey experiment, is based on the uncertainty of the selected action. The second is based on the comparison of the reward the advisor announces for a given state with the one expected by the advisee. In the first case an agent would request advice if the best n options were all within a given vicinity of the best. In QL-agents the estimated quality of each option is directly available when selecting an action, so this calculation presents no problems. In agents that use other algorithms the output is interpreted as a classification since the chosen action is the one with highest output. An action was considered to be unclear when the best n options were within a given vicinity of the best. In the second case an agent would ask for advice if:

$$tolerance * r_j(s) > r_i(s), \qquad (5)$$

where $r_j(s)$ is the advisor's announced reward for state s, and $r_i(s)$ is the advisee's expected reward. The tolerance parameter used in the experiments was 1.0 for Stable agents and 0.9 for Experts.

3.1.4 Where to Get Information?

After having decided that it needs advice, the agent must choose the best source of information. In previous experiments, reported in (Nunes and Oliveira, 2002a, 2003), it was clear that the best advisor was not always the agent with best performance. This can happen for several reasons, but the most common of these is that advisor's and advisee's local environments have different dynamics and respond differently to the same actions in the same state. In this case an agent needs to learn which of its partners it can trust, which it should not and also which has the most successful solution for a given state or class of states. The combination of these types of information can help in the decision of which advisor to choose.

The decision of which peer an agent i should request advice to can take two forms depending on whether the agent needs advice for a full epoch (Novice agents) or for a particular situation. In the first case, the agent will select an agent k among its peers, where:

$$k = argmax_j(trust_{ij}\ ar_j\ rl_i), \qquad (6)$$

where $trust_{ij}$ represents the level of trust agent i has in peer j, ar_j is the average reward of the advice given by peer j and rl_i is the *role discount*. $trust_{ij}$ is calculated by:

$$trust_{n,ij} = \alpha trust_{n-1,ij} + (1-\alpha)1/na \sum_a r_{j,a}/r_{i,a}, \qquad (7)$$

where $r_{j,a}$ is the reward announced by advisor j for advised action a and $r_{i,a}$ is the actual reward received by the advisee i. na is the number of advised actions in epoch n. This update is done at the end of each epoch. The average reward of advice given by peer j is initialized with the score achieved in the epoch where it as saved the advice parameters and updated with the advisee's reward. Whenever the reward in a given epoch is higher than ar_j the parameters of the current epoch are saved and ar_j is set to the current average reward. The role discount is a constant that has a value of 1.0 for all peers that have the same role and a lower value for advisors with different roles (in these experiments the role discount was 0.7). In previous experiments the role discount was calculated in run-time. To do this each agent used an *unsupervised learning* algorithm to keep track of the classes of states it experienced. The role discount was calculated as a normalized measure of similarity between the class representatives of the requesting agent and all its peers. This proved to have reasonably accurate results and was abandoned only because of its computational cost. By accurate we mean that the role discount for agents with the same role was higher than for others and in most cases close to 1.0.

When an agent needs advice for a particular state (s), the choice of advisor is:

$$k = argmax_j(trust_{ij}\ r_j(s)\ rl_j), \qquad (8)$$

where $r_j(s)$ is the estimated reward agent j expects for state s, based on the data stored in the epoch were advice parameters were saved.

Table 1 shows a summary of the *advice-exchange* procedure. This section stated the main questions concerning exchange of information during learning and sketched the most important components of *advice-exchange* along with the reasons why they were integrated in the procedure. Details that are algorithm or problem-dependent will be presented in the following sections.

4 Experiments

In the following sections we will describe two experiments in which we have tested advice-exchange. The first was the pursuit (Predator-Prey) problem, usually considered a toy-problem although some versions are in fact difficult to solve with most learning algorithms. The second was a simulation of an environment in which agents control the traffic in a grid of roads by setting the color and timing of the traffic lights at intersections. This simulation is based on real data, graciously made available by the Lisbon City Hall's Traffic Control Department. The data contains the number of cars that passed in several of the most congested avenues of the city at different days and times during a continuous 15 days period in June 2002. The raw data contains car counts in 5 minutes intervals.

Table 1. Summary of the *advice-exchange* algorithm.

```
While not train finished
   Novice : Select best advisor k (eq. 6)
   While not epoch finished
      1. Get state s for evaluation
      2. Should get advice ?
         Exploration: No, Novice: Yes
         Stable/Expert: eq. 5? Yes
         Otherwise, No (go to 3)
         2.1 Stable/Expert: Select best advisor (k): (eq. 8)
         2.2 Send agent k the current state s and request advice
            Agent k: Load advice parameters,
            Agent k: Evaluate state s,
            Agent k: Produce an advised action (a) for state s
            Agent k: Return a to advisee.
            If integrating advice before acting
               Supervised learning: learn(s, a)
      3. Evaluate state s and produce action (a')
      4. If imitating and advised, use action a, otherwise a'
      5. Receive reward (r)
         5.1 If advised
               If imitating, learn from the result of a':
               EA: If reward > expected reward
                  backpropagation(s,a')
               QL: Regular update based on s, a, r
                  and state after action s' performed
         5.2 If not advised or a = a' learn from reward
               (using regular QL or EA updates)
   End epoch loop: Update performance statistics and *trust*.
End train loop
```

4.1 Predator-Prey

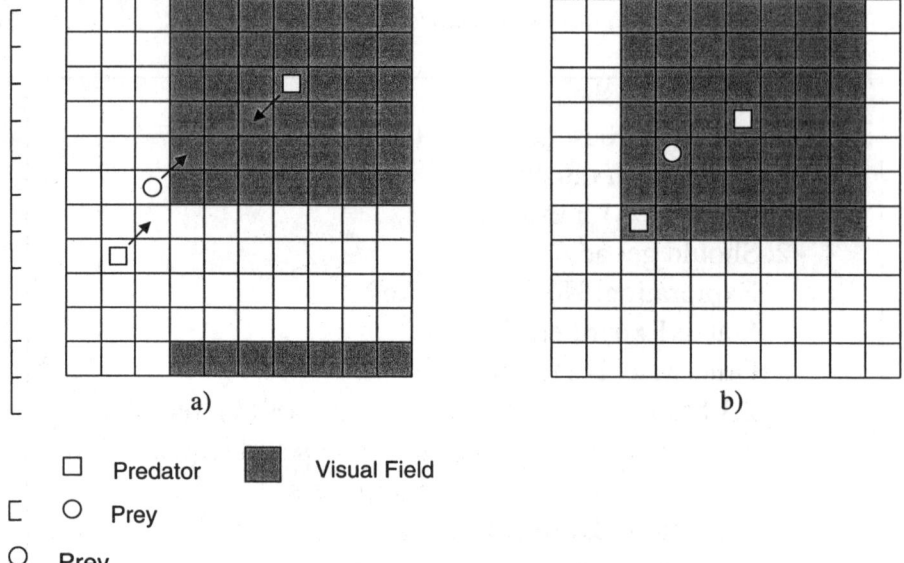

Figure 1. Predator Prey environment at time t (a), and $t + 1$ (b). Arrows represent movements.

This problem was first introduced in (Benda et al., 1985), although the version presented here has been inspired on several variations presented in (Tan, 1993) and (Haynes et al., 1995). One of the grid-worlds in which experiments were performed (the 10x10 version) is depicted in figure 1.

The problem faced by the predator consists in learning to catch a prey in a grid world. A predator is said to have caught the prey if at the end of a given turn it occupies the same position as the prey. The grid-worlds (arenas) are spherical (although they are usually represented in their planar form) and each contains two agents (predators) and one prey. The state of the environment consists on the position of the prey relative to the agent, i.e. if an agent is at position (2,4) and the prey is at (3,6), the state would be (1,2) (see the predator in the bottom-left corner of figure 1 a)). The spherical shape of the grid is taken into account when computing the relative position of the prey as well as when moving. In the experiments reported below an agent's state also contains the position of the prey relative to the partner. Previously published results, using only the prey's position relative to the deciding agent (Nunes and Oliveira, 2003) lead to similar conclusions.

The accuracy of the state representation received by the predator depends on its visual range. The predator perceives the correct position of the prey up to a limit defined by the visual range parameter (3 in this case). If the prey is at a distance greater than the visual range its relative position is disturbed (by Gaussian noise with null average). The further the prey is from the limit of the visual range, the higher is the random noise added to the prey's position.

Each predator has to choose between nine possible actions in each turn, i.e. it chooses either to move in one of the eight possible directions (four orthogonal, and four diagonal), or to remain in its current position. Each predator moves one step in one direction in each turn. Each agent outputs a vector of nine real numbers and the index of the highest output defines the action to be performed.

The prey moves before the predators. To decide in which direction to move the prey detects the presence the closest predator (closest in the sense of being at the shortest Euclidean distance) and moves in the opposite direction, as depicted in figure 1. If there is more than one predator at the same distance one of the predators is picked randomly and the prey moves away from it regardless if it is approaching the other predator. The prey moves only nine out of each ten turns. Any two units can occupy the same cell simultaneously at any time.

In each turn predators are given a reward that as an individual and a global component weighted by $(1 - \beta)$ and β, respectively. The value of β in these experiments was 0.25. The individual component is based on their distance to the prey (d). This reward is equal to 1.0, if the predator has captured the prey, or $0.1(1.0 - d/d_{max})$ otherwise. The constant d_{max} represents the maximum distance between any two positions in the grid world. The global component is the partner's reward. After a successful catch the predator that caught the prey is randomly relocated in the grid-world.

This version of the predator-prey problem does not involve either explicit cooperation or competition between the predators, although cooperative behavior has emerged in some experiments. The same effect was observed using only the position of the prey as state description. Agents use either *Q-Learning* (QL-agents) or *Evolutionary Algorithms* (Holland, 1975; Koza, 1992) (EA-agents) to learn this task. Another type of agents, labelled Heuristic (H-agents), perform a fixed (hand-coded) policy that always tries to reduce the distance to the prey as much as possible. Heuristic agents do not request advice, but they can reply to advice requests. H-agent's policy would be optimal if there was only one agent in each arena considering that an agent will not keep records of the previous movements of the prey.

The scenarios used for these experiments are the following:

1. *Individual*: Four arenas, each with two predators and one prey. In each arena all predators use the same learning algorithm and they do not exchange advice. Two arenas have EA-agents, the other two have QL-agents.
2. *Social Heterogeneous*: Same as previous, except that agents may request advice to any of its seven peers in the same or other arenas.
3. *Social Heuristic*: Same as previous but with an extra arena where two H-agents are performing the same task and may also be chosen as advisors.
4. *Social Homogeneous*: Same as the *Social Heterogeneous* scenario, except that all agents use the same learning algorithm (either QL or EA).

For each of the above scenarios 11 trials were made (x4, or x8, agents of each type), each with different random seeds. Each trial ran for

Learning by Exchanging Advice

9000 epochs and each epoch has 150 turns. For each trial there is a corresponding test which runs for 1000 epochs without learning or exchange of advice. Each agent in the beginning of the test loads the parameters saved during training when it achieved the best score. Partnerships were kept unchanged by this procedure. Agents do not exchange any information except for the one necessary to perform *advice-exchange*.

Two sets of experiments were made differing only in the dimension of the grid. In the first set the grid was 10x10, and in the second 20x20 positions wide.

In this problem advice was incorporated and not imitated. Also, trust was computed as follows: $trust_{ij}$ was initialized to 1.0 and multiplied by a factor slightly higher than 1.0 when advisor j provided advice in epochs where the average reward was higher than the infinite discounted reward. Similarly, it was multiplied by a factor lower than 1.0 when j provided advice and the average reward was not better than the infinite discounted reward. Trust in a given peer was also influenced by other agents. When an agent issued a message of "trust agent X" the receiving agent updated the trust on agent X in the same way as if it had been an advisor for a successful epoch, thus, increasing the trust on this advisor. An agent would issue a message of "trust agent X" to its partner when it had a successful epoch based on the information communicated by a single advisor. In this case "X" would be the advisor's partner. This policy replaced the static role-assignment procedure and allowed agents synchronize their advice requests so that each member of a group would request advice to a different member of the advisor-group. In situations where agents acquire different roles during learning even when they start out in exactly the same circumstances, the run-time definition of roles is important for the success of advice-exchange.

4.2 Traffic Control

The Traffic Control (TC) environment consists of several detached *locations*, each containing one (TC1) or two (TC2) connected crossings. Each location in the TC2 scenario contains 12 lanes over an area of, roughly, 300 x 300m. An environment contains 3 locations, each controlled by a team of agents (QL-agents, EA-agents or H-agents as in the previous experiment). The number of cars generated in each time interval (of 5 minutes) is taken from a set of real data collected in some of the most busy avenues of Lisbon. The TC environment uses real data to calculate, at each time, the number of cars to be inserted. At each 60 seconds interval 1/5 of the cars that must pass through the crossing during the 5 minutes period is introduced simulating a fixed policy traffic-light outside the scope. The insertion of cars is done at different times in each set of adjacent lanes.

This experiment presents several difficulties. Firstly, to find the appropriate timing to end training epochs. Agents must have enough epochs to learn and simultaneously each epoch must be long enough to have a fair estimate of an agent's performance. An optimal situation would be to test an agent's policy for at least one full day, but computational constraints do not allow this. The compromise solution was one simulated hour for each epoch. To diminish the differences in training epochs the data used is restricted to the traffic between 8 a.m. and 8 p.m in regular working days (i.e. holidays and weekends are discarded). All agents start each epoch with no cars in the system.

Another difficulty is that too many mistakes induce heavy traffic jams. In these circumstances some agents are not able to learn because they have no data (when the cars are jammed in a previous crossing for long periods), and huge traffic jams make the simulation too slow to provide results within a reasonable time-frame. To allow the experiments to run with the necessary speed car movements were simplified. Cars do not exchange lanes, nor turn at crossings and the only limitations to their forward movement are the current speed, the position of the car directly in front and the existence of a closed traffic-light ahead. The movement of cars is calculated as proposed in (Nagel and Shreckenberg, 1992) to simulate realistic drivers. The simulation step is of one second and each epoch consists of one hour of simulated-time, i.e. 3600 turns. Agents are asked for a new decision every 20 seconds. In this problem agents imitate the advised action and learn from the result. Training is done for 10000 epochs and test lasts 500 epochs.

It is important, at this point, to state that we do not propose that these techniques are adequate solutions to a real-life traffic control problem. Even though some lessons may be taken that apply to this problem, as well as to others, the objective is not to supply an architecture to solve any specific problem. The choice of this particular test-bed for evaluation is because it has all of the characteristics of the environments these techniques are aimed at. Although the tests are based on real data, the constraints put on the simulation that make it possible to run several hundreds of times faster that real-time, do not allow us, at this moment, to extrapolate the conclusions taken to a real-life scenario.

The quality of the state is composed by a local and a global component, weighted by a factor β (in this case equal to 0.75). The quality $q_{m,i,t}$ for agent i at location m and time t is calculated by:

$$q_{m,i,t} = \beta p_i(C_l/ast_{i,t}) + (1-\beta)p_m(C_g/ast_{m,t}), \qquad (9)$$

where C_l is the patience threshold of drivers for one crossing (10 seconds in this case), $ast_{i,t}$ is the average stopped time of cars currently situated in the incoming lanes of the crossing controlled by agent i. If $ast_{i,t}$ is smaller than C_l it takes the value of C_l so that the first component is always in the interval $[0, 1]$. p_i is a penalty value that is 1.0 if all incoming lanes have an occupation-rate smaller than 1.0, and is multiplied by a fixed value (0.75 in this case) for every full incoming lane. The calculation of the global component is similar, although it uses a different patience threshold (20 seconds) and $atc_{m,t}$ is the average stopped time for all cars in location m. This value includes the time cars may had to wait before entering the scenario. The global penalty value is computed across all lanes in the agent's location.

In this experiment there are two scenarios (*Individual* and *Social*). They are equal in all respects except that in the social scenario agents are able to communicate.

Agents observe a state (s) composed of 10 variables (s_i), where $s_i \in [0, 1]$, for all $i \in [1, 10]$:

- 1–4: Normalized occupation rate of each of the four incoming lanes, i.e. number of cars present over the number of cars required to fill the smallest lane in the location.
- 5–8: Incoming traffic from a given direction (0 for no traffic, 1 otherwise). It is considered that there is incoming traffic if a car has entered the lane in the last 10s or if the lane is full.
- 9: Current color of the agent's traffic-light (0 for red, 1 for green).
- 10: Time since the last change in the traffic-light, normalized at a value of 180 seconds.

Each agent must choose one of two actions. Either set the North-South lanes to green or red (the East-West lanes will automatically switch to the opposite color). An action is requested every 20 simulated seconds. Yellow times (of 5 seconds) are introduced automatically when the light changes from green to red.

Heuristic agents set to green the traffic-light on the lanes with the maximum occupation rate. To prevent quick oscillation of the traffic-light the lanes that have green light have their occupation-rate multiplied by a value larger that 1.0 (1.3 in this case).

4.3 Learning Algorithms

This section contains a summarized description of the learning algorithms used in the experiments. This description is focused on the variations from their standard form that were introduced in this work. *Q-Learning* and *Evolutionary Algorithms* will be given more attention due to their importance in these experiments. For details on the standard versions of these algorithms the reader should consult the referred bibliography or (Nunes and Oliveira, 2004).

Backpropagation (BP) (Rumelhart et al., 1986) is a well known *supervised learning* algorithm, commonly used with networks of differentiable non-linear units connected by weighted links (Artificial Neural Networks, ANN). In this work we use classical, online, *backpropagation* to integrate the information received by *advice-exchange* in agents whose main learning algorithm is also based on ANN and to perform state-to-quality mapping in Connectionist Q-Learning.

Q-Learning (QL) (Watkins and Dayan, 1992) is the most commonly used learning algorithm in the class of RL. Its most simple form, one-level Q-Learning, is based on a table that stores the estimated quality $Q(s, a)$ of performing action a at state s for all state-action pairs. When a reward r_t is received, at time t, the value of $Q(s, a)$ is updated as follows:

$$Q_{t+1}(s_t, a) = (1 - \alpha)Q_t(s_t, a) + \alpha(r_t + \beta Q_{max}(s_{t+1})), \quad (10)$$

where s_{t+1} is the state of the environment after performing action a at state s_t, α is the learning rate and $\beta \in]0, 1[$ a discount factor applied to the estimated quality of the next state ($Q_{max}(s_t)$), which is given by:

$$Q_{max}(s_t) = \max_a(Q(s_t, a)), \quad (11)$$

for all possible actions a when the system is in state s_t.

To incorporate the advice from other agents before acting, a form of *supervised learning* is employed in the update of the quality values. When an agent is advised to use action a as response to state s it will sum a positive value (b_{up}) to $Q(s,a)$ and a negative value (b_{down}) to all other actions available at state s (in the current experiments $b_{up} \approx na|b_{down}|$, where na is the number of actions available for the current state). A similar technique, labelled *Biasing-Binary* (Whitehead, 1991), uses the same absolute quantity both for positive and negative feedback. When incorporation of the knowledge is done after the reward is received, i.e. the agent is imitating and learning from the result, this process is not necessary because QL can learn from the reward even when the action was not its own choice.

When the state and action spaces are continuous, or too large, the implementation of Q-Learning with quality tables is not feasible. In these cases a discretization of the state-space may be necessary, but even this strategy cannot cope with some problems. In the most difficult problems, in which a discretization would either be too coarse or need tables that would be too large to store and too time-consuming to access, the usual solution is to use connectionist Q-Learning (QConn) (Lin, 1992). In this approach the table that stores a mapping of state-action pairs to their estimated quality is replaced by a set of ANN (one for each action) trained with *backpropagation*. Instead of the update described in equation 10 the ANN that corresponds to the executed action is trained using the state (s) as the input example and $r_t + \beta Q_{max}(s_{t+1})$ as the desired response. In the Predator-Prey problem we used standard QL and in the Traffic-Control problem QConn.

Evolutionary Algorithms (EA) (Holland, 1975; Koza, 1992) are a well known learning technique, with biological inspiration. The solution parameters are interpreted as a specimen (or phenotype), and its performance in a given problem as its fitness. After the evaluation of all the specimens the ones with best fitness are selected for breeding. The selected specimens are then mutated and crossed-over to generate a new population, usually of the same size as the previous one.

The variant of EA used in these experiments is based on the work presented in (Glickman and Sycara, 1999), and its main characteristics are:

- The genotype is the set of real-valued matrixes that correspond to the weights an ANN of fixed size.
- Each specimen is evaluated during a certain number of epochs (3 in this case). In the first epoch of evaluation *advice-exchange* is inhibited.
- The selection strategy is elitist (keeps a number of the best specimens in the next generation).
- Mutation is done by disturbing all the values of the parameters with random noise with null average and normal distribution. The variance of the mutation rate is slowly decayed during training.
- The crossover strategy consists on choosing two parents from the selected pool and copying each node of the ANN (along with the weights of all incoming connections) from a randomly chosen parent.

Each agent contains a population of specimens. To incorporate information given by advice an agent will use backpropagation with the advised action as desired response. The selection process will work on the specimens after they have been changed by the *backpropagation* algorithm, which can be interpreted as *Lamarckian learning*. Advice may be replayed at the end of each learning epoch. To do this advice is stored after being used. When the storage-space is filled, incoming advice will replace the oldest advice stored. Only Novice agents may replay advice and this is done when the last epoch was more successful than previous ones. When an agent changes advisor, stored-advice is cleared. Advice-replay allows to incorporate knowledge gathered by advice more rapidly, but it's use must be carefully considered because it is computationally heavy and it may reduce the diversity of the specimens. Other options that may allow a more effective use of advice are being studied.

Since backpropagation of the advised action implies that this is the best action for the given state it should only be done when the agent verifies that this is the case. When advice is integrated before acting the agent must trust the advisor and believe that the advised action is indeed the best response for its state. When an agent is imitating, i.e. incorporation is done after acting, the agent can verify if the action did produce a reward that is similar or higher than what it would expect for that state. If the action proves to be bad advice and produces a relatively low reward, the EA-agent will not incorporate the advice.

5 Results and Discussion

In this section the results of the experiments described above will be presented and discussed.

5.1 Predator-Prey

The results presented in Tables 2 and 3 refer to the average performance achieved in test by the best team (labelled *Best Team*) and the average individual performance of all agents (labelled *Avg*) in several different cases. The team performances are calculated by adding the rewards of both agents in a team at each epoch. The average performance is accompanied by the correspondent standard deviation. The column labelled *Percent* contains the percentual difference between the result on its left and the same result for individual agents of the same type (QL or EA). The measures were taken for each experiment (10x10 and 20x20), scenario (Individual, Social Homogeneous, Social Heterogeneous and Social Heuristic) and learning algorithm (QL and EA).

Table 2. Test results in experiment 10x10 for each scenario-algorithm pair.

Scenario	Alg	Best Team	Percent	Avg
Social Heu.	Heu.	0.3477	-	0.1730(+/-0.0025)
Social Heu.	QL	0.4009	-0.3%	0.1942(+/-0.0074)
Social Heu.	EA	0.4044	+28%	0.1883(+/-0.0116)
Social Het.	QL	0.4018	-0.1%	0.1885(+/-0.0232)
Social Het.	EA	0.4061	+29%	0.1892(+/-0.0148)
Social Hom.	QL	0.4022	-0.0%	0.1843(+/-0.0375)
Social Hom.	EA	0.3263	+3.4%	0.1466(+/-0.0234)
Individual	QL	0.4023	-	0.1908(+/-0.0237)
Individual	EA	0.3156	-	0.1429(+/-0.0184)

Table 3. Test results in experiment 20x20 for each scenario-algorithm pair.

Scenario	Alg	Best Team	Percent	Avg
Social Heu.	Heu.	0.2444	-	0.1217(+/-0.0006)
Social Heu.	QL	0.2387	+19%	0.1177(+/-0.0013)
Social Heu.	EA	0.2412	+3.4%	0.1179(+/-0.0018)
Social Het.	QL	0.2299	+15%	0.1110(+/-0.0020)
Social Het.	EA	0.2292	-1.7%	0.1122(+/-0.0019)
Social Hom.	QL	0.2188	+9.22%	0.0874(+/-0.0097)
Social Hom.	EA	0.2281	-6.2%	0.1118(+/-0.0018)
Individual	QL	0.2004	-	0.0860(+/-0.0049)
Individual	EA	0.2332	-	0.1071(+/-0.0097)

QL and EA-agents have quite different behaviors in these two scenarios. In the 10x10 experiment (table 2) QL-agents are clearly better than EA-agents and H-agents. This is achieved by learning joint strategies in which predators push the prey towards each other. In the 20x20 experiment (table 3) joint strategies are harder to develop and maintain, thus the best result is achieved by H-agents, and the best learning agents are EA-agents. The difference in performance of EA versus QL-agents, in these two experiments, was one of the main reasons for the choice of these variants of the problem. Analyzing the results we can observe that:

- Agents show similar or better team performances in Social scenarios (with the exception of EA-agents in Social Homogeneous scenarios, in experiment 20x20).
- For the agents that show lower individual performances in each experiment (EA-agents in 10x10 and QL-agents in 20x20), improvements on the team performance in Social Heterogeneous and Heuristic environments are clear (ranging from 15% to 29%).
- The best average performances of agents occur in *Social Heuristic* scenarios, even when H-agents show relatively poor performances (as in experiment 10x10).
- Cooperation in Homogeneous environments leads to slightly better results for the agents that have lower individual performances in each scenario.
- The experiments do not allow for a firm conclusion on the comparison of the average results of the best agents in Individual versus Social Heterogeneous scenarios. In experiment 10x10 the average performance of the best agents (QL-agents) shows a slight decrease in Social Heterogeneous scenarios. In experiment 20x20 the average performance of the best agents (EA-agents) shows a slight increase in the same test. Both of these changes are negligible considering the standard deviations of the results.

In summary, the average performance of the best agents is maintained and the agents with lower individual performance learn from their more successful peers, although they seldom surpass their performances. The fact, observed in some experiments, that agents can surpass the performance of their advisors and become Experts themselves, is not as frequent as expected. Nevertheless, it is interesting to point out that QL-agents do achieve better average scores in the Social Heuristic scenario of experiment 10x10 than in any other, even when other agents have lower scores. The hypothesis that we pose for this behavior is that H-agents lead QL-agents to a quick stabilization of the most obvious choices. This can grant QL-agents enough time to explore more thoroughly the possible options to their advisors' individualistic strategy.

5.2 Traffic Control

The results presented in Tables 4 and 5 refer to the scenarios TC1 and TC2, respectively. The column labelled "Avg" contains the average results for each algorithm-scenario pair. Under the label "Std. Dev" are the standard deviations of each of the values for average performance. The column labelled "Best" shows the average test result of the best agent. The columns labeled % show the percentage relative to the best result in each column, the first % column refers to the average results, and the second to the best results. Table 6, as the same structure and refers to the global part of the reward of each team.

In the single-crossing tests (TC1, Table 4), the average results for QL agents is similar in social and individual trials, and EA-agents have a 16% performance increase. The best results for EA-agents have a slight improvement. As in the previous experiment, the agents that have lower performances profit more from communication than others. From the standard deviations we can also observe that the behavior of EA-agents is not only better in average but also more stable.

In the twin-crossing tests (TC2, Table 5), the average results of the best agents (QL) is, again, similar in individual and trials, but EA-agents show a 7% improvement in performance. EA-agents seem able to reach scores higher than QL-agents as can be seen in the Best results column. It is interesting to notice that the same percentual increase in social trials also applies to the best results of EA-agents. The best EA-agent's performance is increased even when none of their peers is able to reach scores at the same level. This seems to indicate that, in some cases, communication can lead a agents to behave better is social domains than the best they are able to do in any individual trial. Nevertheless, in average we can only say that the agents with lower individual performance (EA-agents) can learn to behave as the best (with only 2% difference).

Table 4. Test results in experiment TC1 for each scenario-algorithm pair.

Scenario	Alg	Avg	%	Std. Dev.	Best	%
Individual	HE	0.3424	89%	+/-0.0008	0.34	88%
Individual	QL	0.3832	100%	+/-0.0034	0.39	99%
Individual	EA	0.3059	79%	+/-0.0459	0.37	94%
Social	HE	0.3420	89%	+/-0.0009	0.34	87%
Social	QL	0.3814	99%	+/-0.0059	0.39	99%
Social	EA	0.3652	95%	+/-0.0166	0.39	100%

Table 5. Test results in experiment TC2 for each scenario-algorithm pair.

Scenario	Alg	Avg	%	Std. Dev.	Best	%
Individual	HE	0.2708	93%	+/-0.0013	0.27	68%
Individual	QL	0.2852	98%	+/-0.0093	0.30	75%
Individual	EA	0.2638	91%	+/-0.0347	0.37	93%
Social	HE	0.2710	93%	+/-0.0011	0.27	68%
Social	QL	0.2895	100%	+/-0.0111	0.31	76%
Social	EA	0.2859	98%	+/-0.0352	0.40	100%

Table 6. Global reward results in experiment TC2.

Scenario	Team	Avg	%	Std. Dev.	Best	%
Individual	HEU	0.1041	94%	+/-0.0011	0.11	59%
Individual	QL	0.1024	92%	+/-0.0089	0.12	64%
Individual	EA	0.0865	78%	+/-0.0247	0.16	89%
Social	HEU	0.1044	94%	+/-0.0007	0.11	59%
Social	QL	0.1066	96%	+/-0.0100	0.13	71%
Social	EA	0.1106	100%	+/-0.0274	0.18	100%

If we measure only the global component of the reward we reach similar conclusions. EA-agents improve this component by 22%, in average. QL-agents do not fully take advantage of this rise, but they also show a small improvement (4%).

These tests point in the same direction as the previous ones, advice exchange does improve the average performance of the agents with lower scores, although it may prevent exploration. As is the previous cases not all agents reach the same performance levels and some advisees do not reach the level of performance of the advisors.

6 Conclusions and Future Work

Advice-exchange has evolved, from a simple form of sharing experiences with other agents, to the draft of an architecture that enables cooperation between heterogeneous groups. Along the way, problems such as: disturbances in the learning process (caused by conflicting advice), learning of joint strategies from more successful groups, spurious results that caused agents to trust the "wrong" advisor, and many others, have been addressed by introducing new concepts and tools. The initial objective was to create a process by which agents could improve their learning skills through communication with other agents. The results presented here show that low performance agents can improve their performance by communicating with others.

One of the goals of this work was to create a learning system that could generate better behaviors than individual learning. So far this was only detected in a few specific cases and it is not the rule. In fact, communication can even hurt the learning performance by leading all agents into an area with a local minimum early in the trial.

An important factor to consider is the delay introduced by advice-exchange. If we exclude advice-replay the delay introduced by the process is negligible. Advice-replay may cause considerable delays depending on the number of stored advice. Another important delay factor is communication between processes in different physical locations. In this simulation that factor cannot be measured since all agents are part of the same process. The communication necessary to allow advice-exchange is simple in structure and reasonably low in average, although it tends to be bursty because all agents tend to ask for advice at the same times (when another agent improves its performance making their current solution obsolete). To give a reasonable idea of the learning-time we can say that the Traffic Control simulation, with advice-exchange, is running several hundred times faster than real-time and more than 70% of the computation-time is used in the movement of vehicles and not in the learning procedures.

The objectives of future work are mainly to refine the process into an architecture that promotes better team behaviors. The identification of the type of problem an agent is facing and the adequate control of learning parameters in run-time, based on this information, seems to be a key point to improve performance. Other improvements, based on combination of advice from different sources, rendering unsolicited advice and the use of other learning algorithms, are also scheduled for future work. In some situations it is important to have a global view of a situation, the introduction of team supervisors that learn from a summarized view of the state and may influence the decisions of agents can improve the cooperation capabilities of a team.

The road to new learning systems that interact with complex environments and communicate is still long. Above we have sketched what we believe are some of the fundamental characteristics of agents that can learn by communicating with others as well as from the environment. As was mentioned during the description, the implementation of some of these concepts must be adapted to the type of agents or the environment, but, by creating a set of techniques that promotes communication and effective integration of the exchanged information in the learning process, we can give agents the possibility to interact with others and profit from this interaction.

Acknowledgments

Our thanks to CML (Lisbon City Hall Traffic Control Department) that graciously made available the necessary data to implement the Traffic Control environment, and also to Luís Botelho and Pedro Figueiredo.

References

Banerjee, B., Mukherjee, R., and Sen, S. (2000). Learning mutual trust. In *Working Notes of AGENTS-00 Workshop on Deception,Fraud and Trust in Agent Societies*, pages 9–14.

Benda, M., Jagannathan, V., and Dodhiawalla, R. (1985). On optimal cooperation of knowledge resources. Technical Report BCS G-2012-28, Boeing AI Center, Boeing Computer Services, Bellevue, WA.

Berenji, H. R. and Vengerov, D. (2000). Advantages of cooperation between reinforcement learning agents in difficult stochastic problems. In *Proc. of the Nineth IEEE International Conf. on Fuzzy Systems (FUZZ-IEEE '00)*.

Biswas, A., Sen, S., and Debnath, S. (2000). Limiting deception in groups of social agents. *Applied Artificial Intelligence Journal*, 14(8), 785–797. Special Issue on Deception, Fraud and Trust in Agent's Societies.

Clouse, J. A. (1997). *On integrating apprentice learning and reinforcement learning*. Ph.D. thesis, University of Massachusetts, Department of Computer Science.

Clouse, J. A. and Utgoff, P. E. (1991). Two kinds of training information for evaluation function learning. In *Proc. of AAAI'91*. Anaheim.

Dorigo, M. and Colombetti, M. (1994). Robot shaping: Developing autonomous agents through learning. *Artificial Intelligence*, **71**(2), 321–370.

Glickman, M. and Sycara, K. (1999). Evolution of goal-directed behavior using limited information in a complex environment. In *Proc. of the Genetic and Evolutionary Computation Conf., (GECCO-99)*.

Haynes, T., Wainwright, R., Sen, S., and Schoenfeld, D. (1995). Strongly typed genetic programming in evolving cooperation strategies. In *Proc. of the Sixth International Conf. on Genetic Algorithms*, pages 271–278. Pittsburgh, Pennsylvania, July.

Holland, J. H. (1975). *Adaptation in Natural and Artificial Systems*. University of Michigan Press.

Koza, J. R. (1992). *Genetic programming: On the Programming of Computers by Means of Natural Selection*. MIT Press, Cambridge MA.

Lin, L.-J. (1992). Self-improving reactive agents based on reinforcement learning, planning and teaching. *Machine Learning*, **8**, 293–321.

Litmann, M. L. (2001). Value-function reinforcement learnnig in markov games. *Journal of Cognitive Research*, **2**, 55–66.

Littman, M. L. (1994). Markov games as a framework for multi-agent reinforcement learning. In *Proc. of the Eleventh International Conf. on Machine Learning*, pages 157–163.

Maclin, R. and Shavlik, J. (1994). Incorporating advice into agents that learn from reinforcement. In *Proc. of the Twelfth National Conf. on Artificial Intelligence (AAAI-94)*. AAAI Press.

Nagel, K. and Shreckenberg, M. (1992). A cellular automaton model for freeway traffic. *J. Phisique I*, **2**(12), 2221–2229.

Nicolescu, M. and Matarić, M. J. (2001). Learning and interacting in human-robot domains. *IEEE Transactions on systems, Man and Cybernetics, special issue on Socially Intelligent Agents - The Human In The Loop*.

Nunes, L. and Oliveira, E. (2002a). Advice-exchange in heterogeneous groups of learning agents. Technical Report 1-11/02, FEUP/LIACC/NIAD&R. unpublished.

Nunes, L. and Oliveira, E. (2002b). On learning by exchanging advice. In *Proc. of the First Symposium on Adaptive Agents and Multi-Agent Systems (AISB'02)*. Imperial College, London.

Nunes, L. and Oliveira, E. (2003). Advice exchange between evolutionary algorithms and reinforcement learning agents: Experimental results in the pursuit domain. In *Proc. of the Second Symposium on Adaptive Agents and Multi-Agent Systems (AAMAS/AISB'03)*.

Nunes, L. and Oliveira, E. (2004). Exchange of information during learning. Technical Report 3-02/04, FEUP/LIACC. unpublished.

Price, B. (2003). *Accelerating Reinforcement Learning with Imitation*. Ph.D. thesis, University of British Columbia.

Rumelhart, D. E., Hinton, G. E., and Williams, R. J. (1986). Learning internal representations by error propagation. *Parallel Distributed Processing: Exploration in the Microstructure of Cognition*, **1**, 318–362.

Sen, S. (1996). Reciprocity: a foundational principle for promoting cooperative behavior among self-interested agents. In *Proc. of the Second International Conf. on Multiagent Systems*, pages 322–329. AAAI Press, Menlo Park, CA.

Sutton, R. S. and Barto, A. G. (1987). A temporal-difference model of classical conditioning. Technical Report TR87-509.2, GTE Labs.

Tan, M. (1993). Multi-agent reinforcement learning: Independent vs. cooperative agents. In *Proc. of the Tenth International Conf. on Machine Learning*, pages 330–337.

Watkins, C. J. C. H. and Dayan, P. D. (1992). Technical note: Q-learning. *Machine Learning*, **8**(3), 279–292.

Whitehead, S. D. (1991). A complexity analisys of cooperative mechanisms in reinforcement learning. *Proc. of the 9th National Conf. on Artificial Inteligence (AAAI-91)*, pages 607–613.

Chapter 10

Adaptation and Mutation in Multi-Agent Systems and Beyond

Ladislau Bölöni and Dan Cristian Marinescu

1 Introduction

Reconfigurable and mutable systems are increasingly more popular. As early as 1975, the Microsoft Basic interpreter for Altair contained self-modifying code, introduced to overcome resource limitations (only 4K of space available for the interpreter). A contemporary web browser is a custom application, consisting of a basic framework with multiple extension API's and a large number of plug-ins, codecs, drivers, applets, controls, themes and other add-ons. These extensions are usually developed by third parties, installed/uninstalled dynamically during the lifetime of the application, and frequently changing the behavior of the application in a radical way. Some of the changes in functionality are desired, or at least approved by the user: an example of such an extension is the ability to view new media formats. Frequently, some of the effects are undesirable from the user's point of view: some third party extensions contain *spyware*, pieces of code which report usage statistics and other information about the user. Occasionally, viruses and worms use the very same extension API's.

While web-browsers are the quintessential user-driven applications, reconfiguration and mutability are even more important for autonomous agents. Recently, several agent systems with support for mutability have emerged. Varela and Agha proposed the SALSA language based on the actor programming paradigm (Varela and Agha 2001). The SALSA language is compiled to Java and targets dynamically reconfigurable Internet and mobile computing applications. The SmartApps project (Rauchwerger, Amato and Torrellas 2001) takes an approach of "measure, compare, and adapt if beneficial" for scientific applications, with the restructuring occurring at various levels from the selection of the algorithmic approach to compiler parameter tuning. The Bond agent system (Bölöni and Marinescu 2000a) was one of the first Java based agent systems with support for strong mutability, introducing a mutation technique called *agent surgery*, which describes the mutations as a series of primitive operations on a multi-plane state machine.

Reconfigurable and mutable agents have a special importance in highly heterogeneous systems, such as ad-hoc networks. In such systems, mutation and mobility are strongly intertwined concepts. For instance, the resources available on a desktop computer and a cellphone differ so significantly that agents cannot be migrated from one to the other without being reconfigured, even if both platforms are able to run the same language, such as Java. One solution is to replace the components of the agents with components that satisfy the constraints imposed by the new host. Another choice is to migrate only part of the agent to the new site, using split and merge operations.

Although the ability to change an application at runtime is a programming technique dating back to the beginnings of computer science, there is not yet an universally accepted formal theory or a software engineering model of mutability. The subject of mutability, however is a cross-cutting concern in many fields of computer science. The goal of this chapter is to provide a review of various approaches, and to present original research done by the authors.

To review the contributions from various fields, an organization of the concepts related to mutability (or at least, an understanding of the alternative usages) is necessary. In Section 2, we propose a set of classification criteria for mutable and reconfigurable applications, and propose a taxonomy based on this criteria.

One approach is based on formal modeling of the change of behavior as a result of mutation. The choice of the formal model of agency greatly influences our ability to describe and reason about change. In Section 3 we compare various models in relation to their ability to model change. We present a set of results obtained in the context of the model of a multi-plane state machine of active objects and discuss the advantages and drawbacks of the approach. As a completely random mutation is unlikely to be of any practical use, we are especially interested in *invariants* relative to operations.

In Section 4 we approach adaptability and mutability from the software engineering point of view. Traditional agent oriented software engineering (AOSE) methodologies are not well prepared to handle mutable agents. In fact, the software engineering process, traditionally seen as the steps necessary to transform an initial specification to a final product, needs to be re-thought and evaluated. The methodologies need to be extended and modified to handle the challenging issues raised by mutable agents. We present a proposed set of extensions to the Gaia agent development methodology which supports the analysis and design of agent systems containing mutable agents.

2 A Taxonomy

Mutable and reconfigurable applications can be traced back to the beginnings of computer science. The number of scientific articles dealing with reconfigurability can be counted in the hundreds. Thousands of widely deployed applications are using techniques of mutability. Despite of this, there is no general theory of mutation in applications (and indeed, its desirability and feasibility has not yet been properly investigated). The lack of a common vocabulary of talking about mutable applications makes it difficult to relate the work done by researchers in disjoint fields, such as workflow management (van der Aalst 1999, Han, Sheth and Bussler 1998), user interfaces (Thevenin and Coutaz 1999), scientific computing (Rauchwerger et al. 2001, An, Jula, Rus, Saunders, Smith, Tanase, Thomas, Amato and Rauchwerger 2001) or agents (Varela and Agha 2001, Decker, Sycara and Williamson 1996, Barber, Goel and Martin 2000). Similarities, which might form the core of a general theory, might go unnoticed, because the vocabulary used to describe an adaptive workflow is significantly different from the one used to describe an adaptive user interface.

In spite of the fact that specifications usually do not list mutability as a desirable property of an application, mutable systems have a noticeable presence. Mutability was frequently introduced in small steps in commercial applications, as an answer to general requirements, and resulting in increased flexibility.[1]

The first step towards understanding mutable systems is to attempt to introduce some order in the terminology of the field. The current emphasis on semantics in computer science is a sign that researchers understand the importance of classification (ontologies, taxonomies or even simple terminologies without relations of terms) in the understanding of systems. The next step is to identify some classification criteria which, besides the obvious benefit of categorizing, allows us to identify the most important aspects of mutability. Finally, integrating some of the classification criteria into a taxonomy gives an additional order into the field.

---

[1] A good example is the evolution of web browsers. The ability to dynamically update the browser with plugins, applets, ActiveX controls or client side scripting was not part of the initial design of the World Wide Web. These features, under the influence of business and customer pressure, were introduced step by step, with many false starts and intermediate versions.

2.1 Alternative Names

The requirement to develop software which responds to changes in its local or global environment was handled in many subfields of computer science. As a result, many names were proposed to denote the different concepts, creating a virtual Babel of mutable programs. In this section, we will try to review the different terms used in different fields, without an attempt to completeness.

The adjective *adaptive* appears in many contexts, although with slightly different meanings. *Adaptive workflows* (van der Aalst 1999, Han et al. 1998) are used in terms of workflows, which need to be changed as a result of a runtime event.

Adaptive user interfaces (Thevenin and Coutaz 1999, Pribeanu, Limbourg and Vanderdonckt 2001, Luyten, Vandervelpen and Coninx 2002) were used for describing user interfaces which adapt to the capabilities of the device. Depending on the technique of the adaptation, this can range from a simple data driven, weak mutability process (replacing an image based web-page with a text based one) to component level, hard mutability event (replacing a Swing user interface with a WAP based one). Other terms proposed were plastic user interfaces, migrateable user interfaces, runtime user interface transformations.

Runtime software evolution is used in the recently emerging field of unanticipated software evolution (Gustavsson and Assmann 2002).

2.2 Classification Criteria

We propose five classification criteria targeting the most important aspects of mutability. They are based upon:

- the **amplitude** of change: weak vs. strong mutability;
- the **granularity**: source code, machine code, library, component level mutation;

- **continuity** of interactions: runtime vs. stoptime;
- the **initiator** of the change: externally initiated vs. self initiated change;
- the **mutation technique**: extension API, compositional API, reverse engineering, data driven, hardware and others.

In the remainder of this section, we discuss all these possibilities in detail and provide concrete examples. Most of these criteria apply to agents and to non-agent applications such as interactive programs or client-server systems. To allow for a larger pool of examples, we will occasionally refer to non-agent applications as well. This is justified by the fact that many recent versions of applications traditionally considered interactive, are outfitted with autonomous, agent-like behavior. For example media players such as Microsoft Media Player or RealOne have agent-like subsystems.

Additional criteria exist: for example, the classification based on beneficial vs. malicious nature of the change. Unfortunately, deciding on the malicious or beneficiary nature of a mutation is often non-trivial. Viruses are universally considered malicious, while software updates are considered beneficiary. However, the externally initiated upgrade to the version 2.0 of the Tivo digital video recorders removed previously existent functionality. This action was widely considered malicious by users, but beneficiary to the company. Instead of asking if a certain mutation is beneficial or malicious, we need to ask the question: in behalf of whom a certain agent is acting? This question is a fundamental one in agent research, but it does not relate (only) to adaptation and mutation.

2.2.1 The Amplitude of the Change: Weak vs. Strong Mutability

The first classification criteria refers to the amplitude of the change in the behavior of the system. We call a mutation *weak* if the modified system satisfies the same informal requirement specification as the original one. Examples are mutations which adapt applications to new resource conditions, extend their capabilities for new data formats or give them new abilities which are in the line of their original specification. Patches which fix bugs and vulnerabilities in software are also included here.

In contrast, a *strong mutation* is changing the specification of the system in a radical manner. Examples are agent systems with strong mutability support (Bölöni and Marinescu 2000a), or hardware devices modified with *modchips* (such as the ones transforming the XBox game terminal in a general purpose Linux computer (XBo 2001).

2.2.2 The Granularity of Mutation

The granularity of the mutation refers to the size of the smallest component being changed. *Source code level mutation* refers to performing changes in the source code of the applications. This is a well known technique for interpreted languages such as Lisp or Prolog. Recently, many popular applications targeting the Java platform, such as the aspect oriented programming tool AspectJ (Kiczales, Hilsdale, Hugunin, Kersten, Palm and Griswold 2001) or code instrumentation systems such as JFluid (JFl 2003) are working in this way. *Machine code level mutation* requires transformations in the executable code of fully compiled languages. The mutation can happen either on the executable file residing on the filesystem or on the runtime executable format in the operational memory. *Object level code mutation* includes transformations at the level of compiled but not linked libraries. Examples include *code obfuscators* (Collberg, Thomborson and Low 1997) or *orthogonal persistence systems*. A special case are adaptive libraries such as Standard Template Adaptive Parallel Library (An et al. 2001) when one decides during runtime the actual algorithm to be used.

The most common approach to mutation involves the *component level*. The coarse grain components have independent and well understood specifications. Interaction between components happens through relatively well documented interfaces. Developers of components are usually encouraged to minimize the occurrence of side-effects (which are the most difficult aspect of source code level mutation). Component based mutations for interactive applications include applets, controls, themes, plug-ins etc. The behavior or strategy model used by many agent systems can also be the basis for component level mutation.

2.2.3 The Continuity of Interactions: Runtime vs. Stoptime

An important classification criteria for agent mutation is the ability of the agent to maintain the set of current interactions. In case of **runtime mutation** if the agent continues its execution during the mutation. For example, conversation protocols continue uninterrupted. When the agent or application is stopped and restarted to perform the mutation we talk about a **stoptime mutation**.

There are instances when the runtime mutation is a basic requirement. This is the case of self-healing fault tolerant systems. For interactive applications, such as media players or web browsers, performing updates during runtime is a matter of user convenience. On the other hand many software update systems require not only to stop the application, but also to reboot the computer.

2.2.4 The Initiator of the Mutation

Based on the initiator of the mutation, we classify the mutable systems in three categories. The mutation is *user initiated* when the change happens as a result of direct user action. We assume that the user is fully aware that this action will trigger a modification of the program, such as a software upgrade or installation of a patch. An example of a popular application is the Fortify cryptography patch for the Netscape browser (For 1998). In some cases, the action initiated by the user has undisclosed side effects, as in the case of *spyware* applications. The second case is when the mutation operation is initiated *externally* by an agent. Viruses modifying executables or anti-virus programs removing executable viruses are examples of this approach. Finally, the mutation can be initiated by the agent itself. This is typically the case of agents which can adapt themselves to changing environments such as the SmartApps framework (Rauchwerger et al. 2001), self-healing software or the ubiquitous *self-updating software*. Other examples are *enforced remote updates* such as the case of Tivo digital video recorder, the AOL Instant Messenger, or the Kazaa peer-to-peer file sharing network.

2.2.5 Mutation Technique

Our last classification criteria is based on the techniques used to perform the mutation or adaptation operation. The typical weak mutability applications use *extension API's* through which external components are attached to the main application. The advantage if this approach is that the main application can still retain the control through the application interface. The plugins can run in a controlled environment (such as the sandbox model of the Java applets). Another advantage is that mutation through an extension API is *reversible*, which is not true in general for other techniques.

Runtime *composition API's* allow strong mutation by changing the structure of the application. To qualify as a mutation technique, a component model needs to allow the *runtime* assembly of components. For example, the C++ class model can be seen as a component approach but it is assembled during compile time. Other component models such as Microsoft DCOM, KDE KParts or Gnome's Bonobo does allow runtime assembly. The assembly model for these components is a hierarchical document model, held in container applications. The Java class model on the other hand, allows runtime modification, provided that custom class-loaders and special access methods are used. For the Bond agent framework (Bon 2003), the primitive components are the strategy objects while the assembly model follows the multiplane state machine model of agency.

In absence of an explicit API, mutation can be performed through *reverse engineering*. Often this approach is taken for externally initiated mutations on non-cooperative agents. This is the approach taken by viruses, but also by many legitimate applications, such as code instrumentation (for the purpose of debugging or performance profiling), or some approaches to orthogonal persistence. One can argue that the *reflection capabilities* offered by many modern programming languages (Java, Python, etc) are in fact low level API's for reverse engineering[2].

Data-driven reconfiguration techniques exploit the ambiguity between compiled code and data interpreted as how-to knowledge. Examples are knowledgebase applications such as Prolog programs, Jess/Clips knowledgebases, or applications such as the periodic update of the virus databases of anti-virus programs.

Finally, hardware mutation techniques such as modchips or Flash updates are operating at various levels of the computer hardware.

2.3 A Taxonomy of Mutations

In proposing the five classification criteria, we attempted to make them as independent as possible. Nevertheless, not all combinations represent practical systems. The most frequently encountered combinations are presented in the taxonomy tree of Figure 1. This tree is based on three of the criteria we presented (amplitude, technique and granularity). The remaining two criteria need special treatment. Continuity is a property which depends on the effort of the implementor - runtime mutations are usually preferred, but stoptime approaches are sometimes employed because of simpler and safer implementation. The initiator of the mutation is a property of the individual changes, not of the mutable system. The initiator can be only a human user or a software agent; non-agent applications do not initiate mutations.

---

[2]These are, of course, usable only if code obfuscators were not used.

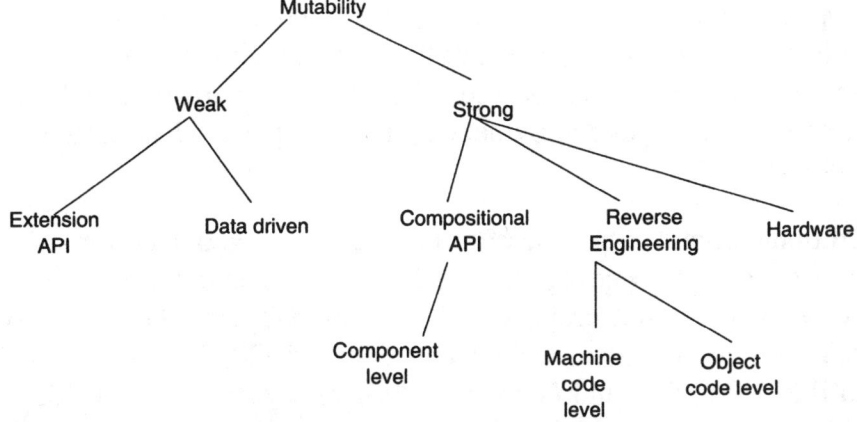

Figure 1. A taxonomy of the most frequently encountered mutability approaches.

2.4 Other Classification Approaches

The classification presented here is a result of a selection of a number of classification methods proposed by various researchers.

Gusstavson and Assmann (Gustavsson and Assmann 2002) propose a classification criteria for runtime software changes based the *technical facet* and the *motivational facet*.

A taxonomy of program transformations for the purpose of code obfuscation is presented (Collberg et al. 1997).

3 A Formal Description of Mutability

3.1 Agent Models and Mutability

A significant number of constructs have been proposed to describe the behavior of agents. Some of these were custom designed to describe agent behavior (such as the BDI model), while others were co-opted from other fields of computer science. Unfortunately, no single model provides a perfect fit from specification to implementation. There is a consensus of researchers and practitioners that during the development and deployment cycle, multiple models need to be employed.

Additional complexity arises from the fact that the perception about the best use of some of the models have changed in the years since they were first proposed. For example, approaches based on modal logic (Levesque, Cohen and Nunes 1990, Kinny, Georgeff and Rao 1996, Rao and Georgeff 1995) were originally considered as implementation models. Due to the computational complexity of these models current approaches employ the concept of modal logic mostly as a specification methodology.

Petri nets, with their well established semantics are arguably the best framework to verify important properties such as liveness (absence of deadlocks) or boundedness. They can readily capture concepts of concurrency, synchronization, contention for resources and so on. However, Petri nets are not considered a good fit as direct implementation models. As specification languages, Petri nets (in the form of colored Petri nets) are highly expressive, but many non-technical people find the Petri net descriptions less intuitive than other models (such as the UML activity or sequence diagrams).

The original statechart model (Harel 1987) is a popular approach for describing the behavior of real-time systems. But the expressivity of the model coupled with a lack of consensus on its semantics makes it rarely used as a basis for formal proofs. Statecharts have been successfully used as execution models for real-time systems and they have also been used for the execution models in agent systems such as SmartAgent (Griss, Fonseca, Cowan and Kessler 2002a).

The UML activity diagrams, inspired both by Petri nets and statecharts have proved to be useful in capturing the semantics of business processes. In their currently standardized form they are not precise enough to be a basis of correctness proofs, although significant research exist towards establishing a precise, unambiguous semantics for UML diagrams (for example, the work of the precise UML group). Although not employable directly as implementation models, tools can be built which are directly generating agents from activity diagrams (for example the DIVA tool generates directly agents for the OpenCybele platform).

Models based on finite state machine based decomposition of active objects represent a trade-off between expressivity and complexity. They can be used to perform formal reasoning, but the results do not cover the activities encapsulated in the active objects. On the positive side, these models can directly serve as implementation models.

Table 1. Models for describing agents, and their suitability for different stages of agent development, spe

Model / suitability...	for specification	for verification	as an implementation model
Modal logic (BDI etc)	good	fair (complexity)	poor (complexity)
Petri nets	fair (complex, not intuitive)	good	fair (complexity)
Statecharts	good	fair	good (specially for real-time systems)
Activity and sequence diagrams (UML 1 and 2)	excellent	poor	poor
Patterns (reactor, proactor, active object)	good	poor	excellent
State machines	good	fair (limited expressivity)	good

3.2 A Multiplane State Machine Model of Agent Behavior

We consider a formal model of agency based on the decomposition of the behavior of the agent into a set of active objects arranged in a multiplane state machine. In order to describe the behavior of the agents in relation their environment, first a model of the environment (the agent's world) needs to be chosen.

The notation used in the following sections is summarized in the following:

K	knowledgebase
S	strategy
G	goal or agenda
$M(Q, q_0, A, \Sigma, \delta)$	finite state machine
$\sigma(G, K, t)$	scheduling function

There are two views of the world we may consider, (a) the *view of an external observer* with a perfect knowledge of the world, who does not need take any action to acquire its knowledge and (b) the *view of an agent*, with an incomplete and imperfect knowledge who needs to take actions to improve its knowledge about the world and these very actions change the state of the world. When we define the goal of the agent, we are thinking in terms of the first view, because we want to modify the true state of the world. However, the actions taken by the agent can be defined only in terms of the second view, the only one available to the agent. For example, a *state transformer function* as in Fagin et. al. (Fagin, Halpern, Moses and Vardi 1995) pp. 154 is expressed in terms of the view of an observer with perfect knowledge. Agent strategies are expressed in terms of the view of the agent.

3.3 Modelling Agent Behavior

We assume that the knowledge of the agent about the world is captured in the *knowledgebase* K of the agents. We usually see the knowledgebase of the agents as being a *model of the environment*; the knowledgebase is an approximation of measurable quantities in the environment. However, a detailed discussion of the representation approaches is outside the scope of this paper.

The **goal** or **agenda** of the agent G is a boolean function applied to the knowledgebase:

$$G : K \to \{\texttt{true}, \texttt{false}\}$$

The goal of the agent is defined on the knowledgebase because the environment is not accessible to the agent. Of course, this allows an agent to satisfy its agenda by self-deceit (modifying the knowledgebase without modifying the environment). Thus, assuring that the knowledgebase is a sufficiently good approximation of the environment is an important part of agent design.

Let $K*$ be the set of possible knowledgebases and $G*$ the set of possible agendas. We define a *strategy* of an agent A as a function which maps the knowledgebase and agenda into a set of *intended actions*

$$S : (G * \times K*) \to \alpha*$$

Now, we are ready to introduce a first model of agents:

Definition 1 *The AM_0 model of an agent is the triplet (G, S, K) consisting of a goal G, a strategy S and a knowledgebase K.*

Definition 2 *We call* **concrete agent building** *the following problem: given an original knowledgebase K_0 and an goal G find a strategy function $S(G, K)$ such that the agenda will be satisfied in a finite time.*

3.3.1 Decomposition in the Plane. Expressing "Change"

From the implementation point of view the AM_0 model is a programmer's nightmare, because of the large monolithic strategy function S. This function is responsible for handling all the events and generating all the actions during the lifetime of an agent.

Definition 3 *The AM_1 model of an agent is the quadruple $(G, S^{(1)}, K, M)$ where G is the goal, K the knowledgebase and $S^{(1)}$ is a set of strategies $S^{(1)} = \{S_1^{(1)}, ...S_n^{(1)}\}$. The finite state machine $M(Q, q_0, A, \Sigma, \delta)$ has the number of states $||Q|| = n$. The current state of the state machine q_i determines the currently active strategy $S_i^{(1)}$.*

Property 1 *AM_0 and AM_1 define the same equivalence classes of agents.*

We let the proof of this property to the reader. We define the higher order agent models AM_n recursively. Repetitive application of the property leads to the conclusion that every agent model AM_n, $\forall n \in \mathbb{N}^+$ defines the same equivalence classes of agents as AM_0.

This property raises two questions: if all models are equivalent to AM_0, why introduce more complicated models? The other question is that the AM_n models can be trivial cases (a single plane with a single state). We don't have guarantees that we can make a decomposition of the agent besides this trivial case.

The answer to the first question is that higher order models capture the natural engineering tendency to assemble solutions from smaller components. The second question is more subtle: one can certainly imagine problems which cannot be decomposed in a meaningful way (for example, none of the model variables has discrete values). What we can say is that most problems in practice are well suited to be decomposed into sub-problems. From an engineering point of view the individual strategies conform to the *active object pattern* (Lavender and Schmidt 1995).

3.3.2 Expressing Concurrency

The AM_n model presented in the previous chapter decomposes the unique strategy of the model AM_0 into a number of strategies active one at a time. In this section we propose a method for further decomposition - we decompose the current strategy of the AM_n model into a number of strategies active concurrently.

Definition 4 *We call an* **m-plane scheduling function** *a function*

$$\sigma : (G* \times K* \times t) \to \{0, 1 \ldots, m-1\}$$

The **scheduling interval of** σ *is a time value* t_{sched} *such that*

$$\forall n \in \{0, 1 \ldots m-1\}, \forall K, \forall G, \forall t,$$
$$\exists \Delta t < t_{sched} \text{ such that } \sigma(G, K, t + \Delta t) = n.$$

This definition basically expresses our notion of a valid scheduling - there is a time interval in which every plane will be scheduled.

Definition 5 *We call a* **multiplane strategy** S^{mp} *with the planes* $S^0, S^1 \ldots S^{m-1}$ *with the associated scheduling function* σ *a function*

$$S(G, K) = \begin{cases} S^0(G, K) & \text{if } \sigma(G, K) = 0 \\ S^1(G, K) & \text{if } \sigma(G, K) = 1 \\ \ldots \\ S^{m-1}(G, K) & \text{if } \sigma(G, K) = m-1 \end{cases}$$

A multi-plane strategy expresses the idea that a strategy function can perform actions dealing with different parts of the world. The AM_1^{mp} agent model uses multi-plane strategies to express its behavior.

Definition 6 *The* AM_1^{mp} *agent model is a quintuple* $(G, S^{(1)mp}, K, M^{mp}, \sigma)$. G *is the goal,* $S^{(1)mp}$ *is a set of first order multiplane strategies, K is the knowledgebase. The current behavior* of the agent is determined by a multi-plane strategy composed of $S^{m_0^0}, S^{m_0^1} \ldots S^{m_0^{m-1}}$.

Definition 7 *An agent is not sensitive to the scheduling function if it is reaching its goal under one scheduling function, it will reach its goal under any scheduling function.*

From a formal point of view, this property is restrictive and difficult to prove. Nevertheless, this is an assumption made by all the non-real-time software currently in production. We assume that the software is not affected by: different processor speeds, scheduling based on the operating system choices and processor loads, external interrupts or garbage collection sessions. All these tacit assumptions are collected in the assumption that the agent is not sensitive to the schedule. While this does not cover the important subclass of real-time agents, it does cover the large majority of agents developed today. In the remainder of this section we will assume that the agents are not sensitive to the scheduling function.

We will denote a special set of states as *error states* and their associated strategies as *error handlers*. We also assume that the transitions in the finite state machine are labelled, and the labels are associated with internal or external transition events which trigger the corresponding transitions. We typically assume two labels with reserved meanings: SUCCESS and FAILURE. We assume that a transition event for which there is no similarly labeled transition is interpreted as a FAILURE transition and their target states are error states. We call a run of an agent *successful* if it contains no FAILURE transition.

3.4 Mutation Operators and Invariance Properties

We introduce a set of *mutation operations* on the multiplane state machines.

O_{as} Add a state.
O_{rs} Remove a state with no incoming or outgoing transitions.
O_{at} Add a transition between two states.
O_{rt} Remove a transition.
O_{ap} Add a plane.
O_{rp} Remove an empty plane.

A *change operation* C is an ordered list of operations $C = O_1 O_2 ... O_n \mid O_i \in \{O_{as}, O_{rs}, O_{at}, O_{rt}, O_{ap}, O_{rp}\}$. The set of operations is *complete*:

$$\forall M_1^{mp}, M_2^{mp} \ \exists C, \ M_1^{mp} \xrightarrow{C} M_2^{mp} \tag{1}$$

We can now propose a set of *invariance properties*.

Property 2 *Adding a new state to the agent does not change the behavior of the agent.*

Property 3 *Adding a new transition to the agent does not change the behavior of the agent in successful runs. It might turn some failed runs into successful runs.*

Property 4 *If we add a new plane to an agent and the output set of the strategies in the new plane is disjoint from the input set of the existing strategies, for all cases where the original agent achieved its agenda, the modified agent will achieve it as well.*

Corrolary 1 *Adding an empty plane maintains the achievability of the agenda.*

Property 5 *Removing FAILURE transitions does not affect successful runs.*

Property 6 *Removing states unreachable from the current state of the agent, or removing transitions going to and from these states does not affect the behavior of the agent.*

Property 7 *Removing states which are reachable only through FAILURE transitions does not affect successful runs.*

One of the difficulties of the model appears when replacing a strategy with an equivalent one. This very simple and frequently encountered operation can be performed using only a series of operations (remove all the incoming and outgoing transitions, followed by the removal of the state, add the new state, re-add the transitions). Unfortunately these series of operations break so many of the conditions of the previous properties that no meaningful invariants can be proved. This dilemma can not be solved only in the terms of the multiplane state machine, but the properties of the associated strategy needs to be considered as well.

We call strategies S_1 and S_2 *equivalent for a run* R, if for all the states of the run, the strategies generate the same action $S_1(G, K_i) = S_2(G, K_i)$, $K_i \in R$. Two strategies are *equivalent for an agent* if they are equivalent for all possible runs of the agent.

We introduce an additional operation:

O_{xs} Replace the strategy S of an inactive state with a strategy S'

For this operation we can prove the following property:

Property 8 *Replacing a strategy with a strategy equivalent for the run does not change the behavior of the agent for the run. Replacing a strategy with a strategy equivalent for the agent does not change the behavior of the agent.*

3.5 How Useful Are the Invariance Properties?

To understand the practical usability of the properties presented in the previous section, we examine three questions. (a) How readily can agents be reduced to the multiplane state machine model?

(b) How likely is that the transformations satisfy the conditions of the properties? (c) How strong are the conclusions of the properties?

Regarding question (a) we can state that a large class of agent systems can be reduced to the multiplane state machine model. Systems like Bond use the A_1^{mp} model directly. SmartAgent (Griss, Fonseca, Cowan and Kessler 2002b, Griss et al. 2002a) is using a model very similar which can be easily equated to an A_1^{mp}. Generally, any agent systems which is assembling agents based on active objects (behaviors, strategies etc) can be readily modeled in A_1^{mp}. On the other hand, systems which does not employ a component model will likely contain a less localized execution trace. These systems can still be modeled as state machines, but the resulting model will necessarily be finer grained, and thus less likely to be useful. The secret of successfully applying the invariance properties is that the individual states correspond to the granularity level of the mutation.

Question (b) is more difficult without actually performing a statistical study of the ways in which mutation is used in agents. The best we can do, is to make some observations based on typical usages. The conditions of the properties are relatively relaxed and easily checkable. The notable exception is Property 4, which requires the computations of the input and output sets of components; this is very difficult or impossible to carry out in an automatic way.

The last question is the strength of the conclusions. Some of these results seem trivial, because they promise a complete lack of change on the global behavior of the agent. Certainly, adding a new state, with no incoming transition does not change the behavior of the agent. We call this result weak, because we know intuitively that this will not be the final state of the changed agent. Nevertheless, results of the presented type has proven to be very useful in many domains of computer science. Invariant transformations are an important component in both static (software) and dynamic (hardware) code optimizations. These techniques perform invariant transformations such as loop unrolling, reordering of independent instructions etc. to achieve faster program execution. In a different example, the extreme programming paradigm (Fowler, Beck et al. 1999, Beck 1999) employs a technique called *refactoring* which leads to improvement in the code structure through a series of invariant operations. In static (non-runtime) software evolution, refactoring a system means to change its structure while retaining its semantics. Refactorings are attractive because they are a means to clean up a system, to facilitate maintenance and testing, and prepare functional changes. They have the advantage that they split software evolution in *harmless operations* (refactorings) whose effects can be checked by program analyzers or regression testing, and *difficult operations* that change semantics, but cannot be easily regression tested. Most of the operations in the properties presented can be thought as the equivalent of the refactoring operations.

Another weakness in the results of properties are the limitations of the multiplane state machine model. Although it supports concurrency, it does not take into account resource contentions. One result of this is the rather strong set of conditions of Property 4 regarding the disjointness of input and output sets. Moving to a model which handles resource contentions (such as Petri nets) and developing a set of similar properties for them would significantly extend the power of the theory (but it would create other problems in the practical applications).

4 A Software Engineering Perspective on Adaptive and Mutable Agents

With the gradual adoption of agent systems in commercial software development it became obvious that the established software methodologies, such as object-oriented analysis and design, are inadequate or insufficient for the analysis and design of agent systems.

The agent oriented software engineering (AOSE) field emerged to fill this gap.

Some proposed methodologies, such as (Wooldridge, Jennings and Kinny 2000), are building upon the existing object oriented methodologies and techniques, e.g., design patterns. There are a number of efforts underway to extend the UML language and the associated software methodology for agent oriented programming (Odell, Parunak and Bauer 2000, Caire, Coulier, Garijo, Gomez, Pavon, Leal, Chainho, Kearney, Stark, Evans and Massonet 2001, Arai and Stolzenburg 2002) or for modeling the knowledge-base of the agents (Cranefield, Haustein and Purvis 2001, Heinze and Sterling 2002). Many methodologies are drawing inspiration from the Belief-Desire-Intention model (Kinny et al. 1996, Padgham and Winikoff 2002). Other approaches are building on techniques for knowledge engineering (Brazier, Keplicz, Jennings and Treur 1995) or on formal methods and languages, e.g., the extensions for the Z language (Luck, Griffiths and d'Inverno 1997). The Tropos methodology (Bresciani, Perini, Giorgini, Giunchiglia and Mylopoulos 2001) is adapting ideas from techniques developed for business process modeling and reengineering (the i* notation (Yu and Mylopoulos 1994)), at the same time retaining the mentalistic notions of belief-desire-intention and related models. Some agent systems have developed their own analysis and design approaches, targeted to the particularities of the agent system such as Cassiopeia.

The introduction of mutable agents creates new problems for the agent analysis and design methodologies. The analysis step needs to take in consideration the possibility of the agent being significantly modified during its lifetime. The design step needs to offer information about which agents should be mutated, at what moment of their life-cycle, and what kind of mutation should be performed. Generally, the methodological discipline is more important for the case of mutable agents.

We now discuss the effect of mutable agents on one of the popular agent design methodologies, the Gaia approach (Wooldridge et al. 2000). The Gaia methodology, with its roots in object oriented approaches such as FUSION, is a good fit for FIPA compliant agent systems such as Bond, *as long as they do not mutate*. In fact, the authors (Wooldridge et al. 2000) explicitly spell out among the applicability requirements that (a) the organizational structure of the system is static and (b) the abilities of the agents and the services they provide are static, do not change during runtime. Several extensions proposed to the Gaia methodology extend the scope of the methodology. The ROADMAP methodology (Juan, Pearce and Sterling 2002) extends Gaia with formal models of knowledge, role hierarchies and representation of social structures. It also extends the permission attributes to allow roles to change the definition or attributes of other roles, although it does not cover the issue of how the modified agents are represented. Our goal is to investigate the feasibility of the removal of these constraints and the changes in the methodology implied by this removal.

We emphasize that no methodology can handle randomly mutating agents. Fortunately, the most frequently encountered operations can be classified in a set of well-understood classes:

- Adding new functionality (roles) to the agent.
- Removing functionality from an agent.
- Adapting the agent to new requirements or a different set of available resources.
- Transferring a functionality from an agent to a different agent.
- Splitting an agent (for instance for the purpose of load balancing).
- Merging agents.

4.1 Adding New Functionality to the Agent

In terms of the Gaia methodology, adding new functionality to the agent is equivalent to saying that the agent will be able to function in new *roles*, while maintaining the previously existing ones. More formally, considering the reconfiguration event e, we can say that if the agent was able to fulfill a set of roles $\mathfrak{R} = \{r_1, \ldots r_n\}$ before the event, and after the event, it will be able to fulfill a set of roles $\mathfrak{R}'$ with $\mathfrak{R} \subset \mathfrak{R}'$.

The corresponding structural definition in the AM_1^{mp} model, employed by the Bond system, is that the extending functionality is an agent surgery operation, which transforms agent A to agent A' and for every run R where the agent A is successful, the agent A' will be successful as well. In (Bölöni and Marinescu 2000b), we demonstrated that elementary surgical operations, such as adding states, adding transitions, and removing transitions labelled FAILURE, maintain this property. In addition, the surgical operation of adding a plane can also maintain this property subject to a set of disjointness conditions.

There is a good mapping between the Gaia concept of roles and the structural implementation. Just as the agent might not be taking on a certain role, although it would be qualified to do it, the agent might not perform certain runs. Thus we can say that through adding new functionality to the agent by agent surgery, the agent acquires the ability to fulfill new roles. The nature of these roles needs to be clearly spelled out.

We also need to maintain the attributes associated with the roles in the Gaia methodology: *responsibilities*, *permissions*, *activities*, and *protocols*. As these attributes are applied to the agent in a cumulative way, an important goal of the analysis process of an agent surgery operation is to check that there is no conflict between the attributes of the agents, that is, the invariants of the agents are properly maintained.

A different question, which needs to be answered by the analysis process, is the opportunity and moment in the agent life-cycle when the new functionality needs to be added to the agent. The answer to this question is not an explicit point in the life-cycle of the agent, but a *trigger*, a specific set of conditions under which the mutation becomes desirable.

While one might argue that the agents can be designed so that they can fulfill all the possible roles needed during its lifetime, this argument ignores the cost associated with having such a multifaceted agent. To put this in a different context, not all the workers of a company need to be qualified for all job descriptions. Nevertheless, it is a frequent occurrence that a worker needs to be sent to additional training so that he or she can fulfill new roles. In many cases, these events can be quite accurately predicted and even planned. Similar considerations apply to agents.

For a concrete example of how the diagrams of the Gaia method can be annotated to handle reconfigurable agents, let us turn to the airline industry for an example. During a flight, the number of (human) agents on the airplane are playing a set of well defined roles: passenger, pilot, stewardess and so on. There are, however, exceptional situations, such as an emergency landing in which cases some of the passengers are required to take on new roles, such as to assist the crew in opening the doors.

The agent model diagram of this situation is shown in Figure 2. In this approach, the exceptional situation is modeled as the mutation trigger. The new state of the agent is modeled in the Gaia agent diagram as a new agent type. Mutated agent types are marked in the diagram with the letter M. The mutation operation is specified using a thick arrow with the "+" label attached (indicating that the mutation retains all the previous roles of the agent[3]).

---

[3]Strictly speaking, the "+" label is not needed on the agent model diagram, because the operation can be inferred from the role inheritance lines. It is however useful on the other diagrams where the role inheritance is not present.

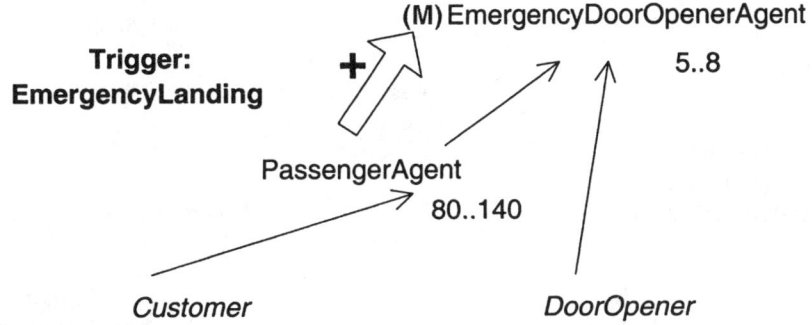

Figure 2. Agent model for an airplane emergency situation

Another diagram which needs to be adapted to handle the needs of the reconfigurable agents is the acquaintance model. While the Gaia acquaintance model does not deal with the details of the interaction, mutations on the agents can frequently change the acquaintances as well. The example presented in Figure 3 also deals with a fictional situation on an airplane. The sudden symptoms of sickness on some passengers and a stewardess triggers a request from a stewardess which makes a passengers step into the role of a doctor. This creates a new interaction pattern, between the doctor, the sick stewardess and the sick passenger. These acquaintance lines would have not existed if the mutation would not have happened.

4.2 Removing Functionality from an Agent

There can be several reasons for removing functionality from an agent. One of them is the course-grain equivalent of *garbage collecting*. At some moment in the agent's life-cycle we might find that some of the roles of the agent will never be activated. The ability to perform roles which will not be activated usually implies some kind of waste of resources. Examples are memory and disk space occupied, network bandwidth by polling for messages which will

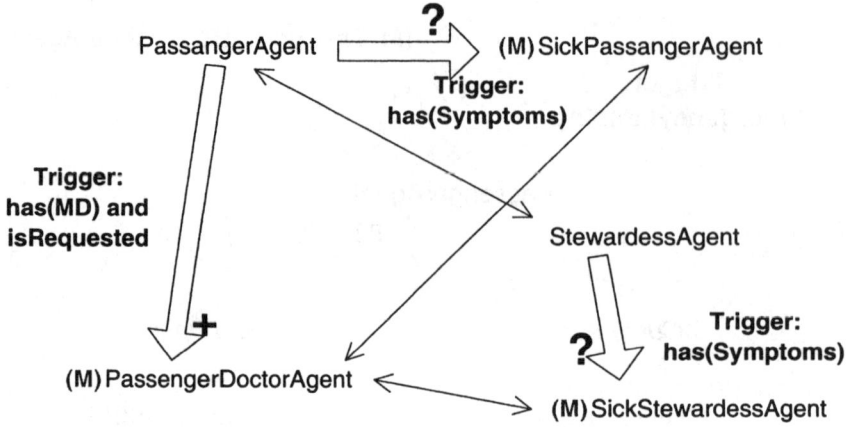

Figure 3. Interaction diagram the situation of a sickness on an airplane

never arrive, processor time for maintaining data structures which will never be queried. It is therefore useful to periodically perform a role-based garbage collection process on the agent. While the process is related to the garbage collection process in programming languages such as Java, there are also some specific differences.

- The garbage collection process happens at the level of components and subsections of the knowledgebase (instead of allocated memory chunks).
- Active components (code) can be also garbage collected.
- The internal structure of the agents can greatly simplify the garbage collection process. For example, for agents based on a multiplane state machine model, the garbage collection process can be reduced to a reachability analysis on the state machines.
- The probability that a garbage collection step will recover some resources is generally lower, then in the case of garbage collecting memory in applications. Moreover, the benefits of role-based garbage collection will tend asymptotically to zero, unless some other mutation operations make additional components unreachable.

The role-based garbage collection can happen in any time during the agent's lifetime. However, the life cycle of agents provides some natural points where the side effects of the garbage collection process are minimal. Such points are: after every transition (for state machine or Petri net based agents), before checkpointing, before moving (for mobile agents) and after every mutation.

Another scenario for removing functionality from agents, is to trigger *specialization* in groups of identical agents. In this case an agent factory generates agents with the ability to perform a set of tasks. The agents are then specialized through removing their ability to perform a certain subset of tasks (Figure 4). The specialization mutation can be either performed under the control of a remote agent, or it can be performed by the agent itself, based on the initial experiences of its life-cycle. This approach is very natural for distributed solution of problems with the "divide and conquer" approach, such as the popular Contract Net protocol (Smith 1980). Another application is the emergence of communication patterns. There is solid evidence that the visual and auditory pathways of mammals are generated using a similar method in the early stages of life. The hardware industry had chosen a similar approach for the zone codes of DVD drives. The DVD drives are manufactured as generic devices, which are capable to play DVDs from any zone. During the first several uses, the DVD drives decides on a particular zone coding, and it permanently removes its own ability to play DVDs from other zones.

4.3 Adapting to New Requirements

Another very important subclass of agent mutations is the adapting to a changing environment, for instance after migration to a new host. Another example is a agent which is using the idle resources of a computer. This agent needs to change between the almost complete control over resources (when the user is not logged in) with only minimal resource allowances (when the user is logged in and working).

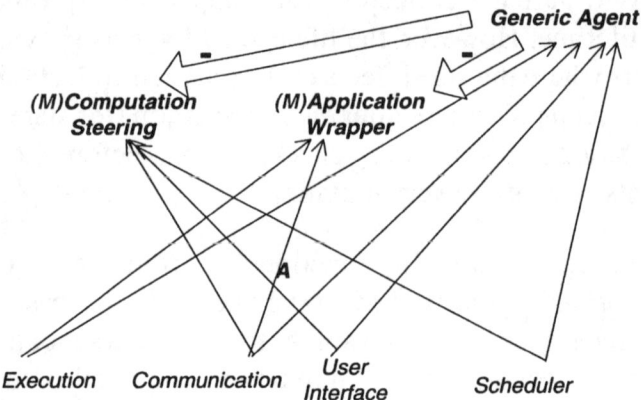

Figure 4. Specialization of generic agents through removal of roles.

Expressed in the terms of the Gaia methodology, the agents implement the same set of roles with a different set of attributes.

- The *responsibilities* and the *protocols* implemented by the role will remain unchanged during this operation[4].
- The permissions associated with the role will be different. The Gaia methodology collects under the concept of *permissions* notions such as resource usage and security permissions. The analysis process assures that the permissions required by the agent after the transformation are satisfied.

The goal of the agent analysis and design is to determine the opportunity and the nature of the agent mutation. The opportunity for migration can be expressed in terms of hard and soft triggers. A *hard trigger* is a boolean function which tells us if an agent cannot fulfill the requirements of its role in a given context. A *soft trigger*, on the other hand, is a cost-benefit analysis of the agent which might suggest a mutation if this leads to an increased performance of an agent.

---

[4]This is a relatively crude approximation which assumes that the functionality of an application is completely specified by the original specification.

Generally, hard triggers are leading to changes with implementations with lower resource usage, while soft triggers are biased towards implementations with higher performance and associated higher resource (permission) requirements.

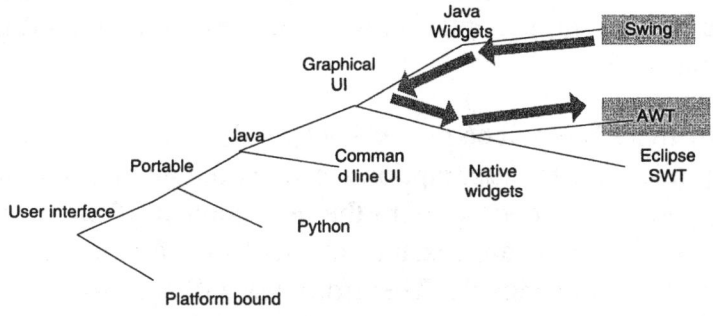

Figure 5. Design decision tree

The adaptation scenarios can be described in the terms of a design decision tree. In the Gaia approach (as in many other software engineering approaches) the designer moves from an initial, very high level specification, to an increasingly more specialized choice. These choices form a *design tree*, which will be assumed to be a separate one for every role the agent can play. For the example given in Figure 5, the choices are all valid approaches to create a user interface of the program. During the design process, a set of decisions are made to determine the permissions (in the resource usage sense) of the program. The leaf nodes typically correspond to the actual implementation of a program, but of course, not all the potential choices are actually implemented. For an agent which does not require reconfiguration, only one branch of the design tree is explored. Once the decision is made, the design tree is not used in the actual operation of the agent, although it might be kept, implicitly in the design documentation.

For a reconfigurable agent, more than one leaf node is fully instantiated. We need to emphasize that this involves the same analysis and programming steps as for any other agent. However, in this case, the design tree has a practical utility during the lifetime of the agent. Let us assume an agent which is executed with a Swing based, fully graphic user interface, and needs to be migrated to a Personal Digital Assistant. A simple check of the permissions of the role will tell us that the current implementation will not work and a reconfiguration operation is needed.

The role, therefore will be moved backwards in the design tree, at every step releasing the assumptions made at the given point, until the assumptions do no conflict with the new context of the agent. Normally, however, this happens at an abstract specification level, which on its own, is not runnable. Therefore, we will start to move toward the leafs of the decision tree again, this time however making decisions according to the new context. Our goal is to reach a fully implemented leaf node which conforms to the current set of permissions. The required transformation will be, therefore, one which transforms from the original design choice into a new design choice.

4.4 Splitting and Merging Agents

Splitting and merging agents are operations which are surprisingly easily implementable in many agent systems. Moreover, very compelling application scenarios can be found to justify them. The reason why this technique is not more frequently used in applications is because there is no accepted software engineering methodology to specify them. Also, splitting and merging is not a mechanically intuitive concept such as agent mobility[5].

---

[5] Agent mobility, in practice involves: stopping an agent, serializing it, transferring data and code over the network, signalling back that the transfer was successful, destroying the agent in the original location, restarting it on the new location. It is quite obvious that this is a complex, and not entirely intuitive process which has little to do with movement in the mechanical sense. However, the power of the metaphor made people accept the notion of mobile agents easier because most of us can visualize it.

The software engineering process for splitting and merging agents involves most of the notions presented in the previous sections. We need to identify the *triggers* of the split and merge mutation. The agent model identifies agent types involved into the split and merge operations.

In Figure 6 we present an agent model for an agent which represents a military unit[6]. The military unit can split into two components, the reconnaissance unit and the cover unit. This implies a distribution of the roles between the two split units. We should note that some of the roles (such as the communication role) will be are replicated in the two split units.

We have introduced two new elements in the agent model, the split operator and the merge operator. Each of them are labelled with the trigger of the split and merge operations respectively.

5 Conclusions

Mutable and reconfigurable software is increasingly more popular. It is driven by the needs of users to have their applications adapt to the changing global or local environment. These requirements appear in an amplified form for software agents. The software environment of agents is much more dynamic then the relatively controlled environment in which server applications or desktop programs live.

Reconfigurable agents are bringing additional challenges to the already complex problems of modeling agents and reasoning about them, and the emerging field of agent based software engineering.

---

[6]Many readers might consider that a military unit is better modeled using an agent society. This, however, is a matter of choice. From a point of view of battlefield simulation, small military units can be more conveniently modeled as a single agent. Many researchers pointed out the significant similarities between modeling individual agents and modeling agent societies.

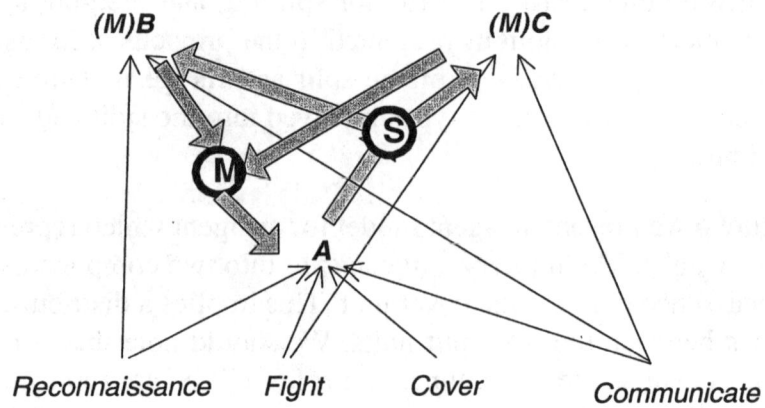

Figure 6. The agent model for splitting and merging operations

We find it interesting, that the implementation and deployment of mutable and reconfigurable systems proceeded despite the lack of appropriate theoretical foundations.

In this chapter we provided a review of the field of mutable and reconfigurable agents from various perspectives. We attempted to organize the various implementation methods by providing a classification for mutable and reconfigurable agents - at the same time tying it to the larger field of reconfigurable software. We discussed the various approaches which can provide a formal model for reasoning about mutable agents. We presented some results based on a multi-plane state machine model, and discussed their strengths and limitations. Finally, we discussed the challenges posed by mutable agents for the agent oriented software engineering and have presented an approach for extending Gaia, one of the popular AOSE approaches towards handling mutable agents.

References

An, P., Jula, A., Rus, S., Saunders, S., Smith, T., Tanase, G., Thomas, N., Amato, N. and Rauchwerger, L.: 2001, STAPL: An adaptive, generic parallel C++ library, *Proceedings of the 14th Workshop on Languages and Compilers for Parallel Computing (LCPC)*.

Arai, T. and Stolzenburg, F.: 2002, Multiagent systems specification by UML statecharts aiming at intelligent manufacturing, *Proceedings of the first international joint conference on Autonomous agents and multiagent systems*, ACM Press, pp. 11–18.

Barber, K. S., Goel, A. and Martin, C. E.: 2000, Dynamic adaptive autonomy in multi-agent systems, *JETAI* **12**(2), 129–147.

Beck, K.: 1999, *Extreme Programming Explained: Embrace Change*, Addison Wesley.

Bölöni, L. and Marinescu, D. C.: 2000a, Agent surgery: The case for mutable agents, *Proceedings of the Third Workshop on Bio-Inspired Solutions to Parallel Processing Problems (BioSP3), Cancun, Mexico*.

Bölöni, L. and Marinescu, D. C.: 2000b, A component agent model - from theory to implementation, *Second Intl. Symp. From Agent Theory to Agent Implementation in Proc. Cybernetics and Systems, Austrian Society of Cybernetic Studies*, pp. 633–639.

Bon: 2003, Bond webpage, URL http://bond.cs.ucf.edu.

Brazier, F., Keplicz, B. D., Jennings, N. R. and Treur, J.: 1995, Formal specification of multi-agent systems: A real-world case, *First International Conference on Multi-Agent Systems (ICMAS'95)*, AAAI Press, San Francisco, CA, USA, pp. 25–32.

Bresciani, P., Perini, A., Giorgini, P., Giunchiglia, F. and Mylopoulos, J.: 2001, A knowledge level software engineering methodology for agent oriented programming, *in* J. P. Müller, E. Andre, S. Sen and C. Frasson (eds), *Proceedings of the Fifth Inter-*

national Conference on Autonomous Agents, ACM Press, Montreal, Canada, pp. 648–655.

Caire, G., Coulier, W., Garijo, F. J., Gomez, J., Pavon, J., Leal, F., Chainho, P., Kearney, P. E., Stark, J., Evans, R. and Massonet, P.: 2001, Agent oriented analysis using Message/UML, *AOSE*, pp. 119–135.

Collberg, C., Thomborson, C. and Low, D.: 1997, A taxonomy of obfuscating transformations, *Technical Report 148*, Department of Computer Science, University of Auckland.

Cranefield, S., Haustein, S. and Purvis, M.: 2001, UML-based ontology modelling for software agents.

Decker, K., Sycara, K. and Williamson, M.: 1996, Intelligent adaptive information agents, *in* I. Imam (ed.), *Working Notes of the AAAI-96 Workshop on Intelligent Adaptive Agents*, Portland, OR.

Fagin, R., Halpern, J. Y., Moses, Y. and Vardi, M. Y.: 1995, *Reasoning about knowledge*, MIT Press.

For: 1998, Fortify for netscape, URL http://www.fortify.net .

Fowler, M., Beck, K. et al.: 1999, *Refactoring: Improving the Design of Existing Code*, Addison Wesley.

Griss, M. L., Fonseca, S., Cowan, D. and Kessler, R.: 2002a, SmartAgent: Extending the JADE agent behavior model, *Proceedings of the Agent Oriented Software Engineering Workshop, Conference in Systemics, Cybernetics and Informatics*, ACM Press.

Griss, M. L., Fonseca, S., Cowan, D. and Kessler, R.: 2002b, Using UML state machines models for more precise and flexible JADE agent behaviors, *Proceedings of the Agent Oriented Software Engineering Workshop, AAMAS*, ACM Press.

Gustavsson, J. and Assmann, U.: 2002, A classification of runtime software changes.

Han, Y., Sheth, A. and Bussler, C.: 1998, A taxonomy of adaptive workflow management, *CSCW-98 Workshop – Towards Adaptive Workflow Systems*.

Harel, D.: 1987, Statecharts: A visual formalism for complex systems, *Science of Computer Programming* **8**(3), 231–274.

Heinze, C. and Sterling, L.: 2002, Using the uml to model knowledge in agent systems, *Proceedings of the first international joint conference on Autonomous agents and multiagent systems*, ACM Press, pp. 41–42.

JFl: 2003, JFluid web page, URL http://www.sunlabs.com/projects/jfluid.

Juan, T., Pearce, A. and Sterling, L.: 2002, ROADMAP: Extending the Gaia methodology for complex open systems, *Proceedings of the first international joint conference on Autonomous agents and multiagent systems*, ACM Press, pp. 3–10.

Kiczales, G., Hilsdale, E., Hugunin, J., Kersten, M., Palm, J. and Griswold, W. G.: 2001, An overview of AspectJ, *Lecture Notes in Computer Science* **2072**, 327–355.

Kinny, D., Georgeff, M. and Rao, A.: 1996, A methodology and modelling technique for systems of BDI agents, *in* W. V. de Velde and J. W. Perram (eds), *Agents Breaking Away: Proceedings of the Seventh European Workshop on Modelling Autonomous Agents in a Multi-Agent World, (LNAI Volume 1038)*, Vol. 1038 of *LNAI*, Springer-Verlag, p. 56.

Lavender, R. G. and Schmidt, D. C.: 1995, Active object: an object behavioral pattern for concurrent programming, *Proceedings of Pattern Languages of Program Design,*.

Levesque, H. J., Cohen, P. R. and Nunes, J. H. T.: 1990, On acting together, *Proceedings of the Eighth National Conference on Artificial Intelligence*, American Association for Artificial Intelligence, pp. 94–99.

Luck, M., Griffiths, N. and d'Inverno, M.: 1997, From agent theory to agent construction: A case study, *in* J. P. Müller, M. J. Wooldridge and N. R. Jennings (eds), *Proceedings of the ECAI'96 Workshop on Agent Theories, Architectures, and Languages: Intelligent Agents III*, Vol. 1193, Springer-Verlag: Heidelberg, Germany, pp. 49–64.

Luyten, K., Vandervelpen, C. and Coninx, K.: 2002, Migratable user interface descriptions in component-based development, *Proceedings of the 9th International Workshop on Design, Specification, and Verification of Interactive Systems*.

Odell, J., Parunak, H. and Bauer, B.: 2000, Extending UML for agents, *in* G. Wagner, Y. Lesperance and E. Yu (eds), *Agent-Oriented Information Systems Workshop at the 17th National conference on Artificial Intelligence*, pp. 3–17.

Padgham, L. and Winikoff, M.: 2002, Prometheus: A methodology for developing intelligent agents, *Proceedings of the first international joint conference on Autonomous agents and multiagent systems*, ACM Press, pp. 37–38.

Pribeanu, C., Limbourg, Q. and Vanderdonckt, J.: 2001, Task modelling for context-sensitive user interfaces, *Lecture Notes in Computer Science* **2220**, 49–??

Rao, A. S. and Georgeff, M. P.: 1995, BDI agents: from theory to practice, *in* V. Lesser (ed.), *Proceedings of the First International Conference on Multi–Agent Systems*, MIT Press, San Francisco, CA, pp. 312–319.

Rauchwerger, L., Amato, N. M. and Torrellas, J.: 2001, SmartApps: An application centric approach to high performance computing, *Proc. of the 13th Annual Workshop on Languages and Compilers for Parallel Computing (LCPC), August 2000, Yorktown Heights, NY.*, pp. 82–92.

Smith, R. G.: 1980, The contract net protocol: High-level communication and control in a distributed problem solver, *IEEE Transactions on Computers* **29**(12), 1104–1113.

Thevenin, D. and Coutaz, J.: 1999, Plasticity of user interfaces: Framework and research agenda.

van der Aalst, W. M. P.: 1999, Generic workflow models: How to handle dynamic change and capture management information?, *Conference on Cooperative Information Systems*, pp. 115–126.

Varela, C. and Agha, G.: 2001, Programming dynamically reconfigurable open systems with SALSA, *ACM SIGPLAN Notices* **36**(12), 20–34.

Wooldridge, M., Jennings, N. R. and Kinny, D.: 2000, The Gaia methodology for agent-oriented analysis and design, *Autonomous Agents and Multi-Agent Systems* **3**(3), 285–312.

XBo: 2001, Xbox on linux, URL `http://xbox-linux.sourceforce.net` .

Yu, E. S. K. and Mylopoulos, J.: 1994, From E-R to "A-R" - modelling strategic actor relationships for business process reengineering, *in* P. Loucopoulos (ed.), *Proceedings of the 13th International Conference on the Entity-Relationship Approach*, Springer, Manchester, UK, pp. 548–565.

Chapter 11

Intelligent Action Acquisition for Animated Learning Agents[*]

Adam Szarowicz, Marek Mittmann, Jaroslaw Francik

Generation of animated human figures especially in crowd scenes has many applications in such domains as the special effects industry, computer games or for the simulation of the evacuation from crowded areas. Current systems allow for partially automatic generation of scenes involving a few interacting characters but expensive manual labour is still necessary in order to enrich the characters' behaviour repertoire. In this chapter we explore the possibility of applying reinforcement learning to acquire new high-level actions for animated characters. The chosen algorithm is the deterministic version of Q-learning. This allows for easy definition of the task, since only the ultimate goal of the learning agent must be defined. Generated actions can then be used to enrich the animation produced by an animation system. Results achieved when training agents with forward and inverse kinematics control are also demonstrated and compared.

1 Introduction

In recent years we have seen a substantial increase in graphics processing power especially with the appearance of cheap hardware graphic accelerators. This has improved the quality of computer

---

[*] The work was partially supported by the British Council/KBN (Polish State Committee for Scientific Research) grant, project number 239/2002.

graphics available in home and office applications such as CAD programs or computer games. It has also boosted the level of special effects in film post-production. However a noticeable gap exists between the possible quality of character animation and the level of intelligence in which the characters are equipped by game developers or even professional animation packages. Therefore creation of animated scenes involving interacting characters remains a problem in such applications as film post-production and special effects, computer games or event simulation in crowded areas.

Ideally, animated characters used in such applications should have some autonomy and awareness of their environment, which is a feature lacking in current animation packages. Additionally a substantial plan repertoire of high-level everyday actions such as drinking coffee or opening a door is a key point in the creation of realistically looking animation sequences. However manual creation of actions for different characters is a tedious and expensive task. Automation in action creation would eliminate the need for human labour, shorten the task of generation of the crowd scenes and also reduce the costs.

The work presented in this paper is primarily aimed at rapid creation of scenes with multiple characters interacting in a simulated environment, where fine detail of the characters is not important. Applications for such a system would be found in film special effects and post-processing, computer games and also simulation of evacuation from crowded areas.

This chapter is divided into a number of sections. First we review the systems built to generate animation sequences with intelligent characters. The review comprises discussion of general-purpose animation architectures with special focus on learning, physics-based dynamic simulation systems, crowd creation systems, and q-learning. We then briefly present our system FreeWill aimed at producing crowd scenes involving many interacting characters. This is followed by a discussion of how the system can be extended to produce better and more appealing animation sequences and the proposed solution achieved through application of machine learning is outlined. The results are then presented and discussed.

2 Current State of the Art in Automatic Character Animation

2.1 General Animation Architectures

One of the dominant approaches taken by researchers trying to build intelligent characters is represented by research groups trying to build animation systems by creating a small number of relatively sophisticated characters, animated using kinematic techniques and/or predefined motion. The characters are usually equipped with a high degree of autonomy, can learn and perform complex actions.

A good example here is C4, an architecture recently proposed by a research group working at the MIT Media Lab (Isla et al, 2001, Burke et al 2001). C4 was designed to allow creation of autonomous and semi-autonomous creatures. The work is informed by the behaviour of living animals (in particular dogs). The main concept is a 'brain' divided into numerous systems communicating through an internal blackboard. The whole system is highly modular which is its main architectural principle. C4's emphasis on learning especially in the form of training a home pet (a dog) makes the system biased towards solutions allowing to handle this type of interaction. This includes dedicated data structures and a hierarchy of purposefully defined objects. It is not obvious whether the learning system is flexible enough to allow unsupervised learning of more advanced activities.

An extension of C4 which includes much greater learning capabilities is described in Blumberg et al (2002). The applied learning algorithm is a modification of the reinforcement learning technique. However the task of the learning engine is not to learn the necessary motor skills but rather is defined on a higher level, "with respect to a motivational goal of moving in a certain way" and happens in real-time during the interaction with the system. Thus the authors define their learning system as being more abstract than the traditional approaches and additionally aim to use learning as a means of increasing online interactive capabilities of their characters and not as a design tool. The whole system was built to mimic a dog training technique called the "clicker training". To generate motion the system uses a so-called pose-graph whose motions are derived from source animation amended by an interpolation technique (Downie, 2000). Thus the animation is realistic and transitions can be generated in real-time but the actions must be prepared by an animator and pre-programmed into the system.

Another example of applying reinforcement learning to animation includes Yoon et al (2000), where RL techniques were used to create motivational and emotional states for a human character. This system incorporated such concepts as motivation driven learning (where the source of the reinforcement signal for learning was the creature's motivational module), organisational and concept learning but not motor learning. Similarly as before the learning happens on a higher level and only affects the character's behaviour in an indirect way.

Other interesting work presented by the MIT Synthetic Character Group includes a study of user interaction with high-level control over synthetic wolves, whose emotional state is maintained by a computer (Tomlinson et al 2002), or a behaviour-based reactive autonomous cinematography system (Tomlinson et al 2000). In the former the intelligence was put not into a virtual character but rather into a virtual camera and lights in order to enhance the emotional expressiveness of the animation. A summary of the systems created by the group can be found in Downie et al (2002).

The common feature of all systems created by the researchers from the MIT Synthetic Character Group is that the motor systems are always built upon animation material prepared by animators rather than creating character motion from scratch (Downie et al 2002).

Another example of this approach to the creation of autonomous characters is represented by John Funge's cognitive architecture (Funge, 1998, Funge, 1999, Funge et al, 1999). Funge's research interests concern building cognitive systems dedicated for animated characters. He strongly supports the idea of adding autonomy to computer-generated creatures and he defines an autonomous character as "a character that, during the course of a computer game or animation, can decide how to behave on its own." (Funge, 1999). Funge argues that virtual characters should maintain a cognitive model that is an internal model of their world and additionally an explicit representation of some other knowledge about the character's world (Funge, 1998, Funge et al, 1999). Therefore he describes autonomous cognitive characters as characters having some domain knowledge – knowledge about the world's dynamics. At the same time he claims that generally the character should not be omniscient, which happens when the characters gain access to the 'true' world model. Funge formalises his ideas using the situation calculus.

Funge emphasises that the cognitive layer is only one among many models underlying an animated character (Figure 1). He recognises that a developer trying to build a virtual scene will be faced with many problems apart from those of artificial intelligence. What he proposes in not only a cognitive architecture but also a whole process of building a virtual character.

However Funge concentrates on building a single character without addressing the problem of how to exchange information or coordinate its behaviour with other creatures. Also the process of building the whole system, especially how to merge all different aspects of the cognitive engine is not addressed in detail and the learning aspect in only mentioned without deeper consideration.

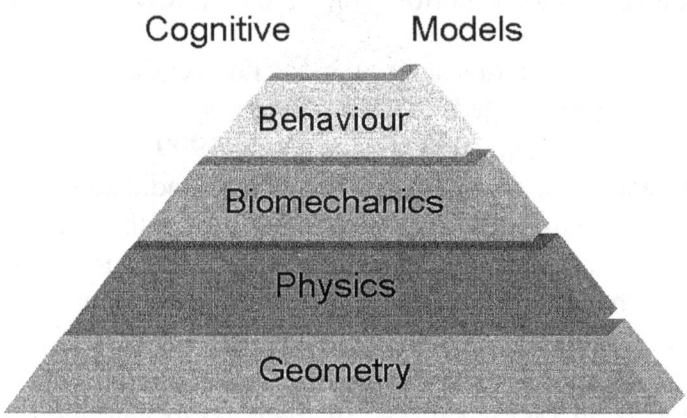

Figure 1. Pyramid depicting different models used to build a virtual character (after Funge, 1999)

Another example of a complex animation system is that proposed by Terzopoulos and his collaborators (Terzopoluos et al 1994, Terzopoluos et al 1996). They created a marine world inhabited by realistically looking and behaving animals (mainly fish) to explore vision and navigation systems. They also employed machine learning to acquire complex motor skills for the simulated fish. The virtual characters are able to learn low-level motions and also high-level behaviours. In their approach the researchers use physics-based simulation and create a dynamic model of the fish with muscles and springs. Such an approach to motion control however makes the simulation computationally demanding and so Grzeszczuk (Grzeszczuk, 1998, Grzeszczuk et al, 1998) proposed an application of neural networks to emulate dynamics. They claim that

that using this approach physically realistic animation can be generated one or two order of magnitude faster than when using numerical simulation (Grzeszczuk et al, 1998). Still the system can only be applied to characters with relatively small number of degrees of freedom.

An application of genetic algorithms to behaviour learning for animated characters was presented by Wan and Tang (2002).

2.2 Physics-Based Controllers

An approach similar to that described above is creation of (usually bipedal) characters with very complex motor skills using dynamic simulation. In this approach the emphasis is put on creation of one realistically modelled character simulated using the motion equations and physics-based controllers. Terzopoluos, Grzeszczuk, and Tu (Tu and Terzopoulos, 1994, Grzeszczuk and Terzopoulos, 1995) present a system for animating dynamically simulated fish and snakes. However similar approach applied to dynamic simulation of human figures requires that the characters have many degrees of freedom thus making it computationally expensive. Despite that there is a lot of research being done in this field, some examples are presented here. Hodgins et al (1995) propose controllers for three different athletic behaviours. Apart from dynamic simulation they also use state machines, techniques for reducing disturbances to the system introduced by idle limbs and inverse kinematics. Van de Panne and others (van de Panne et al, 2000) propose a limit cycle algorithm for the animation of a walking biped and a dynamic motion planner for simplified characters (Acrobot, Luxo). Anderson and Pandy (1999) investigated realistic simulation of human gait using a 23 degree-of-freedom model, similarly Laszlo et al (1996) applied the limit control cycle technique to a 19 degree-of-freedom model of a human.

There have been few attempts to build dynamic controllers which could control more than one specific motion. Examples of those are the ones proposed by Pandy and Anderson (1999) who tried to create a controller applicable to both jumping and walking behaviours and also the work presented by Faloutsos and his colleagues (Faloutsos et al, 2001a, Faloutsos et al, 2001b, Faloutsos, 2002), who additionally applied Support Vector Machines to automatically learn preconditions of different dynamic actions.

2.3 Crowd Simulation

A similar branch of research also concerning creation of intelligent characters is automatic generation of animation sequences involving many computer characters. This is often referred to as crowd scenes. One of the first papers to be published in this area was Reynolds (1987) where an algorithm for simulating flocking behaviour was presented. Tu and Terzopoulos (1994), in the paper mentioned before, apply flocking algorithms to fish equipped with a vision system. Hodgins and collaborators did some research into modelling motion of many agents with significant dynamics (Brogan and Hodgins 1997, Hodgins et al (1995) and also on adapting similar behavioural patterns to different creatures and environments (Hodgins and Pollard, 1997, Pollard and Hodgins, 1998). Metoyer and Hodgins (2000) presented a framework for rapid crowd motion prototyping, where simplified bipeds are plying American football. Additionally their agents can learn high level behaviours from real data using a memory-based learning algorithm. Two interesting recent architectures are ViCrowd (Raupp Musse and Thalmann, 2001) and ALOHA (O'Sullivan et al 2002). Both architectures support simulation of human crowds in real time. ViCrowd attributes intelligence to groups of characters rather than individual agents. The system allows for control of the animation using scripted behaviours, external interaction or reactive rules and events. Crowds, groups and characters are equipped with intentions, beliefs and knowledge. A dedicated scripting language has also been created for control of the simulation.

ALOHA puts greater emphasis on time efficiency of the simulation, thus optimising the geometry of the characters as well as motion and behaviours. The system can additionally be coupled to a voice generation module (Cassell et al, 2001).

3 Overview of Other Concepts

3.1 Q-learning

To perform fully automated acquisition of high-level animation actions it is desired that a user only define a goal for the learning task without intervening in the way the action is performed. Reinforcement Learning (RL) methods (Sutton and Barto, 1998, Mitchell, 1997) fulfill these criteria. RL is a Machine Learning technique whereby autonomous software (the agent) learns by trial and error which action to perform by interacting with the environment. Models of the agent or environment are not required. At each discrete time step, the agent selects an action given the current state and executes the action, causing the environment to move to the next state. The agent receives a reward that reflects the value of the action taken. The objective of the agent is to maximize the sum of rewards received when starting from an initial state and ending in a goal state. One incarnation of RL is Q-learning (Watkins, 1989). The objective in Q-learning is to generate Q-values (quality values) for each state-action pair. At each time step, the agent observes the state s_t, and takes action a. The choice of actions in early stages is usually random (any action may be selected from the possible actions set) and becomes more informed as the agent learns more about the environment (agent prefers actions which give higher rewards thus exploiting its knowledge). After executing an action the agent then receives a reward r dependent on the new state s_{t+1}.

The reward may be discounted into the future, that is rewards received n time steps into the future are worth less by a factor γ_n than rewards received in the present. Thus the cumulative discounted reward is given by

$$R = r_t + \gamma r_{t+1} + \gamma^2 r_{t+2} + \cdots + \gamma^n r_{t+n} \quad (*)$$

where $\gamma \in [0,1)$. The Q-value is updated at each step using the update equation (*) for a deterministic Markov Decision Process (MDP) as follows

$$Q(s_t, a) \longleftarrow r_t + \gamma \max_{a'} Q_{n-1}(s_{t+1}, a') \quad (**)$$

A sequence of actions ending in a terminal state is called an epoch.

Q-learning can be implemented using a look-up table to store the values of Q for a relatively small state space. Neural networks are also used for the Q-function approximation (Bertsekas and Tsitsiklis 1996, Haykin 1999).

Reinforcement Learning has been applied to create successful board games implementations (Schraudolph 1994, Thrun 1995), with unmanageable state spaces. Backgammon is the most successful example (Tesauro 1994). Reinforcement Learning has also been used in robotics to control one or more robotic arms (Davison and Bortoff 1994, Schaal and Atkeson, 1994), recent applications of RL to character animation have been presented before.

3.2 The Agent's Senses: Collision Detection and Avoiding

Collision detection problems are crucial for any automatic animation. This is how the physical limitations for the movement can be determined. Even the internal, biomechanical conditions could be reduced to the analysis of collisions between bones in joints, however such approach seems to be impractical.

More generally looking, the collision detection appears to be the main source of information on the agent's environment. The agent's perception (or awareness) of the ambient world is essential. As the agents are not omniscient they need some kind of senses. The ability to detect collisions may be considered to be the sense of touch; however the important difference is that this technique may be used not only in relation to the objects actually being in contact with the agents, but also to the objects that potentially could be within their reach. This is how it is used in q-learning based action planning, as described in the previous subsection: numerous states are taken into consideration before a limited final subset is chosen.

The desired collision detection algorithm should avoid testing all the polygonal faces in both objects for overlap. Instead, much more efficient solutions are based on spatial volumes (volumes that entirely enclose objects). Tightness of fit between an object and its bounding volume is crucial for the precision of the collision test. The simplest, yet widely used bounding volumes are spheres. However when object are in close proximity this approach is imprecise or it requires strong spatial subdivision techniques applied in a hierarchical manner. Another solution is axis aligned bounding boxes method (AABB). This method also produces relatively large bounding boxes if the underlying objects are not axis-aligned, and that is the case for example with the avatar's arms. A good solution may be the OBB (oriented bounding boxes) approach (Arvo and Kirk, 1989) as a good trade-off: they are a snug fit and the method is not very computationally intensive. When applying the separating axis theorem proposed by Gottschalk (1996) it is enough to make just 15 tests to determine if the boxes overlap. An extended OBBTree algorithm (Gottschalk et al., 1996) introduces hierarchical box subdivision for even higher precision.

3.3 Agent Architectures

The growing popularity of agent-based architectures and methodologies brought new discoveries into the field of autonomy, distribution and interaction (Rao and Georgeff, 1995, Wooldridge and Jennings, 2000, Wood and DeLoach, 2001, Mylopoulos *et al*, 2001, Winikoff *et al*, 2001) and gave a new opportunity to apply recent advances in AI to the problem of automatic animation generation. However there has been a very limited application of those systems into the field of computer animation. Although intelligent characters are often called agents, not many of the advances made by the agent community have explicitly been incorporated into the intelligent character architectures. One such example may be Flake et al, (2001).

An example of a system trying to solve this problem is FreeWill (Forte and Szarowicz, 2002, Szarowicz et al 2002) upon which we based our implementation. FreeWill proposes an architecture suitable to create intelligent and realistic animation by incorporating elements found in both animation-driven systems and distributed (multiagent) solutions. Each avatar participating in the animation consists of an intelligent agent implemented as a modified SAC agent (Winikoff et al, 2001) together with a body layer, which is responsible for handling the visible part of the agent (see Figure 2).

Details of the FreeWill architecture and its relation to both animation and agent-based systems can be found in Szarowicz and Forte (2003). FreeWill's main task is to produce an animation sequence with many interacting avatars. This is done by allowing the avatars to perform both reactive and proactive actions. Thus the level of sophistication of the resulting animation relies heavily on the number of high-level proactive actions present in the agent's plan library. It is worth noting that the avatar control is done using the kinematics approach and not dynamic simulation. The avatars can be controlled in two modes – using forward and inverse kinematics control. Because each high-level action (such as opening a door) has to be pre-processed and scripted into the system a natural question arises – 'is it possible to create such actions automatically?' For such a mechanism to be feasible it must be possible for an animator or film director to define only the high-level goal of a given action and then be presented with a generated solution (or family of solutions), which could be used to enrich the final animation. Reinforcement learning seems to be a good candidate technique for such automation.

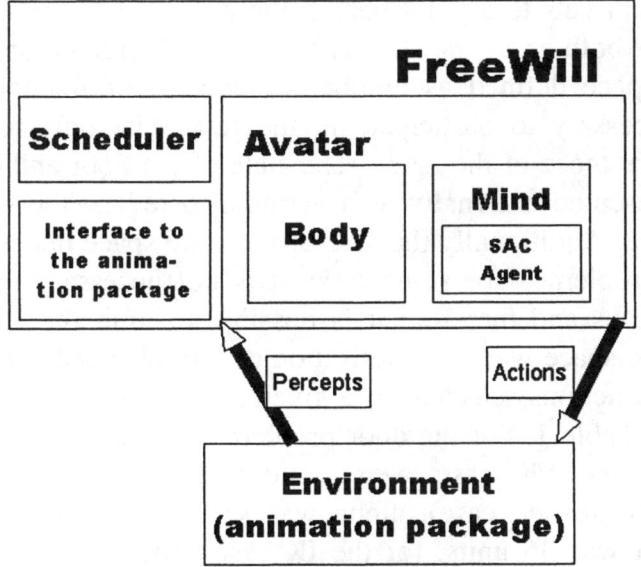

Figure 2. The FreeWill avatar

4 Implementation of the Q-learning

We have decided to select for implementation the deterministic version of the Q-learning algorithm. This is because, as our virtual world is fully simulated (as opposed to real robotic environments), it is also deterministic and therefore the implementation can be simpler. The implementation was done in a few steps:

a) First the agent was assigned a number of simple actions, which are gathered in the Table 1.

b) Next the state space was defined. For every action there are two possible different simulations – one for inverse kinematics control and one for the forward kinematics. We have simulated two different actions: a task of getting through a locked door and a task of lifting a static object (a teapot). Therefore the simulation goals were 1) the agent gets to a point behind the door and 2) the teapot gets lifted. For both cases the state space was a discretisation of a continuous space defined as number of degrees of freedom for the joints necessary to participate in the task. The only represented states were those of the agent. The state of the door and the teapot was represented externally as a variable to reduce the size of the state space. Additionally the size of the state space depends on the type of problem – for some tasks it is not necessary to consider some actions and therefore it is possible to limit the state space. The action space was a discretisation of arm displacements and rotations for actions selected for a given simulation from the set presented in Table 1. For the door problem the selected low-level actions for the FK case were actions 1,2,3,6 (first case) and 1,2,3,4,5,6 (second case), alpha was set to 20 degrees, step displacement was 35 units, for the IK case actions 1,2 were chosen with hand displacement of 5 units in each dimension. For the teapot problem the FK actions were 1,2,3,4 (first case) and similarly but with decreased alpha (10 degrees) and increased number of states along every state space dimension for the second case, the selected IK actions were movements of the hand in a 3 dimensional space (action 1), the hand displacement for every dimension was set to 8 units.

Table 1 Low-level actions used to train the avatar.

Forward kinematics control	Inverse kinematics control
1. Rotate arm up/down by Δalpha	1. Move palm by (Δx, Δy, Δz)
2. Rotate arm forward/backward Δalpha	
3. Rotate forearm by Δalpha	
4. Rotate hand along Z axis by Δalpha	
5. Rotate shoulder along Z by Δalpha	
6. Move forward/backward by Δx	2. Move forward/backward by Δx

c) The simulation was started and the Q-values were modified according to the update equation was (**), the action selection equation was given as follows:

$$P(a_i | s) = \frac{k^{\hat{Q}(s,a_i)}}{\sum_j k^{\hat{Q}(s,a_j)}}$$ (***), where $P(a_i | s)$ is the probability of selecting action a_i given that the agent is in state s, k > 0 is a constant which determines how strongly the selection favours actions with high $\hat{Q}$ values (Michell, 1997). In the case of our problem k = 2.

For most of the time (approx. 95%) the agent was placed in random starting states after the end of every epoch (an epoch is a set of actions ending in an absorbing state – that is when the agent receives a rewards of any value). Thus the exploration/exploitation policy was defined by equation (***). In this mode the maximum number of random actions the agent was taking in every epoch was limited to several hundred (i.e. if the agent didn't reach an absorbing state by then it was reset to a new random state). For the last 5% of the epochs the agent was placed in the starting position for the task (in front of the door or at the table in a neutral position) thus optimising the path through the state-space for that particular problem.

The Q-table was represented as a lookup table and the values were initialised to 0. The χ was set to 0.95

d) Positive rewards were given to the agent whenever it managed to reach the goal successfully, whereas negative rewards were used to prevent it from performing illegal moves, e.g. outstretching a joint or colliding with an obstacle.

5 System Implementation and Results

5.1 The Framework

The experimental apparatus consists of an application communicating with 3DStudio Max software package, which was used to model the three-dimensional scene (Figure 3). The characters were created using Character Studio.

The main part of the framework is an external application written in C++ that processes all simulations. It controls the scene and acquires information about object interactions through the COM interface, which is exposed from within the 3DStudio package by means of a dedicated script (a max script). This script contains definitions of functions for manipulating the objects and defines the appearance and initial positions of objects placed in the scene. Since the communication through COM interfaces is not fast enough, some critical components have been moved to a plug-in, that directly communicates with the max script engine. This solution is very efficient, but inconvenient to implement, so only necessary functions have been implemented in this way. The final sequence of actions is saved as a script controlling the avatar and the scene objects. Animation created with that script can then be rendered.

Intelligent Action Acquisition for Animated Learning Agents

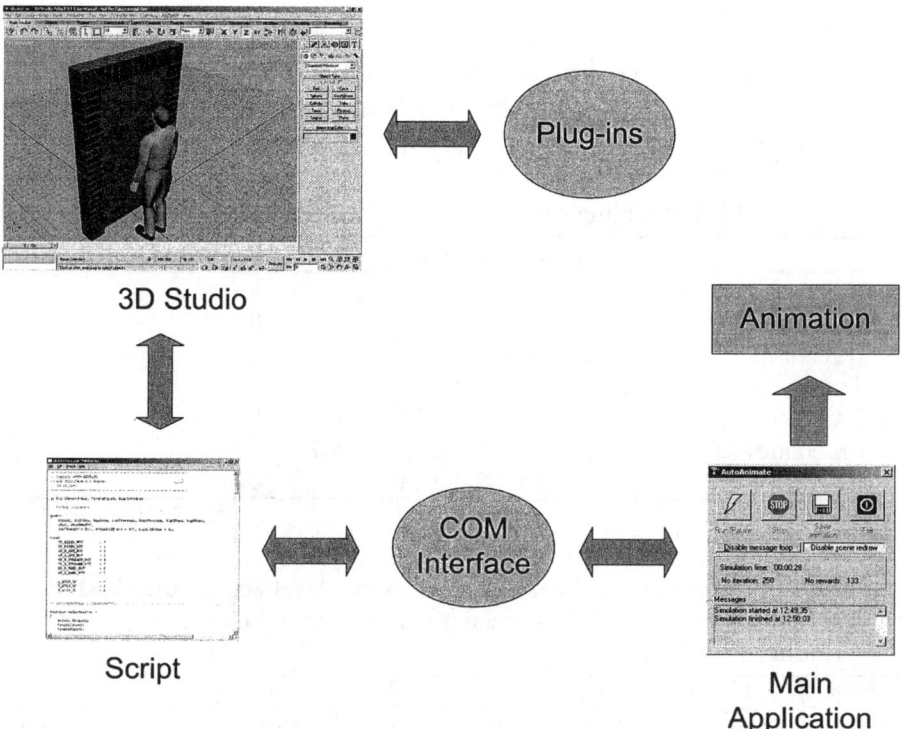

Figure 3. The architecture of the experimental apparatus

Recently, the framework described above has been replaced by a new architecture called KINE+ (Francik, 2003). It made the application independent of the 3DStudio Max concerning the simulation task. Moreover, KINE+ is a part of a wider FreeWill+ project, an extension of FreeWill described in section 3.3. It is a flexible environment for multiple animated, embodied, communicative and collaborative agents. The new system is based on a multilayered architecture dealing with increasing level of abstraction in each part of the agents's design, and applies behavioural animation (Reynolds, 1987, Tu and Terzopoulos, 1994, Musse and Thalmann, 2001) as its main design principle. The potential of automatically acquired actions may be used in two ways. They may be applied just to enrich agents' plan libraries. Much more promising is the approach in which new actions are acquired *ad hoc*, whenever they are needed. It drastically extends the time of animation generation, making it useless for on-line solutions. For off-line animation production this technique may however break many inherent limitations.

Table 2 Resulting action sequences for the teapot problem

	Action sequence	Sequence length
Door:		
FK control (1st case)	1 1 6 6 6	5
IK control	2 4 6 6 6	5
Teapot:		
FK control (1st case)	3 4 0 4 4 2 1 8 7	9
FK control (2nd case)	4 0 0 0 0 2 2 6 1 2 4 1 8 7	14
IK control	1 1 2 4 4 2 2 0 6 0	10

Table 3 Size of the state space, number of low-level actions used and required number of epochs for each simulation (* simulated state space, ** sufficient state space).

	FK			IK		
	State space	Actions	Min. no. epochs	State space	Actions	Min. no. epochs
Door	12 672	8	400	1 296 000* (6720)**	8	900
	2 141 568	12	1500			
TeaPot	12 936	10	15000	2240	8	10000
	120 960	10	40000			

5.2 Results

Four example resulting action sequences for the door and teapot problems are presented in Table 2. Table 3 summarises the results achieved for all modelled problems. Convergence and average reward (sum of rewards received by the end of an epoch) graphs for the teapot simulation are depicted in Figures 3-6. Sequences of shots from resulting animations are presented in Figures 7, 8 and 9.

Intelligent Action Acquisition for Animated Learning Agents 373

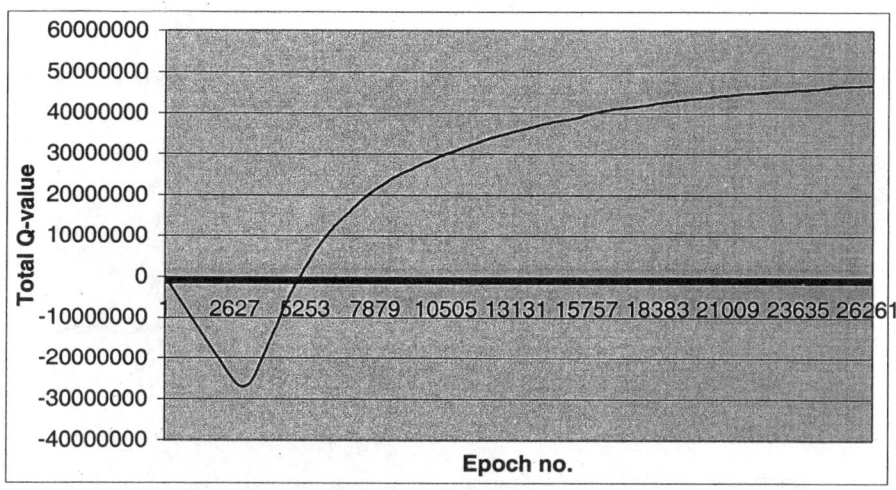

Figure 3. Convergence graph for the FK teapot problem (first case)

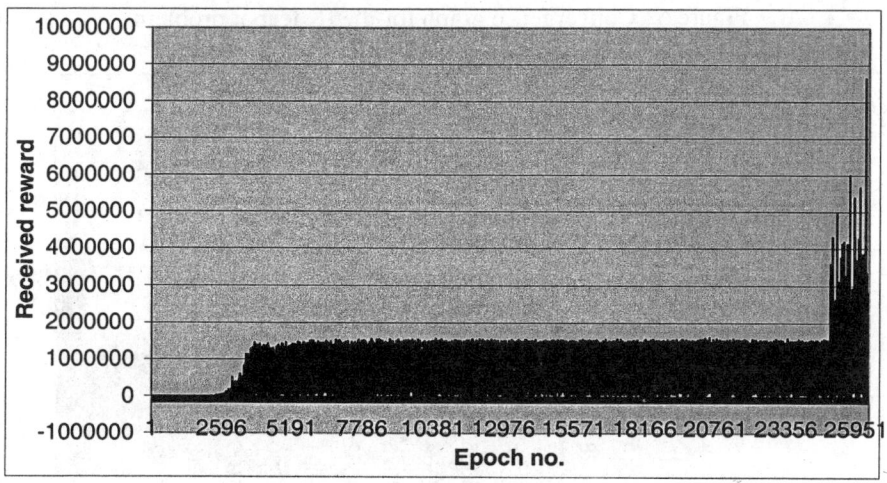

Figure 4. Reward received by an agent (the peak at the end is caused by removing the limit on maximum number of random actions)

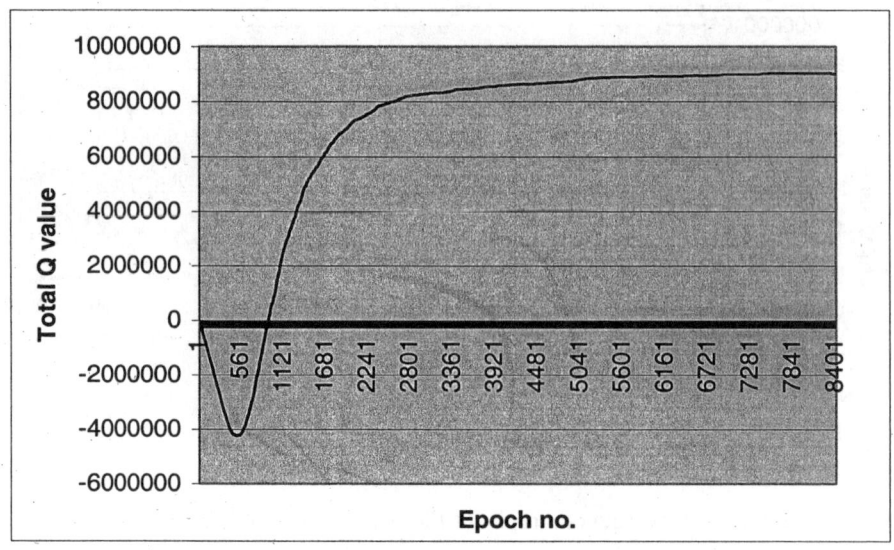

Figure 5. Convergence graph for the IK teapot problem

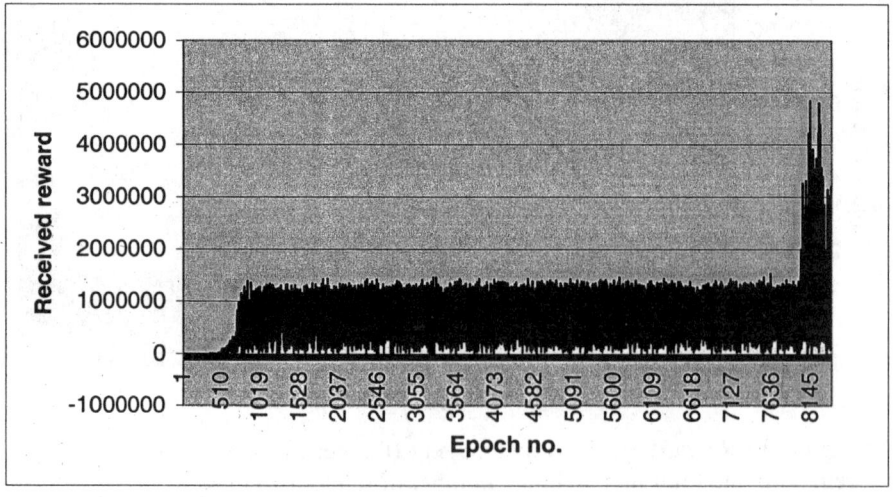

Figure 6. Reward received by an agent (the peak at the end is caused by removing the limit on maximum number of random actions)

Intelligent Action Acquisition for Animated Learning Agents

Figure 7. An avatar learning to get through a locked door. The FreeWill ver. 2 system and 3DStudio MAX with Biped Plug-in.

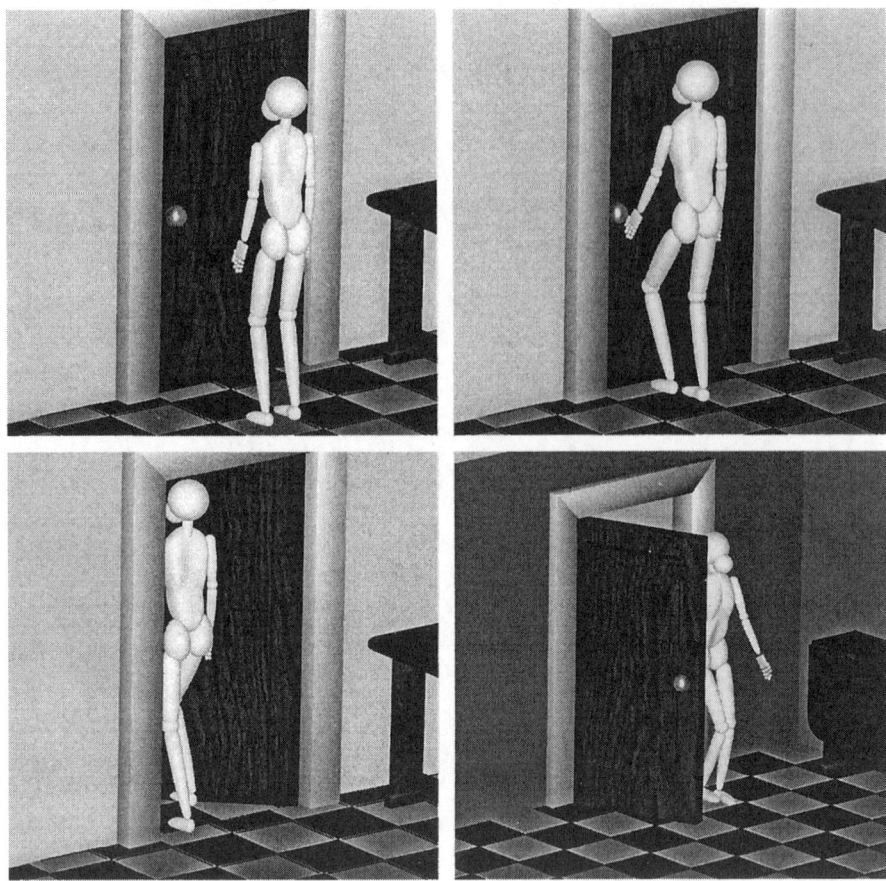

Figure 8. Getting through the door task. Screenshots of an animation derived from the best solution found. The FreeWill ver. 2 system and 3DStudio MAX with Biped Plug-in.

Intelligent Action Acquisition for Animated Learning Agents

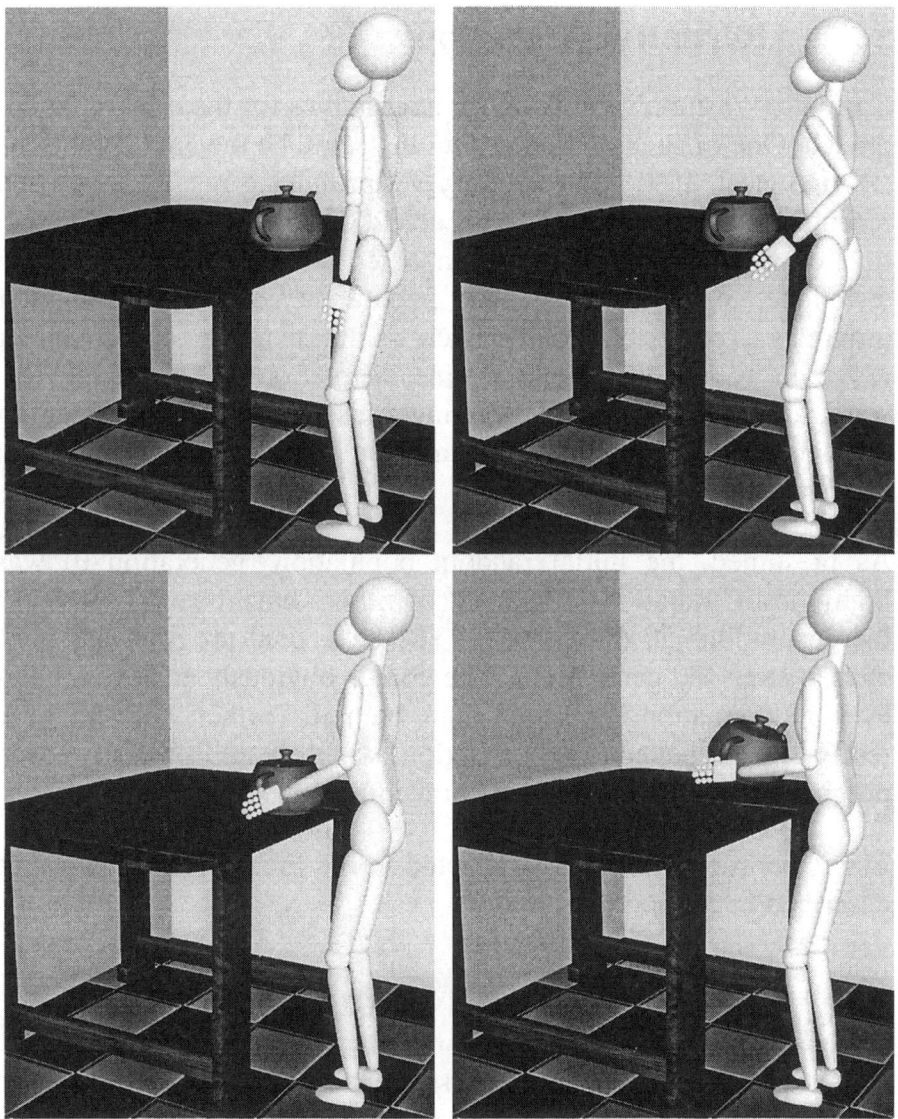

Figure 9. Lifting a teapot. Screenshots of an animation derived from the best solution found. The FreeWill ver. 2 system and 3DStudio MAX with Biped Plug-in.

5.3 Alternative Algorithms

Currently we are also looking for alternatives for the q-learning approach. One of such possible solutions could be the ants algorithms (Dorigo et al. 1991). Introductory yet promising results have been recently obtained by Ewa Lach (Lach 2003) on the basis of application of genetic programming, an extension of genetic algorithm technique, in which the genetic population consists of computer programs (Koza, 1992). In the Lach's solution the agent's computer programs were constructed of nine instructions, like moving forward, turning, moving and swinging arms etc. The goal was getting through a locked door, just the same as in the previous q-learning experiments.

As presumed, the initial random population (generation 0) was highly unfit. Rapid improvement of fitness have been observed in the generations 50 through 116. The best, final program has been chosen after 480 generations. The goal is obviously achieved, however the animation is still not quite natural. Further work includes research on a better fitness function, which would prefer solution programs that effect in more natural animation sequences.

The experiment has been conducted in the FreeWill framework as described in the previous section.

6 Conclusions and Summary

In this paper a way of learning sequences of low-level actions to achieve a goal of an animated agent has been explained. The agent has been controlled using both forward and inverse kinematics and the learning algorithm applied was Q-learning. This algorithm proved to be sufficiently effective to learn new actions by a virtual agent with several degrees of freedom. Another benefit given by this technique is that automatically generated sequences can easily be scripted and parameterised and used in other animation tools. The created animation sequences are faithful enough to be applied in a crowd scene generating software such as FreeWill or other systems presented in Section 2. An application of explicit action learning as presented in this chapter would allow generating a library of complex motions for groups of avatars, which would otherwise have to be added using manual scripting, hand animation performed by skilled artists or expensive motion capture based techniques. Easy integration of the above architecture with professional animation packages and use of bipedal avatars make it especially fit for simulation of systems with multiple human-like characters with articulated bodies. Additionally our results can also be applied in the field of robotics, provided that the robot can already perform more basic actions such as walking. The experiments show that although scaling up is easier for the forward kinematics (the representation of states is more consistent), the inverse kinematics control solution is easier to program (fewer dimensions in the state space, less different low-level actions) and generally generates more natural actions. Action learning using inverse kinematics mode of control provides therefore a feasible learning mechanism with results offering good trade-off between quality of motion and computational (in the case of physical simulation discussed in Section 2.2) or labour (in the case of more traditional techniques for creating complex actions) costs.

However adding more degrees of freedom to the presented technique will eventually create a very substantial state space with long simulation times and therefore a more compact representation is required. Therefore our next step will be an application of neural networks for state approximation. Other learning techniques (such as genetic programming) will also be applied to the constructed framework to compare results achieved from different methods.

So far we have only experimented with avatars interacting with static objects. A bigger challenge would be to try to learn interaction between agents – e.g. passing an object. The experiments presented in this paper provided good foundation for attempting that challenge.

Acknowledgments

The authors would like to thank Dr. Peter Forte for his invaluable input to the FreeWill project. Thanks also go to Paul Honey for creating the skin for the animated avatars, Ewa Lach for the explanation of results obtained by applying genetic programming to the action acquisition problem and Dr. Paolo Remagnino for his feedback on Reinforcement Learning.

References

Anderson F. C. and Pandy M. G., Three-Dimensional Computer Simulation Of Gait, Bioengineering Conference Big Sky, Montana, June 16-20, 1999

Arvo J. and Kirk D., A survey of ray tracing acceleration techniques. In Introduction to Ray Tracing Course Notes, Proceedings of ACM SIGGRAPH'88, 1989.

Bertsekas D.P. and Tsitsiklis J.N., Neuro-Dynamic Programming, Athena Scientific, 1996

Blumberg B., Downie M. Ivanov Y. Berlin M. Johnson M. P. Tomlinson B., Integrated learning for interactive synthetic characters, ACM Transactions on Graphics, Vol. 21, Iss. 3 July 2002, pp. 417-426

Brogan, D.C. and Hodgins, J. K. Group Behaviors for Systems with Significant Dynamics, Autonomous Robots 4(1), pp. 137-153, 1997

Burke R., Isla D., Downie M., Ivanov Y., Blumberg, B., Creature smarts: The art and architecture of a virtual brain. In Proceedings of the Computer Game Developers Conference, 2001

Cassell, J., Vilhjálmsson, H., Bickmore, T., BEAT: the Behavior Expression Animation Toolkit, Proceedings of SIGGRAPH '01, pp.477-486, August 12-17, Los Angeles, CA, 2001

Davison D. E. and Bortoff S. A., Acrobot software and hardware guide, Technical Report Number 9406, Systems Control Group, University of Toronto, Toronto, Ontario M5S 1A4, Canada, June 1994

Dorigo M., Maniezzo V., Colorni A., Positive Feedback as a Search Strategy, Tech. Rep. no 91-016, Politecnico di Milano, Italy, 1991

Downie M., Behavior, Animation and Music: The Music and Movement of Synthetic Characters, M.Sc. Thesis, The Media Lab. MIT, 2000

Downie M., Tomlinson B., Blumberg B., Developing an aesthetic: character-based interactive installations, Computer Graphics Vol. 36, Issue 2, May 2002

Faloutsos P., Composable Controllers for Physics-Based Character Animation, Ph.D. Thesis, Department of Computer Science, University of Toronto, 2002.

Faloutsos P., van de Panne M., Terzopoulos D., The virtual stuntman: dynamic characters with a repertoire of autonomous motor skills, Computers and Graphics, Volume 25, Issue 6, pp. 933-953, December, 2001

Faloutsos P., van de Panne M., Terzopoulos D., Composable Controllers for Physics-Based Character Animation, ACM SIGGRAPH 2001, Los Angeles, California, 12-17 August 2001

Flake S., Geiger C., Küster J. M., Towards UML-based Analysis and Design of Multi-Agent Systems, in Proceedings of International NAISO Symposium on Information Science Innovations in Engineering of Natural and Artificial Intelligent Systems (ENAIS'2001), Dubai, March 2001

Forte P., Szarowicz A., The Application of AI Techniques for Automatic Generation of Crowd Scenes, The 11th International Symposium on Intelligent Information Systems, Advances in Soft Computing, Physica-Verlag: pp209-216, Sopot, Poland, 2002

Francik J., A Framework for Program Control of Animation of Human Avatars. Studia Informatica, Vol. 24, No. 4 (56), 2003, pp. 55-65.

Funge J. D., Making Them Behave: Cognitive Models for Computer Animation, PhD thesis, Department of Computer Science, University of Toronto, 1998

Funge J. D., AI for Games and Animation. A Cognitive Modeling Approach, A K Peters Natick, Massachusetts, 1999

Funge J. D., Tu X., Terzopoulos D., Cognitive Modeling: Knowledge, reasoning and planning for intelligent characters, Computer Graphics Proceedings: SIGGRAPH 99, Aug 1999

Gottschalk S., Separating axis theorem. Technical report TR96-024, Dept. of Computer Science, UNC, Chapel Hill, 1996.

Gottschalk S., Lin M. C. and Manocha D., OBBTree: A Hierarchical Structure for Rapid Interference Detection. Proceedings of ACM SIGGRAPH, New Orleans, Lo, 1996, pp. 171 – 180.

Grzeszczuk R. and Terzopoulos D., Automated Learning of Muscle-Actuated Locomotion Through Control Abstraction, Proceedings of SIGGRAPH 95 ACM SIGGRAPH, pp. 63-70, 1995

Grzeszczuk R., PhD Thesis, NeuroAnimator: Fast Neural Network Emulation and Control of Physics-Based Models, Dept. of Computer Science, University of Toronto, May 1998

Grzeszczuk R., Terzopoulos D., Hinton G., NeuroAnimator: Fast Neural Network Emulation and Control of Physics-Based Models, proceedings of SIGGRAPH 98, Computer Graphics Proceedings, Annual Conference Series, pp. 9-20, Orlando, Florida, 1998

Haykin S., Neural Networks a Comprehensive Foundation, Prentice Hall, 1999

Hodgins, J. K., Wooten, W. L., Brogan, D. C., O'Brien, J. F., Animating Human Athletics, Proceedings of Siggraph '95, In Computer Graphics, pp 71-78, 1995

Hodgins, J. K. and Pollard, N. S., Adapting Simulated Behaviors For New Characters, SIGGRAPH 97, Los Angeles, CA, 1997

Isla D., Burke R., Downie M., Blumberg B., A Layered Brain Architecture for Synthetic Creatures, pp. 1051-1058, in Proceedings of Seventeenth Joint Conference on Artificial Conference IJCAI-01, 4-10 August, Seattle, USA, 2001

Koza, J.R., On the Programming of Computers by Means of Natural Selection Artificial System, MIT Press, 1992

Lach E., Genetic Programming in the Animation of Human Avatars. 3rd International PhD Students' Workshop on Control and Information Technology IWCIT'03, Gliwice, Poland, pp. 43-48.

Laszlo J., van de Panne M., Fiume E., Limit Cycle Control and its Application to the Animation of Balancing and Walking, Proceedings of SIGGRAPH 1996, (New Orleans, LA, August 4-9, 1996), in Computer Graphics Proceedings, Annual Conference Series, ACM SIGGRAPH, pp.155-162, 1996

Metoyer, R. A., Hodgins, J. K., Animating Athletic Motion Planning By Example. Proceedings of Graphics Interface 2000, pp. 61-68, Montreal, Quebec, Canada, May 15-17, 2000

Mitchell T.M., Machine Learning, McGraw Hill, 1997

Musse S. R, Thalmann D., Hierarchical Model for Real Time Simulation of Virtual Human Crowds, IEEE Trans. on Visualization and Computer Graphics, 2001, Vol.7, No2, pp.152-164.

Mylopoulos J., Kolp M., Castro J., UML for Agent-Oriented Software Development: The Tropos Proposal, in Proceedings of the Fourth International Conference on the Unified Modeling Language, Toronto, Canada, October 2001

O'Sullivan C., Cassell J., Vilhjálmsson H., Dingliana J., Dobbyn S., McNamee B., Peters C., Giang T., Levels of Detail for Crowds and Groups, Computer Graphics Forum, Vol. 21(4) pp 733-742, November 2002

Pandy M. G. and Anderson F. C., Three-Dimensional Computer Simulation Of Jumping and Walking Using the Same Model, in Proceedings of the VIIth International Symposium on Computer Simulation in Biomechanics, August 1999

Pollard, N. S. and Hodgins, J. K.. Adapting Behaviors to New Environments, Characters, and Tasks. Yale Workshop on Adaptive and Learning Systems, 1998

Rao A. S., Georgeff M. O., BDI Agents: From Theory to Practice, in Proceedings of the First International Conference on Multi-agent Systems ICMAS95, June 12-14, 1995

Raupp Musse, S. and Thalmann, D., A Hierarchical Model for Real Time Simulation of Virtual Human Crowds, IEEE Transactions on Visualization and Computer Graphics, V. 7, N.2, pp. 152-164, April-June, 2001

Reynolds, C. W., Flocks, herds, and schools: A distributed behavioral model, Computer Graphics, SIGGRAPH '87 Conference Proceedings, vol. 21(4) pp25-34, ACM SIGGRAPH 1987

Schaal S. and Atkeson C., Robot juggling: An implementation of memory-based learning. *Control Systems Magazine*, 14, 1994

Schraudolph N. N., Dayan P., Sejnowski T. J, Temporal difference learning of position evaluation in the game of Go. In J. D. Cowan, G. Tesauro, and J. Alspector, editors, *Advances in Neural Information Processing Systems 6*, pp. 817-824, Morgan Kaufmann, San Mateo, CA, 1994

Sutton R.S.and Barto A.G., Reinforcement Learning: an introduction, MIT Press, 1998

A. Szarowicz, P. Forte, "Combining Intelligent Agents and Animation", AIxIA 2003 - Eighth National Congress on AI, September 22-26, Pisa, Italy, 2003

Szarowicz, A., Amiguet-Vercher, J., Forte, P., Briggs, J., Gelepithis, P.A.M., Remagnino, P., The Application of AI to Automatically Generated Animation, in AI2001: Advances in Artificial Intelligence, 14th Australian Joint Conference on Artificial

Intelligence, AI'01, Springer LNAI 2256, pp 487-494, Adelaide, Australia, 2001

Terzopoulos D., Tu X., Grzeszczuk R., Artificial Fishes: Autonomous Locomotion, Perception, Behavior, and Learning in a Simulated Physical World, Artificial Life, 1(4) pp.327--351, 1994

Terzopoulos D., Rabie T., Grzeszczuk R., Perception and Learning in Artificial Animals, Artificial Life V: Proc. 5th Inter. Conf. on the Synthesis and Simulation of Living Systems, Nara, Japan, 1996

Tesauro G., TD-Gammon, a self-teaching backgammon program achieves master-level play. *Neural Computation*, 6(2) pp.215-219, 1994

Thrun S., Learning to play the game of chess. In G. Tesauro, D. S. Touretzky, and T. K. Leen, editors, *Advances in Neural Information Processing Systems 7*, MIT Press Cambridge, MA, 1995

Tomlinson B., Blumberg B., Nain D., Expressive autonomous cinematography for interactive virtual environments, Proceedings of the Fourth International Conference on Autonomous Agents, Barcelona, Spain, pp.317–324, 2000

Tomlinson B., Downie M., Berlin M., Gray J., Lyons D., Cochran J., Blumberg B., Leashing the AlphaWolves: mixing user direction with autonomous emotion in a pack of semi-autonomous virtual characters, Proceedings of the ACM SIGGRAPH symposium on Computer Animation, San Antonio, Texas, pp.7-14, 2002

Tu X. and Terzopoulos D., Artificial Fishes: Physics, Locomotion, Perception, Behavior, Proc. of ACM SIGGRAPH'94, Orlando, FL, in ACM Computer Graphics Proceedings, p.43-50, 1994

van de Panne M., Laszlo J., Huang P., Faloutsos P., Dynamic Human Simulation: Towards Agile Animated Characters, Proceedings of the IEEE International Conference on Robotics and Automation 2000, pp. 682-687, San Francisco, CA, 2000

Yoon S.Y., Blumberg B. M., Schneider G. E., Motivation driven learning for interactive synthetic characters. In Proceedings of Autonomous Agents 2000

Wan T. R. and Tang W., Simulating Virtual Character's Learning Behaviour as An Evolutionary Process Using Genetic Algorithms, Journal of WSCG, Volume 10, Number 3, 2002

Watkins, C.J.C.H., Learning from delayed rewards, PhD thesis, University of Cambridge, Psychology Department, 1989

Winikoff M., Padgham L., and Harland J., Simplifying the Development of Intelligent Agents, in AI2001: Advances in Artificial Intelligence, 14th Australian Joint Conference on Artificial Intelligence, LNAI 2256, pp. 557-568, Adelaide, December 2001

Wood M. and DeLoach S.A., An Overview of the Multiagent Systems Engineering Methodology, in Agent-Oriented Software Engineering. P. Ciancarini, M. Wooldridge, (Eds.) LNAI Vol. 1957, Springer Verlag, Berlin, January 2001

Wooldridge M., Jennings N.R., Kinny, D., The Gaia Methodology for Agent-Oriented Analysis and Design, Autonomous Agents and Multi-Agent Systems, Vol. 3, No. 3, pp. 285-312, 2000

Chapter 12

Using Stationary and Mobile Agents for Information Retrieval and E-Commerce

Samuel Pierre

The deployment and widespread use of Internet generated a renewed interest in distributed architectures. However, a great number of services offered with these architectures require high-speed network connections. Moreover, there is a widespread proliferation of portable computers and devices (e.g., laptops, palmtops, etc.) which are equipped with low processing capacity. Thus, there is a need for a new *service engineering* that is able to deal with both the demands of high-speed network connections and the capacity limits of the new portable information devices. This new service engineering could very well benefit from multi-agent systems and mobile agents, especially when it comes to the realms of information retrieval and e-commerce. This chapter presents a synthesis of agent technologies and some of its novel applications. Section 1 introduces the basic concepts related to agent technologies. Section 2 presents an agent-based architecture for information retrieval. Section 3 provides the implementation details of this architecture, whose performance is assessed in Section 4. Section 5 suggests a new architecture for product search in e-commerce. Also, Section 6 concludes by highlighting the salient features of agent-based architectures.

1 Basic Concepts and Background

Multi-agent systems (MAS) emerge from a number of domains such as distributed systems, artificial intelligence, knowledge representation, user modeling, and telecommunication networks. This section first defines the concepts of *agents* and *multi-agent systems*. Next, agents' communication and cooperation mechanisms are presented. Finally, the differences between mobile agent and mobile code are highlighted.

1.1 Agent and Multi-Agent Systems

Often, *agent* has been used as a multipurpose term, and it is associated with a great number of definitions. From a general perspective, an agent may be defined as a person who works for, or manages the business of an individual, a group, or a country (Pelletier et al., 2000). This general definition highlights two fundamental aspects:

- an agent is active;
- an agent responds to a request from somebody or something.

As far as multi-agent systems are concerned, the concept of agent varies greatly. Nevertheless, it is possible to find a middle ground among these various concepts. In fact, the term agent refers to an artificial actor that remotely fulfills one or more tasks. It is also a virtual entity, equipped with a partial representation of its environment, which is able to respond to such an environment. It can also communicate with other agents within a multi-agent universe and its behavior results from its observations, its knowledge and its interactions with other agents. Hence, they are "software entities", real or abstract, of various levels of complexity. Moreover, they are autonomous, and triggered by a user's direct request in order to accomplish a specific task tailored to a given context.

An agent may interact with other agents. It may also perceive its environment and react in real time according to the changes which occur. Thus, an agent behaves similarly to a software program which evolves in a complex and dynamic environment. In the context of intelligent services, an agent is somewhat of a user's personal assistant. It must be involved in a cooperative process with the user for whom it must accomplish some tasks (Lander and Lesser, 1997). When certain complex tasks are assigned to an agent, it must be able to clarify this complexity to the user. In fact, agents interact in users' environments in order to reduce their workloads or to provide necessary assistance. This assistance is even more efficient when the agent is able to assess users' interests, habits and preferences. The term *agent* is used in a great number of domains; hence, its sense depends on the context in which it is used.

In this chapter, we describe the agent as a software entity which:

- acts on the behalf of another entity (a human or another agent);
- is autonomous, and designed to accomplish specific tasks;
- reacts to external events emerging in its environment.

A *multi-agent system* (MAS) is a computer system characterized by its software and hardware environments, by its main elements (its agents) and by the action and interaction mechanisms which exist between these elements and their environments. The MAS environment can be described by a detailed description of the hardware (memory, computers, networks, etc.) and software (databases, operating systems, knowledge base, etc.) components available to the system.

Many different specifications have been put forward for the design of a relevant MAS architecture. Such an architecture must:

- include mechanisms to insert and delete agents in the agents' society;

- allow the construction of an agent society, that is a group of agents which collaborate in order to reach a specific objective (the agents are not supposed to know one another);
- offer mechanisms to solve conflicts.

To a lesser extent, an MAS must include mechanisms which allow the integration of heterogeneous and reusable agents, particularly for online distributed applications. However, due to security reasons, it remains risky to interact with unknown agents.

1.2 Cooperation and Communication Mechanisms

In an agents' society, agents share information as a means of communication, in order to demonstrate their competencies and to meet common objectives. The design of MAS requires methods to facilitate communication (Pelletier et al, 2003). The expression "*communication among agents*" refers to the mechanisms which allows messages to be sent and received, as well as the communication means (protocols) which support the agents' cooperation. This section presents the technical aspects of communication and cooperation among agents.

1.2.1 Communication Among Agents

Generally, an MAS architecture includes three layers: typology, communication, and cooperation. The *typology* layer defines the agents and analyzes their possible activities. The *communication* layer indicates the information exchanged among the agents. The *cooperation* layer stipulates the transactional models among the agents.

In order to minimize ambiguity and inaccuracies and to communicate efficiently, heterogeneous agents must share a common language, a communication standard. ACL (*Agent Communication*

Language) is a language used to communicate among agents. Agents share messages in order to accomplish a task, according to the communication protocols. KQML (*Knowledge Query and Manipulation Language*) is an archetype of these protocols. It derives from the three following elements:

- *performative statements*, which represent the surface layer of the language;
- *semantics*, which represents the deep layer of the language;
- *ontologies*, which represent the context.

In this chapter, the expressions KQML language and KQML protocol are used interchangeably. When we refer to KQML as a communication standard, we refer specifically to the features of the protocol. When highlighting the exchange of messages among agents, we tend to refer to the communication language. KQML and ACL are the most widely known inter-agent communication languages. Their specifications are similar to those of message passing protocols.

An *ontology* is a common language and vocabulary that serves as a framework to decode exchanged messages. A common ontology allows intelligent agents to share and reuse their knowledge. We say that an agent commits to an ontology when its visible actions are consistent with the definition of this ontology.

The first inter-agent communication approach is the *procedural approach*. According to this approach, communication is modeled by the exchange of information maintained by procedural directives. Script-based languages, for example Tcl (Tool Command Language) and Telescript, rely on this approach. The advantage of the procedural approach is that it allows the transfer of programs which are immediately executable. However, breaking down a program into various procedures remains a challenge as its contents and structure must be understood.

We then devised the *declarative approach*, which models communication through declarative exchanges (definitions, hypotheses, etc.). A declarative language must be expressive and scalable. That is the case of the language ACL. The declarative approach includes two types of communication models: blackboard communication and communication through messages (Pelletier et al., 2003).

Blackboard communication provides access to a common data structure. The blackboard is an area where information, partial results and hypotheses are polled for the use of each agent. The knowledge and hypotheses displayed on this board are exchanged among the agents. The agents' task scheduling is conducted through an agenda system that is directly linked to the blackboard. In fact, agents are provided with two sources of information: the public board communication and their private, or local, knowledge database. Agents are thus provided with their own resources as well as a partial representation of their environment.

Communication through messages, as its name indicates, is accomplished through messages exchanged among agents. Hence, agents benefit from a local knowledge base, as well as the other agents' knowledge through messages they exchange, thus ensuring knowledge dissemination.

1.2.2 Cooperation Among Agents

Cooperation among agents implies sharing information, tasks and resources. Communication includes the set of techniques which allow agents to distribute tasks, information and resources in order to accomplish a common activity. In an MAS, cooperation among agents is conducted through various methods. The best known methods are: communication, negotiation, arbitration and specialization.

Communication is necessary for cooperation. Communication may be direct (point-to-point) or selectively broadcasted (multicasting).

It is a rudimentary element to ensure tasks dispatching and action coordination. Once communication is established, the operational features of the systems are activated, enabling communication among the agents. Interpretation mechanisms, which constitute the reasoning features of the agent, must be secured. The agent's knowledge enables it to interact with other agents to accomplish a task, hence the importance of this cooperation method.

At some point, conflicts may arise while information is shared. Two approaches are used to solve conflicts: arbitration and negotiation. *Conflict solving through arbitration* mainly consists of specifying behavior rules which restrain the population of agents. Each agent receives a specific task and breaks it down into sub-tasks. The protocol calls these agents and prompts it to auction these sub-tasks. The agents qualified to accomplish these sub-tasks (i.e., agents equipped with the required resources and knowledge) may bet on them. According to the resolution strategy favored by the protocol, the agent with the highest bet wins the sub-task. Thus, the agent becomes responsible to carry out each sub-task won for a given activity.

In the case of *conflict solving through negotiation*, agents with a specific expertise gather to solve a given problem. Many negotiation strategies exist. One of them is based on the game theory: the agents build a game tree to coordinate their activities. Each agent specifies its choice at each step of the plan represented by the tree, until a compromise is reached.

Specialization is the process by which agents progressively adapt to their tasks. Specialization is beneficial to the entire population provided it enhances the community's capacities to transform a problem into a similar problem that has already been solved. Agents may specialize through learning. The learning approach allows an agent to adapt more easily to changes which occur in its environment. Knowledge may be acquired by various means: feedback, resources, direct enquiries with other agents, observations of other agents or an agent's own experiences.

1.3 Mobile Agent and Mobile Code

The expression *mobile agent* includes two distinct concepts: that of agent and that of mobility. The mobility of an agent refers to its capacity to migrate from one machine to another, through a network, in order to intervene on behalf of its host or another software entity (Karmouch and Pham, 1998; Kotz and Gray, 1999; Hagen et al., 1998). The concept of agent mobility stems from characteristics borrowed to the migration process, which consists of an inter-computer transfer process. A process is an abstraction of an operating system which includes the code, the data and the executable state of the program.

Code mobility refers to its capacity to modify dynamically its execution place, or the code fragment source during the execution of an application. The code may be interrupted, moved, and restarted on a remote machine (i.e., the mobile agents) or an application may use code located on a remote machine and transferred only during the execution (code on demand or applets). Therefore, mobile code technology renders possible the development of a number of innovative services.

The other mobile code option remains the client-server model with remote calls, implemented with technologies such as CORBA (Common Object Request Broker Architecture) or COM (Component Object Model). In fact, mobile computing paradigms are mostly based on the client-server model and may be grouped into three categories: mobile adaptation, extended client-server model and data access by mobile clients.

Mobile adaptation allows the dynamic allocation application and system resources according to the changes required by mobility.

The *extended client-server model* requires a separation of the applications on the server as well as their optimization for mobility. This allows the transfer of certain functionalities from the server to the client, or vice-versa, before disconnecting. As for *data access by mobile clients*, it is linked to the replication of the data stored on the server and the consistency of the data in the client's cache. Universal access and personalization (data preservation) are two reasons why mobile agents are favored.

Agents are said to be *mobile* when they have the capability of migrating among various physical nodes of a network (Emako Lenou et al., 2003; Glitho et al., 2002). This main characteristic does not depend on the intelligence that agents are equipped with. In current systems, mobile agents make little use of the abstractions developed in the field of artificial intelligence. Rather, they tend to use migration processes. However, the concept of mobility is different from process migration which requires an operating system that is more complex, or even totally dedicated.

However, migration process has contributed to the introduction of code and object mobility. Mobile agents build upon those notions and now implement two types of mobility, regardless of the operating system: *weak mobility and strong mobility* (Pierre, 2003).

Weak mobility refers to traditional remote mobile code. It is used to send a program to a given site before it is activated; the program remains on a given site for its entire lifetime. In this case, only initialization data can be transferred.

Strong mobility (partial or total migration) refers to a program, a mobile agent, which is able to dynamically change its execution site. Thus, a mobile agent can initiate its execution on Site A, migrate to Site B, where it continues to perform the very action it was conducting immediately before it migrated. Static variables and the execution stack are transferred. Hence, although an applet only exploits code mobility between the server and the host (a client servlet towards the server), a mobile agent requires a transfer of code, data, process and authority between machines (Baumann et al., 1998).

2 Multi-Agent Architecture for Information Retrieval

Applications which use mobile agents can be classified into three categories: those where a very specialized agent performs a simple task on one or many servers, those where the agent conducts a complex task over a long span of time with little movement, and, finally, those where the agent perform a complex task on a certain number of servers. This latter category comprises most e-commerce and information retrieval applications (Cabri et al., 2000; Brewington et al., 1999; Dagupta et al., 1999; Varshney, 2001; Varshney et al., 2000).

In order to perform complex tasks, the agent must have access to algorithm code which can be complex and often requires a large quantity of data. Hence, the execution speed and the code size are critical factors. However, these algorithms and this data would benefit from being shared with other applications or supported by the server itself, for local access. It is also futile for the agent to transport code that is already present or that duplicates operations already performed by the server. This section addresses the design of multi-agent architectures which solves these problems. It is particularly relevant for information retrieval applications, but it can also be used with other types of applications (Bah et al., 2002; Glitho, 2000; Glitho et al., 2000; Goutet et al., 2001).

2.1 Mobile Agent Information Retrieval

The quantity of numerical information exchanged has grown exponentially with Internet. This growth generated such a massive amount of information dispatched and disseminated, that the relevance of the information found in this gigantic pool of data that is Internet, may remain questionable. It is relatively difficult for non-experienced Web users to find, scan and sort relevant information within reasonable delays, and without considerable efforts. Hence, it is necessary to set up mechanisms and computer tools adapted for information retrieval (Chamam et al., 2003).

In information retrieval, the underlying principles and the models used to access information are probabilistic. Indexing is necessary to efficiently access large databases, but also to organize and limit the set of accessible elements. In most information retrieval systems, the basis of indexation stems from the contents to be indexed. In order to maintain these bases, indexing robots (Web crawlers) are regularly set up. These robot programs (ro "bots") visit servers, explore public data to gather a list of URLs. The indexing derivation process may be a simple extraction, it may be extracted through inference, or analyzed and assigned the indexed items (Goutet, 2001).

Although maintaining multilingual indexing bases provides many challenges, the set of hyperlinks to return to the user remains problematic due to the large quantity of information available on Internet. In semantic-based systems, linguists classify documents according to possible keywords. Moreover, natural language-based search engines use human editors to optimize results. *AltaVista*, *InfoSeek*, *Lycos* and *Excite* use heuristics to establish the priority order of Web pages.

Generally, heuristics single out documents with large quantities of keyword occurrences. Such an approach can prove to be very inefficient as the most relevant documents do not always contain the keyword used to launch the search. This inefficiency is due to some intrinsic aspects of natural languages such as synonymy (e.g., car, vehicle, automobile, limousine) and polysemy (e.g., hit, browse, surf). Other problems such as the lack of standards to organize information on the Web and the ontology needed to describe the contents of data sources limit the search arenas investigated by various Web search services. The ontologies required to describe the contents of documents are not limited to static and intelligent information retrieval agents, they are also relevant to mobile agents.

Compared to the client-server approach, mobile agent-based information retrieval offers enhanced flexibility and efficiency. Mobile agents move to the source of information to process and filter data on site while using asynchronous, cooperative, and communicative mechanisms with mobile or stationary agents. Mobile agents may also select search strategies and plan their migration route according to their knowledge of the user's past preferences and behavior, thus favoring a more efficient search by solely returning the links associated with relevant information. Mobile agents allow to bypass the constraints of bandwidth, latency and unreliable links. To sum up, information retrieval based on mobile agents should allow rapid and effective information retrieval regardless of its location and its asynchronous operations ought to reduce the network load.

Moreover, information retrieval based on mobile agents is bound by certain requirements such as autonomy and coordination. In fact, it needs to locate the source of information and identify the services offered by this source. If the information and the resources necessary to its execution are unavailable, the agent should adapt to this situation by planning a new itinerary. Certain solutions to resource constraints consist of partitioning the agent's logic processing brains on a neighboring information source node. This requires effective synchronization mechanisms, especially as the number of agents and information sources increases.

Certain inter-agent communication models use direct coordination that is typically based on client-server communication. When the information sources are widely distributed, these direct models signal their location through complex strategies and by leaving residual information on the nodes visited. Thus, communication depends on the network reliability, and, consequently, reduces the advantages of agent mobility. One could select a rendez-vous model, a type of time synchronization of the running environment where only the co-resident agents of a same execution environment can communicate. This approach solves the agent location problem without eradicating the risks of missing a rendez-vous.

Other indirect coordination models use blackboards for inter-agent communication. Blackboards remain on each site and they are used to exchange messages locally with other agents. A mobile entity who wishes to communicate exploits tuple spaces associated with a blackboard without specifying the stored data, but solely by searching signs or occurrences of an element through its contents. Thus, this spatio-temporal approach splits up inter-agent communication and turns out to be an interesting safety component for mobile agent-based systems on Internet.

2.2 Characterization of the Architecture

Considering the limits of an application based on a single mobile agent (transports the entire code during every move), we suggest the separation of a single agent into many integrated agents within a multi-agent architecture, most of them remaining mobile. This section presents this architecture, starting with its objectives and specifications (Goutet et al., 2001; Goutet, 2001).

Mobile agents are best known for two frequently mentioned qualities: they favor the reduction of the network load and they are network-aware, i.e., they know its topology and its configuration. Although they are often taken for granted, in fact, these characteristics are rarely implemented in current mobile agent systems. Moreover, the classical approach seeks, on the contrary, to use consecutive layers of protocol to maximally mask the physical characteristics of the network from the applications.

Hence, we seek to design an architecture that will minimize the network load. More particularly, one of the main weaknesses of the client-server approach and classical fixed networks in general is that they do not support geographical location and generate bandwidth waste by processing identically local links and intercontinental links which can be associated to a large number of hops. This can often cause rush hour overloads and generate important delays in packet transportation. Although this currently has a tendency to change, more particularly in the highly competitive field of WAP (Wireless Application Protocol) cellular telephony, much work remains to be done.

Basically, there are two types of agents: passive or reactive agents, and active agents. Although the former solely respond to messages from their environments (user, system or another agent), the latter act from their own will. They can trigger actions following an internal event even without external messages. Generally, active agents are primary actors, and they use passive agents to accomplish their assigned tasks. However, contrary to objects, passive agents are permanent and they retain an internal state which allows them to respond and adapt to the messages they receive. Consequently, the response to two identical messages may vary according to the agent's past history.

Multi-thread agents may be simultaneously active and passive. In our case, active agents are mobile, and to remain sufficiently light, they cannot carry a great quantity of knowledge to other agents. One of the objectives of our architecture is to provide these agents with a simple way to find an agent which is capable to provide the services they need.

First of all, we must define "service". For a number of reasons, we selected the interfaces as the agent's characteristic which represents a function or a set of functions an agent is capable to perform for others. The characteristic selected must represent the agent's functions as closely as possible and it must be independent from other parameters. It must be possible and easy to find an agent according to this characteristic. Finally, an agent must be able to embrace many functionalities.

The Grasshopper platform offers many search possibilities, according to the agent's name, description or class. However, these characteristics are either unique, such as the agent's name or class, or too vague and dynamic, such as the description. Hence, it is relevant to consider the interface to represent a set of functions an object contains in Java or other object-oriented languages. Knowledge of this interface is even necessary to communicate with another agent. Instead of being constrained to a single standard interface, we chose to select the interface as a representation of a service offered by an agent. The corresponding search function, which is lacking in Grasshopper, is included in this architecture at the level of the register, as shown in Figure 1 (Goutet, 2001).

In our architecture, the register is a specific agent which focuses on the necessary functions for which the platform is not responsible. As mentioned above, one of these functions is the search for other agents. In order to do so, the register keeps a list of the agents present on the platform, as well as their interfaces. When they arrive on the machine, passive agents sign in with the register for each interface they wish to present. They sign out upon their departure or when they are deleted. The platform may advise the register of the agent's arrival or departure, but the agents themselves select the interfaces by which they want to be found. Hence, the registration process cannot be totally automated nor ignored in the agent's code.

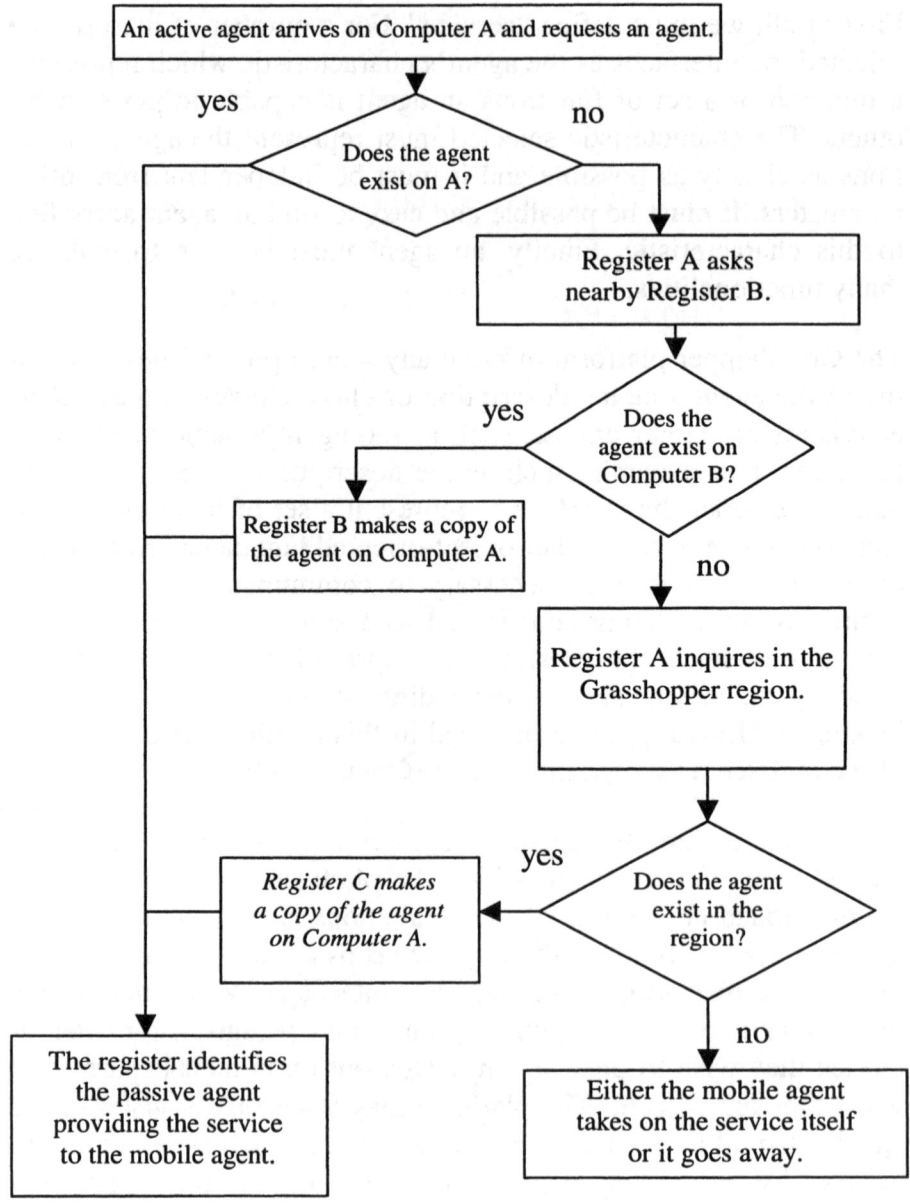

Figure 1. Agent search algorithm

If the register does not find an agent in its own database, it first inquires with nearby registers in order to benefit from its location and reduce the network load. In case of a negative response, it then seeks within the safety region to which it belongs. In successful cases, it copies the found agent on the local machine and transfers the request to the mobile agent. Hence, remote links and machines, those that do not belong to the same sub-network, are solely used as a last resort and network resources are thus preserved. Another advantage of this feature is that it liberates the mobile agent from the code which prepares and conducts the search, as well as the associated error processing code, a code which may represent a considerable size of the agent. Figure 2 shows an example generated by the developed applications: the communication process between an active mobile agent and a passive agent.

In order to minimize the total network load and to improve the performance of applications, we must provide the system with a "location" element. Many approaches are possible. Grasshopper already provides a certain implicit "regionalization" of space by introducing the concept of "region". They assemble the agencies with the same safety and property characteristics, thus providing certain information regarding geographical proximity. They also provide agents with search functions. As this is insufficient (the network of an international organization could be spread over many countries with the same safety policies), the register will thus be responsible for saving, and providing other agents with, topological information about the network, more particularly nearby agencies and agents. The concept of proximity may vary from one register to another. It reflects the proprietary system policies regarding network use. In fact, larger "neighborhoods" imply a wider use of the local network and could be favored by an administrator working with underused broadband links.

As the register is permanently in contact with the agency, it happens to be in a good position to act as a "bodyguard" to complement the system safety functionalities. It can observe the agents movements in order to, for example, prevent access to undesirable agents or parasites who seek to use the system resources without returning anything in exchange. Figure 3 illustrates the relationships among the different components of the architecture.

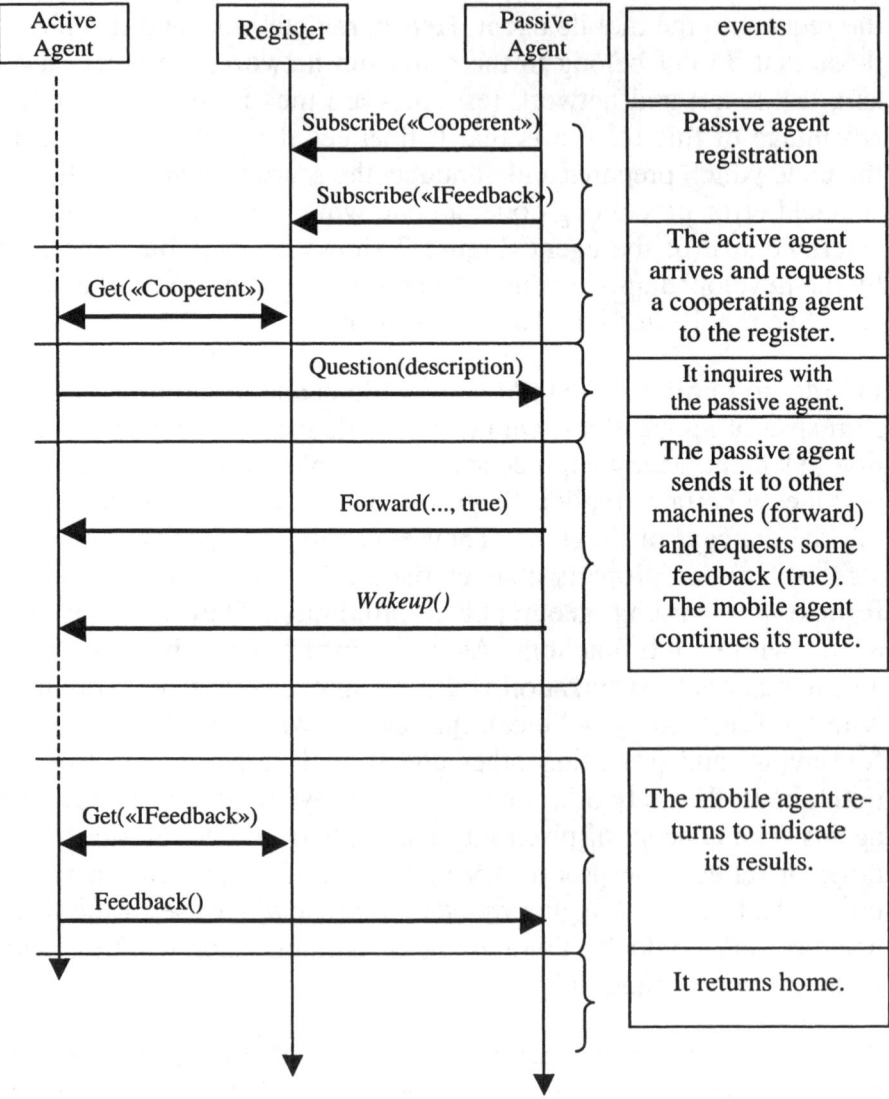

Figure 2. Communication between an active mobile agent and a passive agent

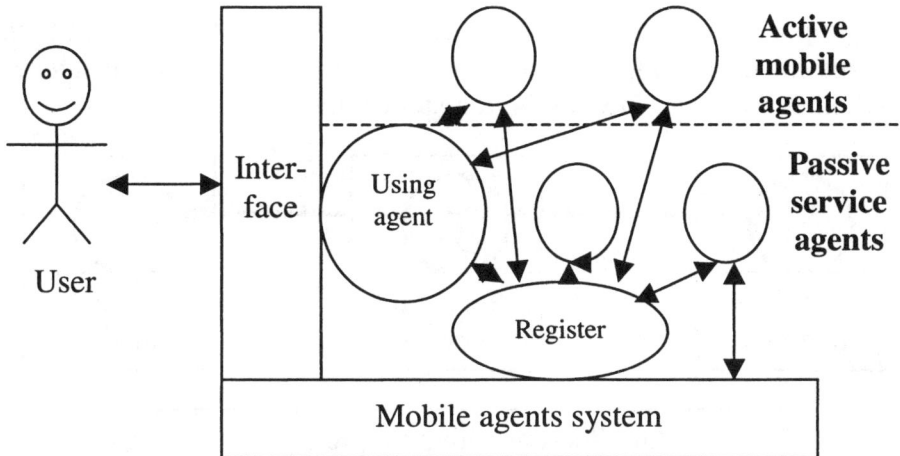

Figure 3. Relationships among the components of the architecture

Grasshopper keeps a record of each agent in cache memory for ulterior usage, but it then becomes very difficult, if not impossible, to load a different version of an saved previously agent. The mobile agent must often deal with the discrepancies which exist between the interfaces provided by the server and its own needs, which loads on the code transported every time. The architecture we suggest allows to reuse this code through the encapsulation of a separate agent which is added to the server interface. Figure 4 shows how the suggested architecture reduces the network load.

By reusing the code, the bandwidth is economized, and the administrator's load is also reduced. Nevertheless, this implies that the system is capable to efficiently manage up to several thousand agents while protecting the host from malicious agents or parasites which seek to exploit the system resources without providing anything in return. For example, a picture or information server aims at providing access (through their agents) to a maximum number of people. Thus, we must prevent a single agent from remaining on the server database for many hours, or even days, consuming memory and processing time to the detriment of other users. To do so, as an extension of the platform, the register happens to be in a strategic position. Grasshopper (as most mobile agent systems) recognizes all of the agents which arrive, remain or leave an agency, whether they sign in with the register or not.

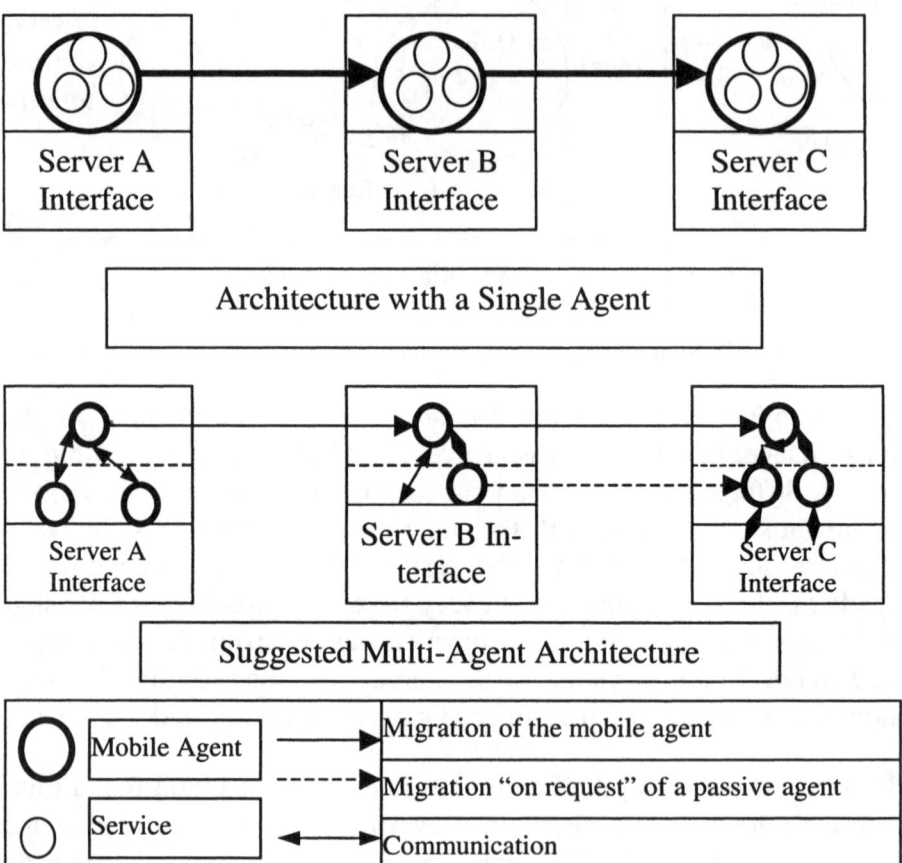

Figure 4. Architecture to reduce network load

Two types of knowledge are considered in this architecture: the knowledge of its network topology and the knowledge of its contents (agents, places and data). The network is represented by a set of places (agencies) organized into zones. These zones must be well-limited and independent from the safety or search regions to represent a proximity relationship among the agencies. This leads to a set of addresses, agencies and zones which are linked through an "is in" relationship indicating that an agency belongs to a zone or that a zone is included within another zone.

An agency may belong to two zones without affecting the algorithms presented subsequently. Thus, we obtain a graph which accommodates more simple heuristic search techniques than a graph illustrating the set of physical links of the network. Figure 5 illustrates such a graph.

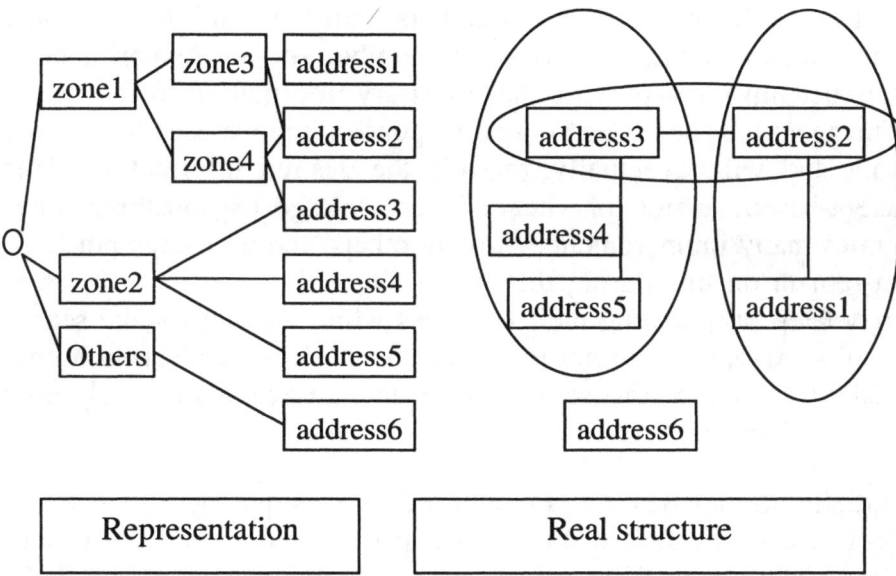

Figure 5. Network representation

A resource – a place, an agent, a file, a database – is represented by an "address". This address includes the address of the machine, as well as the address of the agency where the agents is headed, the name of the resource (with the complete route in case of a file). This address comes with information designed to guide the agent in its search for a useful resource to accomplish its task. In order to reduce the load of the agent, it must transport a minimum of information while the relevant data is stored into other agents.

2.3 Experimental Number Application

This application aims at helping users find a correspondent to obtain phone information by searching through a database of possible correspondents (Goutet, 2001). This section presents the principles, the design selections, as well as the modifications affected to the initial application.

2.3.1 Principles of the Application

Instead of dialing a series of numbers until they find the right one, users dial a unique experimental number (or an "experimental" Internet link) and provide the necessary information to the agent. The agent then searches for a correspondent predisposed to take the call, and will, eventually, provide the desired information. This agent-based version of vocal server of large organizations integrates many improvements. Among others, the user does not have to remain on line during the search. Once the agent is sent, users may keep on pursuing their activities while waiting for the search results. Also, they are not forced to take the communication immediately. Moreover, this service offers an interface that is much more personalized and user-friendly.

Initially, the application is composed of a unique agent that transports the user's information and a list of predisposed correspondents. The agent visits the machines of each user listed until the right person is found. One of the improvements to the initial application renders the list of correspondents dynamic. Hence, the agent can be sent back towards other correspondents during the process, even those who were not included in the initial list, as they were unknown from the user. This first modification allows the application to move into the realms of mobile agents and opens up possibilities of other modifications discussed further.

2.3.2 Design Choices and Modifications to the Initial Application

One important decision regarding the design of this application is related to the telephony-agent interface. The field selected was that of Internet telephony where a phone call is transmitted through an IP network. Many protocols are currently being experimented to harmonize IP and real time. H323 and SIP (Session Initiation Protocol) are the two main protocols. Glitho (2000) considers these protocols fail to meet the desired objectives: to support a large inventory of services and providers, achieving the rapid and simple creation, management and personalization of services, ensuring network independence and cooperation with existing services. To compensate for these weaknesses, Parlay was introduced as a supplementary layer placed above these protocols to offer service providers a simple standard interface (Desrochers et al., 2000). This technology is client-server oriented. Thus, we may link mobile agents solely to the SIP (the agent could integrate and SIP client), to Parlay, or to both.

In the initial version of the application, the agent traveled through a series of predetermined places. The addition of a dynamic list of destinations opens the door to new possibilities. In fact, the agent's goal is to find a phone correspondent, either a certain type of information represented by an IP address (in the case of IP telephony), a Boolean response (responds/does not respond) and eventually other possible extensions.

The current directory of phone services, which is more or less useful to retrieve needed information, includes answering machines or services, repertories, voice servers, etc. It is natural to consider adapting these tools to IP telephony and more particularly to our application, especially given that it is actually simple, as the agent transports a description of the call objectives in a text format which is more condensed and comprehensible than voice by machines. Each one of these tools or services can be implemented through an agent that will communicate with the mobile agent through a simple interface described with the architecture. The role of the mobile agent is thus reduced to traveling through a dynamic list of addresses, each address corresponding to a human or virtual agent. The scope of the application is itself widened, allowing, for example, to search a user assigned to many devices, to find a person in an organization who is assigned to certain functions or who may provide certain information.

2.4 Internet Picture Retrieval Application

This application searches through a network to find pictures requested by a user who wishes to use "clip arts", produce birthday cards, etc. In the developed version (Goutet et al., 2001; Goutet, 2001), the pictures and their description are obtained through HTML pages. The agent travels to each server which hosts a picture database and scans HTML pages provided by the server as the database interface. Then, the agent returns to its starting point with pictures that correspond to the user's request. More precisely, the agent travels through a list of destinations, beginning by an initial list of known sites. When it arrives at a site which contains a picture database, it goes through the HTML pages and seeks information according to the method described further. It then uses the criteria provided by the user to jot down the results of this search. It returns once it has gathered a sufficient number of results or once it has completed its route.

A database with a JDBC (Java DataBase Connectivity) interface would have been an interesting addition to this prototype, but due to security reasons, very few sites offer direct Internet access to their databases. In order to design a realistic application, or even an application that may be used immediately, we selected the interface that is the most often used on Internet: HTML pages, eventually dynamic HTML pages. Our agent should be able to extract the necessary information (i.e., the addresses and descriptions of pictures) from HTML pages as generally and as simply as possible.

3 Implementation of the Information Retrieval Architecture

Most of the architecture presented in the previous section was implemented (Goutet et al., 2001; Goutet, 2001). This section presents the list and the structure of the agents composing the sub-set of the architecture that we have already implemented. New agents may be added without major modifications.

3.1 Generic Classes and Interfaces

The expression "generic class" includes classes used in every application, and which are, consequently, not specific to any agents nor any algorithms selected other than those chosen when the architecture was designed. More particularly, they are: the *Class Address*, which may find a network resource and the *Class Link* which may add to this address priorities or information about its route position to the agent. Figure 6 describes these two classes.

Also, the *Class Job* allows to represent an agent's task through a string of characters which describes the task and assigns it to an agent. It is particularly useful for passive agents who may receive simultaneous requests from many agents, but it can also be used by active agents, in order to organize tasks, for example. This class may be considered as a simple internal representation of communication through KQML messages as it would contain the message sender as well as the message itself.

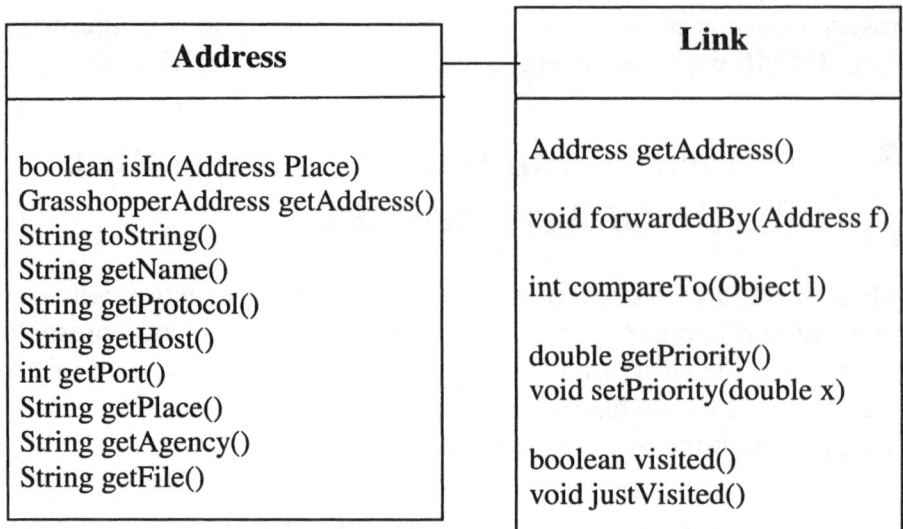

Figure 6. Interfaces of *Address* and *Link Classes*

One of the characteristics of the *Class Address* is that it can be built from a wide inventory of Java objects representing a great number of concepts, hence a large number of constructors (not illustrated on the figure). Grasshopper class descriptions also include "GrasshopperAddress" which includes an address under Grasshopper in host/agency/place format, or an "identifier" used to identify a unique object in Grasshopper.

One of the particularities of the architecture we suggested is that we do not search an agent from a description or a name, but from the interface it presents. This is motivated by the fact that we must know the agent's interface in order to create a communication proxy towards this agent in Grasshopper. Another reason, which is mostly conceptual, is that (aside from a few exceptions) an active agent arriving at a new agency is not seeking a specific agent but, rather, an agent that is capable to accomplish a certain task that corresponds to a specific interface. The most important interfaces are: *IRegister, Cooperant* and *IResearcherAgent*.

IRegister is the interface offered by the Registers. It must allow the addition and the removal of services, as well as communication between a mobile agent and the agent providing the required service. It includes the following functions:

- *void subscribe(Identifier agent, String style)* which adds a "style" service provided by the agent "agent". The reverse operation is performed by the function *void unsubscribe(Identifier agent, String style)*.
- *boolean question(Identifier researcher, String destination, String description)* is the utmost messaging function, sending a "description" message from a "seeking" agent to a "destination" service. In this case, we are referring to a service, rather and a specific agent. The register is responsible for locating an agent offering this service.
- *Identifier get(String destination)* allows an agent to have the identifier of an agent that provides a "destination" service without having to explicitly send a message.
- *boolean get(String destination, GrasshopperAddress place)*: here, service is requested on another machine "place". The agent providing the service (if it exists) will migrate to, or copy itself on, the machine place, or even provide a communication proxy and remote service if it cannot travel or if the network policy prevents it from doing so.

The general interface of passive agents is *Cooperant*. It must allow communication, search and operations from the register to the agents. This interface integrates the following functions:

- *void question(Identifier researcher, String description)* is the search and communication function. Previous results and messages are then sent directly to the "searching" agent who requested them.
- *boolean go(GrasshopperAddress place)* is the function used to request an agent to move to another machine (here "place") to provide a service or in cases where a machine stops at home.

- *AgentInfo getInfo()* provides the agent's Grasshopper information, more specifically the identifier that will be used subsequently to find it.

IResearcherAgent is the interface of *HuntGroup* (the mobile agent). It is mostly adapted for information retrieval agents, and more specifically for the mobile agent of the HuntGroup application developed. It may be easily modified or inherited for other applications. It contains the following functions:

- *void forward(Address [] stops, boolean notify)* and *void forward(Link[] stops, boolean notify)* are used to refer the agent to new addresses. The Boolean parameter "notify" is used to tell the mobile agent that we wish to be informed of the search results in order to complete and update its information. In this case, once it has accomplished its task, the agent returns to show its route.
- *void wakeUp()* wakes up the agent who is awaiting an event or a result.
- *String getCallDescription()* provides the search description to the agent.
- *void setResponse(boolean x, boolean notify)* is used to respond to the agent. The parameter "notify" is synonymous to the functions "forward".

3.2 Agents

The main role of the register is to record the agents offering a specific service on the same agency and a search function for these agents. If the searched agent is not found on the machine, the register can search it and copy it on nearby, known machines. This operation requires a certain knowledge of the network. It may also be configured differently according to the local network policies. The register offers the *IRegister* interface. It remains a stationary agent, even if it can be copied on a machine that is similar to the original one. Figure 7 shows the structure of the register implementation.

Using Agents for Information Retrieval and E-Commerce 415

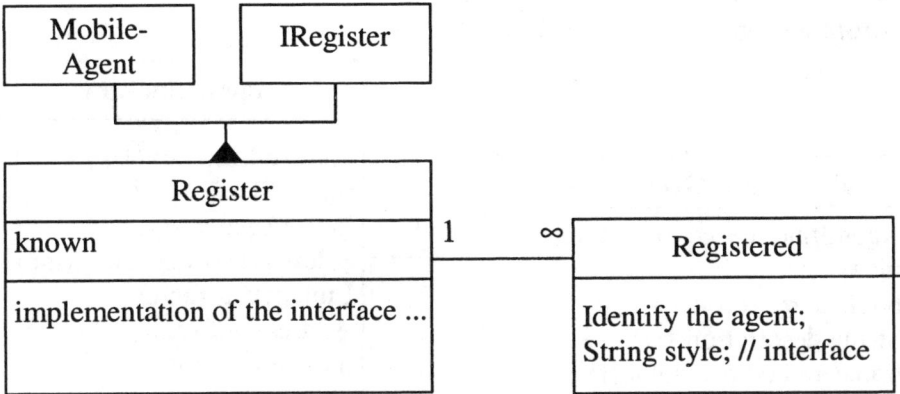

Figure 7. Structure of the register

The "GUIAgent" is the user interface. It is generally used to provide a unique, personalized interface to many applications. It implements the "IUserGUI" interface and remains static due to its graphic and personal character. Note that the graphic classes (Frame, etc.) are often difficult to transport. Moreover, we must also consider that the machines themselves are often plagued by their own limitations (for example, "Personal Java", designed for small devices such as palmtops, does not support the "swing" library).

The *HuntGroup* is the active mobile agent that is at the heart of the developed applications. It implements the *IResearcherAgent* interface and transports the itinerary and the results (including the solutions and its "knowledge" of the network). It ensures the mobility functions in the developed applications, leaves the more specialized functions to passive agent and transports the fewest elements possible to execute the application. Figure 8 shows the structure of the agent *HuntGroup*.

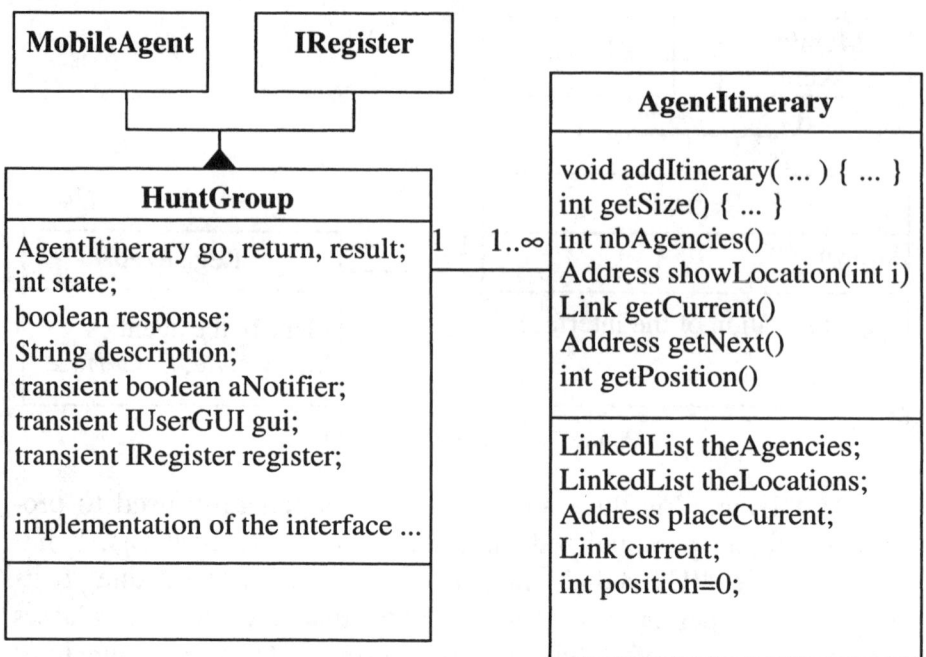

Figure 8. Structure of the mobile agent *HuntGroup*

Certain variables of the *HuntGroup* are labeled "transient" as they are not transported. In fact, they are used to access the GUI and the local register. As those change with every movement, it would be useless, or even dangerous, to transport them. We also chose to provide the agent with many itineraries. This is due to the fact that in *AgentItinerary*, the destinations are organized by priority order, according to a given task, rather than by "chronological" order. If the agent must accomplish many successive tasks, it then requires a sub-itinerary for each of these tasks, including the favored destinations of the task. It will thus save many successively used itineraries. *AgentItinerary* also includes the itinerary travel functions whose body will vary according to the version developed. Both lists *theLocations* and *theAgencies* respectively preserve the agent's destination addresses and the geographical proximity zones.

The last important agent of our architecture, *KnowAgent*, records knowledge. It is semi-stationary (it moves only on request, and preferably through clones) and implements the *Cooperant* interface. It is thus a passive agent. Its role consists of preserving knowledge and providing it to mobile agents as needed. For this purpose, we will use the information retrieval techniques described above. This task is attributed to the class *KnowManager*, while the class *KnowAgent* looks after communication and task management. A *Vector* class represents a number vector which may be high but which contains many null values, as it is often the case in information retrieval. These values will be preserved in a hash table. Figure 9 shows the structure of this agent.

Numerous agents may be added to provide the services required by the applications. They are generally passive, semi-stationary agents implementing the *Cooperant* interface or one of its inherited interfaces. We also implemented an *Answering* agent replacing human users to respond to the agents and a *Researcher* agent whose goal is to find information through HTML files.

3.3 Implementation and Testing Environment

The Grasshopper mobile agent platform was chosen for the implementation. Developed by the German organization IKV, the initial version was initially available in August 1998. Java was selected as the development language of the platform, particularly due to its portability. Grasshopper meets the OMG (Object Management Group) standard on mobile agents, i.e., the MASIF (Mobile Agent System Interoperability Facility) standard. The latter was designed to ensure interoperability between different mobile agent platforms. Grasshopper is one of the distributed agent environment (DAE). It is composed of regions, places, agencies and two types of agent: stationary and mobile.

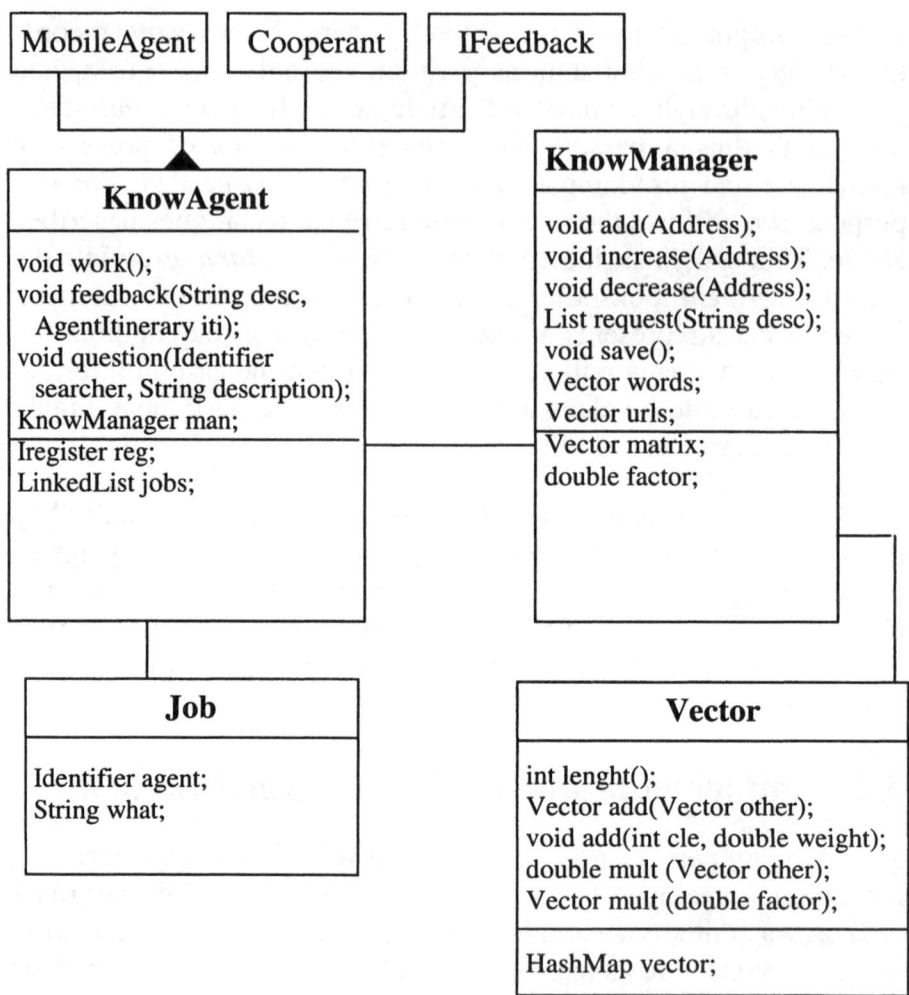

Figure 9. *KnowAgent* structure

In order to measure the data transported over the network, we used three Windows NT 4.0 workstations equipped with an Intel Pentium II 400 processor. These machines were supported by a local-area network with 100 Mbps Ethernet. Trip lengths were measured with Java classes simulating the agents' movement within a network, in order to automate the measures without requiring a distributed testing environment.

4 Evaluation of the Information Retrieval Architecture

Performance measures are mostly linked to the network load, as this is the variable we seek to optimize through this architecture (Goutet et al., 2001; Goutet, 2001). We note that the execution time remains inferior to a second, which is totally acceptable for applications where the human response time remains the limiting factor. The variations induced by the various implementation choices and by the information retrieval algorithms will be examined.

4.1 Transportation Measures

This first series of measures aims to compare different versions of the "HuntGroup" application in terms of code size moved through the route. First of all, it consists in isolating the parameters of the mobile agent system used (in this case, Grasshopper). When Grasshopper regions find agencies and agents, or to apply a common security policy, the agencies must register with the region when connecting to the network. This operation can be relatively frequent in the case of mobile devices (cordless phones, palmtops, etc.). However, such operations are not necessary for the system to function. Hence, we measured the cost of the agency registering with the region, that is 14 K in the direction of *the agency towards the region*, and 24 K in the other way, for a total of 38 K. Compared with the agent traveling costs, this value is relatively high, although moderate considering the smaller "mobility" of the agencies (addition frequency/agency deletion).

Next, we considered the different versions of the "HuntGroup" application. The basic version offers a single agent which holds and transports the GUI and all of the corresponding addresses possible. The size of this set of classes is 16.4 K. Table 1 summarizes these results.

Table 1. Measures of the effect of the cache and the Grasshopper region

	Direction	One Way (Kbytes)	Return
1st expedition with the region	1->2	26.5	36
	2->1	14.4	26
	Total	40.9	62
2nd expedition	1->2	8.4	17.5
	2->1	7.4	18
	Total	15.8	35.5
Without region: (1st expedition)	1->2	-	24
	2->1	-	12.8
	Total	-	36.8

These measures indicate a mean gain of a factor 2 due to the Grasshopper cache, but also while the regions are not used. Hence, it is relevant not to use the regions, especially as the registers take on the same functions (agent search, security) in the developed architecture. The measures displayed in Table 1 were obtained without using the Grasshopper regions.

An improved version of this agent consists in separating the graphic interface from the agent, although this does not create a significant difference for the transportation costs of the agent, as the graphic classes are present on each Java machine and contain very little data. A third version provides intelligence (information retrieval algorithms) to the agent. This increases the class size to 33.3 Kbytes, without taking into account the amount of knowledge carried at each move of the agent, justifying the use of a multi-agent architecture to keep both intelligence and the right performances, such as those presented in Table 2.

The versions of the application used from now on are based on the architecture presented in the previous sections. The difference between the different version resides in its use of the network knowledge within the *AgentItinerary* class. A first, "simple" version, travels to its destinations according to a priority order, while the "local" version priority is established in terms of the current zone destinations. The "complex" version selects its subsequent destination according the priority of each known zone. Each one of these version is more complex than the previous one. Hence, the latter is the "heaviest" to transport. However, from this complexity, we expect a reduction in the number of trips necessary, thus, yielding a reduction in the total network load. Figure 10 shows the differences between the last two algorithms.

Table 2. Comparison of transport costs of different versions

	Direction	One Way (Kbytes)	Return
1st "simple" expedition	1->2	18	19
	2->1	3.2	8
	Total	21.2	27
2nd "simple" expedition	1->2	5	5.7
	2->1	0.5	5.7
	Total	5.5	11.4
1st "local" expedition	1->2	18	19
	2->1	3	9
	Total	21	28

Table 2 illustrates that there is very little difference between the different versions for the first expedition as well as the subsequent ones. Other measures, not reported in this chapter, confirm these results. Moreover, the use of the Grasshopper and the Java machine caches (loading the object *Class*), allows to preserve the algorithm code. One could wonder whether the agent size would vary much during its route, as it accumulates knowledge and results. In order to assess this, an agent route load (one way) was measured with a larger number of initial destinations. No significant differences were found with an itinerary of 20 addresses over 2 zones, instead of 3 addresses over 2 zones, and an increase of 2 Kbytes for an itinerary of 20 addresses and 20 zones.

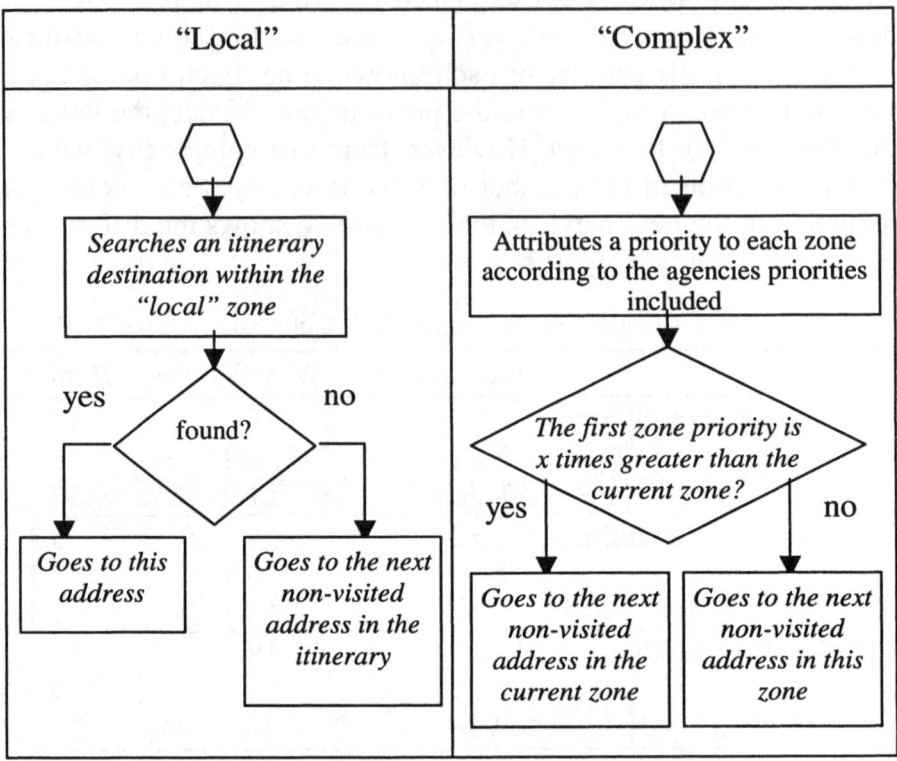

Figure 10. Itinerary follow-up algorithms

4.2 Information Retrieval Scenarios

The three algorithms are then compared by simulating their movements and learning over a scenario that includes a few organizations, administrations and access providers in three cities (City1, City2 and City3). Figure 11 illustrates such a case. We measured the number of agent movements for each version for a given series of requests, and considered both "local" (within a sub-network) and "regional" movements. Figures 12 to 16 show the evolution of the number of movements for the three version of the developed itinerary routing algorithm: "simple", "local" and "complex".

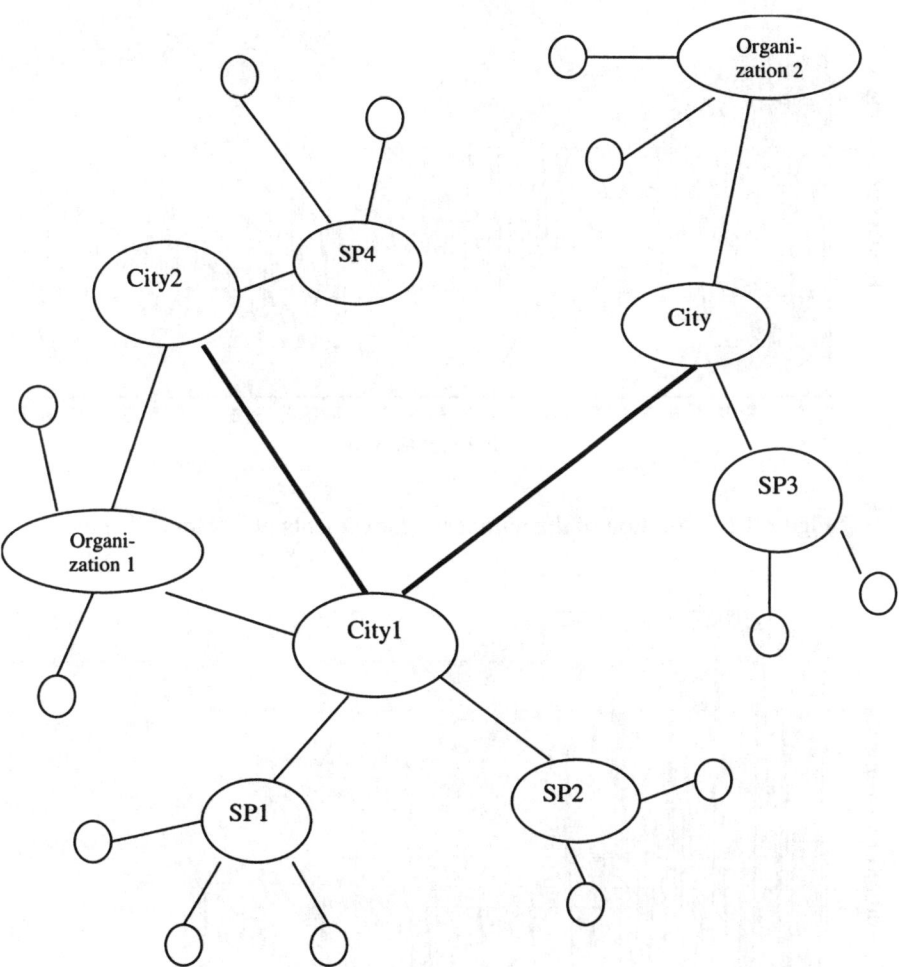

Figure 11. Measure scenario

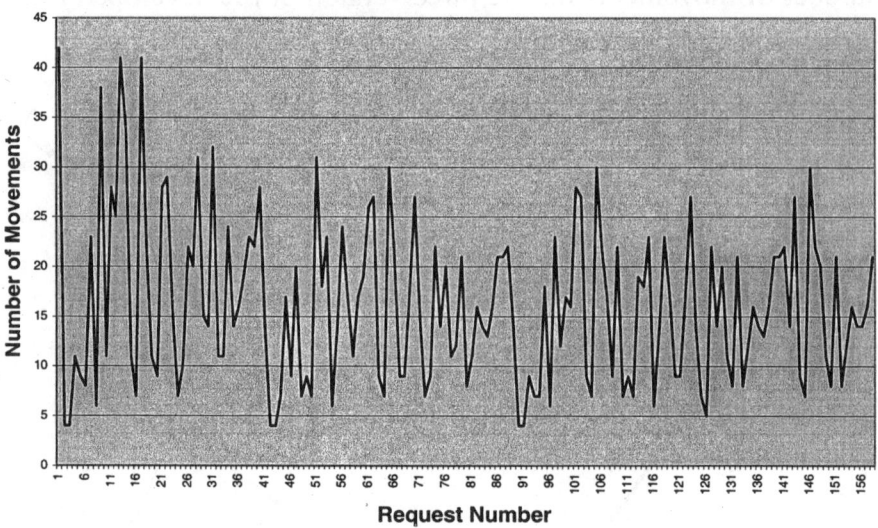

Figure 12. Evolution of the number of movements of a "simple" agent

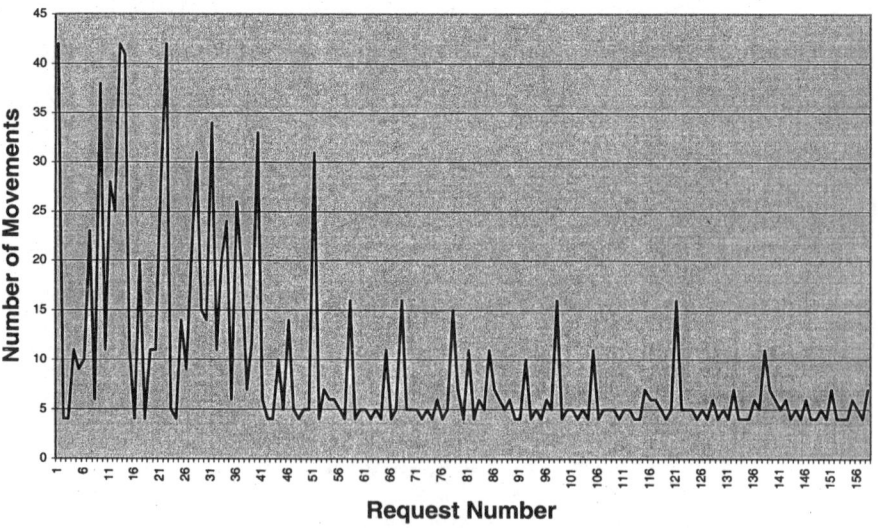

Figure 13. Evolution of the number of movements of a "local" agent

Using Agents for Information Retrieval and E-Commerce 425

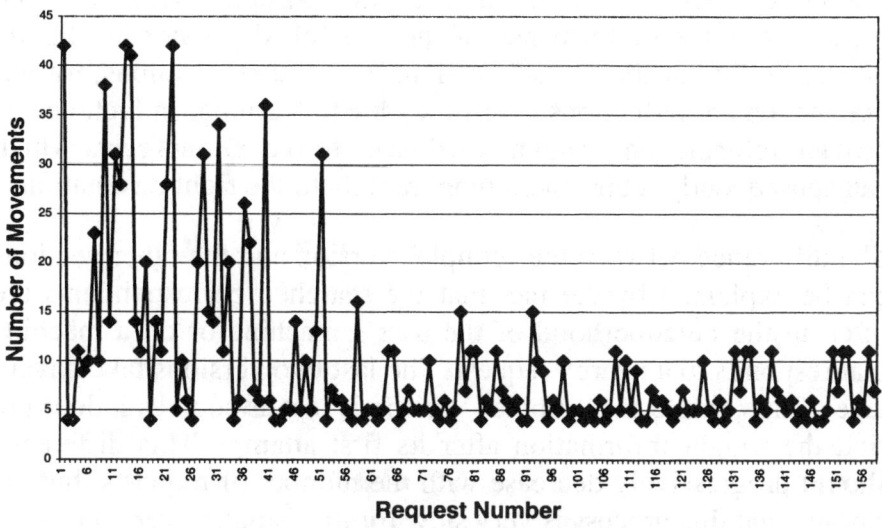

Figure 14. Evolution of the number of movements of a "complex" agent

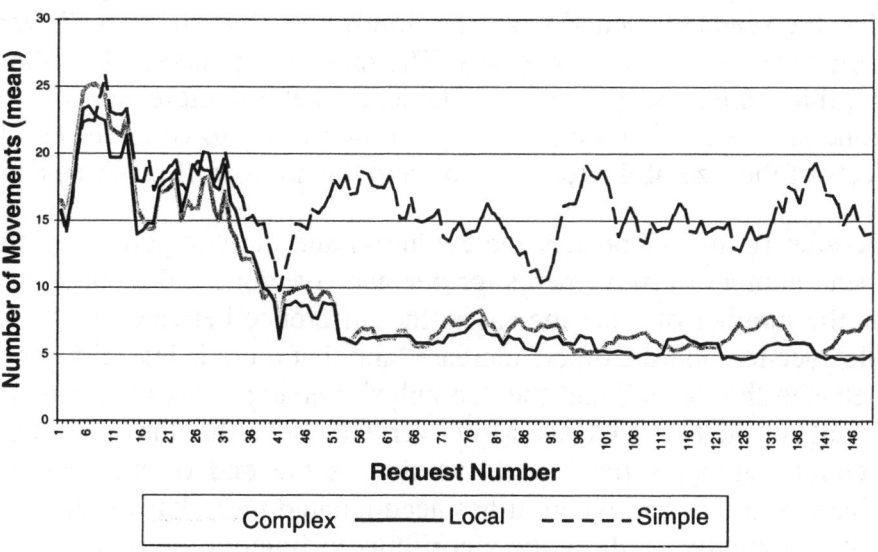

Figure 15. Mean comparisons of the number of movements

We notice that the performance of the "simple" version is the worst, while the performance of the "complex" version is slightly inferior to that of the "local" version. For all three versions, we notice movement reductions over time, due to learning. In fact, agents provide information through "feedback" to the *KnowAgents* which can subsequently return them more rapidly to the right destination.

The difference between the "simple" version and the other versions can be explained by the fact that the searched correspondents are often in the neighborhood of the user's machine or on a machine that responds to a nearby request. The last two versions take advantage of this, while the "simple" version is lost as soon as it does not find the sought information after its first attempt. This difference should progressively decrease with the number of requests, but we can see that this process is very slow for the "simple" version.

The performance of the "complex" version are slightly deceiving, given its higher level of "complexity". This is explained by the fact that the selected scenario is very simple, and that the "local" version finds the information easily. The more complicated algorithms are thus unnecessary, even unsuitable, as they are more sensitive to zone sizes (in this case). It could be interesting to observe the effects of the size of the test network and the number of machines.

Figures 16 and 17 present the evolution and the comparison of the mean number of movements for one and five zones. We notice that as the number of zone increases, the difference between the "simple" version and the others increase, and that there is less difference between the "local" and the "complex" version. This confirms the results of the first analysis. Moreover, we notice that the "complex" version surpasses the "local" version at the end of the learning phase, at a moment where it has accumulated more knowledge, but not sufficiently to allow the algorithms to function optimally. This suggests that the "complex" version would be more recommended for real, dynamic environments.

Using Agents for Information Retrieval and E-Commerce   427

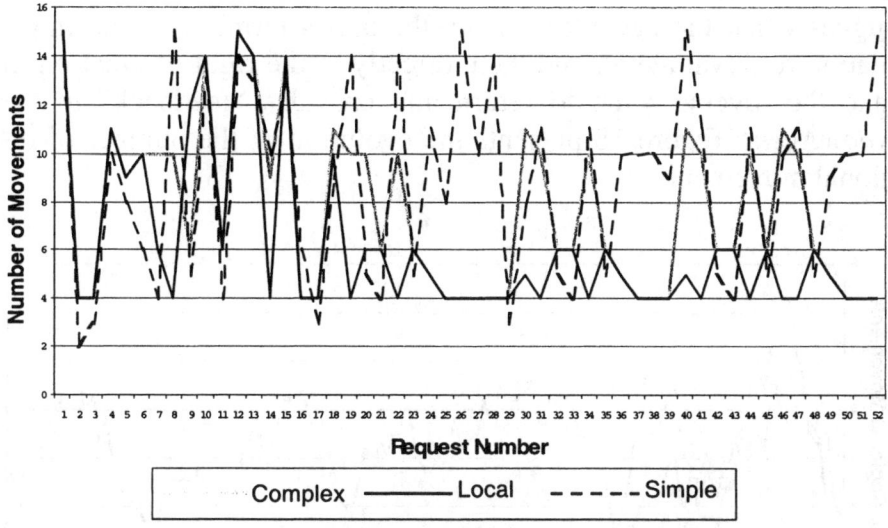

Figure 16. Mean number of movements for a single zone

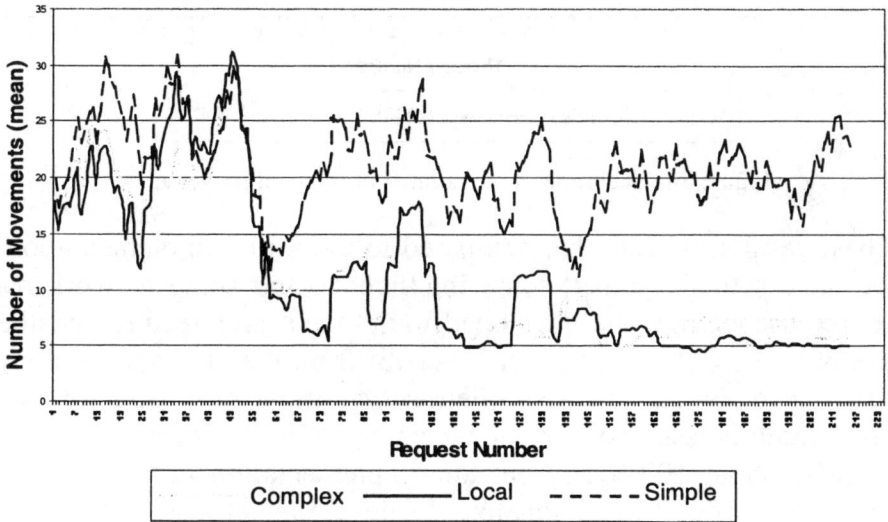

Figure 17. Mean number of movements for five zones

We notice that the inferior limit value for these measures is 4. This suggests that the agent goes from the user's terminal to the information retrieval agent, and then, directly to the right destination. It take the reverse route to return and provides "feedback" to the *KnowAgent*. Figure 18 presents the evolution of the number of regional movements.

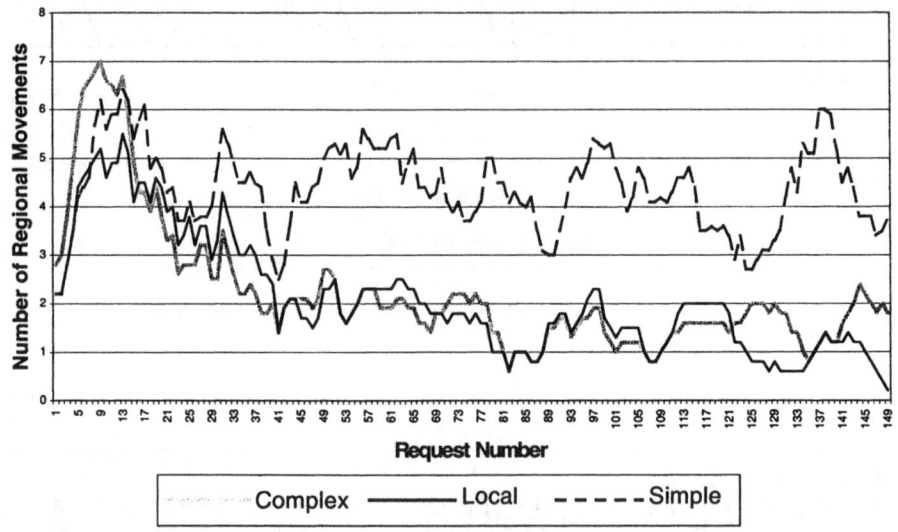

Figure 18. Evolution of the number of regional movements

These results indicate that compared to the initial implementation, we have actually gained some intelligence and some network resource usage from the developed multi-agent architecture and the use of routing algorithms which benefit from the network topology knowledge. Nevertheless, a client-server implementation for this application is less costly (establishing and SIP communication only requires about 500 bytes) but this implementation is less flexible and more difficult to personalize. Its development is also less scalable, as all the data would have to go through a central served located in a single city, thus requiring a greater number of connections and "regional" data. Although a distributed client-server implementation would offer a better performance, it remains less personalizable and less flexible than a mobile agent implementation.

5 Multi-Agent Architecture for Product Retrieval

E-commerce is one of the most promising fields where mobile agents may bring significant value added (Dagupta et al., 1999). The agent may search a specific product for a client by visiting a certain number of commercial sites. This section presents an architecture where many agents cooperate to find a product on a series of virtual commercial sites (Chamam et al., 2003; Chamam, 2003).

5.1 Description of the Problem and General Scenario

Before buying a product online, the client generally wishes to find a store that offers a "good" price for this product, or even the best price on the market. Although the price is an important decision factor, it is not always the sole criteria for the client who is mindful of the quality/price ratio. Establishing this ration depends very much on the agent's (profile) own criteria and his level of satisfaction of the agent regarding the product. This implies that the agent be equipped with more or less complex intelligence.

To clearly illustrate, we will consider solely the price criteria, and imagine that the client only wishes to buy the item that is the least expensive, where k "good" prices for a commercial product sold by a certain number of virtual stores (e-shops). One (or some) agent(s) is (are) then sent from the client's machine to visit e-shops and find a given product. In this section, we will specifically address the total search time and the mean induced network load by certain mobile multi-agent solutions. Two types of situations may occur:

1. The client is looking for the best price.
2. The client is looking for k prices $(p_1,\ldots, p_k)$ lower than a threshold price p_s. These prices are called *admissible prices*.

In this section, we will only address the first situation described above.

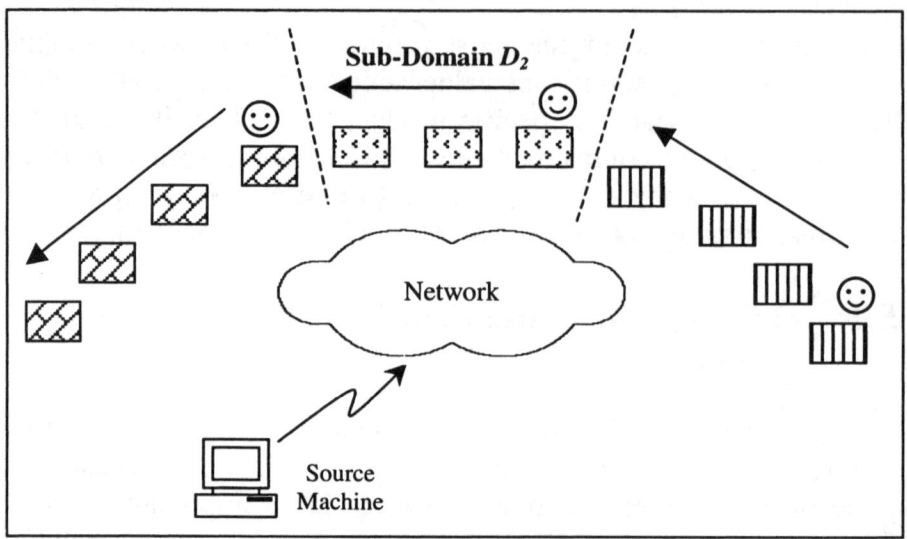

Figure 19. Search with numerous agents

As illustrated in Figure 19, the client starts from a known list of N e-shops, the *domain*, that will be visited, and where the products with admissible prices will be found. m agents are then sent over the network to find the best price or one/some admissible price(s), according to the client's needs. The case $m = 1$ is a specific mono-agent case, where a single mobile agent visits all of the N sites. In the general case where $m \neq 1$, each agent A_i visits a subset of N_i sites $D_i = \{s_1,\ldots, s_{n_i}\}$ of the domain, called the agent's sub-domain. This sub-domain is affected to each agent by a dispatching module (*Dispatcher*).

To examine a multi-agent solution, we will consider a practical price search scenario: client A wishes to buy one (or many) computer(s) and wants to visit N e-shops who sell such products. The sites are published over the network (Internet) and they are accessible to the agents. Each e-shop describes the characteristics of its products in XML format and publishes the information available on its Website. The information related to the computers is the following: the make, speed, hard disk capacity, memory, monitor, price, etc. Using XML standardizes the presentation of this data and facilitates its interpretation by the machine, or by its agent. The site also offers some information about the company selling the products: its name and address.

In this particular example, the client wishes to purchase a computer with certain specific characteristics (for example, a 1000 MHz Pentium III, with a 20 Gbytes hard disk, 512 Kbytes memory, and a 19" monitor whose price is less than $ 1,200). The user inserts those characteristics in a graphic interface and launches a program. One (or many) agent(s) is(are) then sent over the network to find a computer with these features. It (They) hit(s) various e-shop sites and consult(s) their public catalogues. If an item that matches the client's criteria is found, the agents must be warned so that they interrupt their expedition and return to the client's station. At this point, an order could be placed, but we will limit this example to price searches.

5.2 Solution and Suggested Algorithms

For all of the solutions presented in the remainder of this chapter, the following terminology will be used:

T: total mean time of the search. It is the mean time required to perform all of the price search operations.

l: latency time. The time required for an agent to migrate from Site A to Site B. We suppose that this time is identical for all pairs of sites.

tt : processing time. The time an agent spends on a given site. It includes the agent's processing time, the time spent in queue and the agent's execution time on the site (mainly to search and sort prices). Actually, this time depends on various factors, such as the server's performance, the memory, the complexity of the task to perform on the site, the network traffic, the saturation level of the queues, etc. However, to simplify our examples, we suppose that this value is identical for all agents and for all sites.

m : number of agents involved in the search. Each agent is sent towards a sub-domain assigned by a dispatching module on the source machine.

N_i : number of e-shops in the sub-domain of agent A_i.

N : total number of e-shops, $N = \sum_i N_i$.

p_a : agent's size, in kilo-bytes (Kb). We suppose that all agents have the same size.

p_m : size, in kilo-bytes of a message sent by an agent to another agent or to the source machine. The exchanged messages will mainly contain information on the prices found. We suppose that all of the messages have the same size.

t_m : time required for an agent to send a message to another agent or a platform (blackboard). It includes both the message processing time and the latency time.

t_{mm} : time required for an agent to send a *multicast* message to all of the other agents or from a platform (such as the blackboard example discussed further) to the agents.

Moreover, the agents will be equally distributed over the domain. Each agent will have $N_i = E(N/m) + i$ sites in its sub-domain ($i = 0$ or 1 according to the remainder of the Euclidean division of N over m). For example, if 10 sites must be visited by 3 agents, 2 agents will have 3 sites in their sub-domain, and a 3rd agent will visit 4 sites.

In the case of a search for the best price, the agents involved must visit all of the sites included in their respective sub-domains to ensure they have the *k* best prices for all of the domain sites. The algorithm to search the best price for each agent is presented in Figure 20.

```
BEST PRICE ALGORITHM
BEGIN
  price_minimal = very_large_number;
  WHILE ($D_i$ is not empty) DO
    S := first element on list $D_i$;
    Migrate to S;
    Search (product);
    IF ((product found) AND (Price (product)) < price_minimal ))
    THEN price_minimal := Price (product);
    Di = Di − {S};
  End WHILE
  Migrate (address_source);
  Deliver (price_minimal);
END
```

Figure 20. Algorithm to search the best price

5.3 Architecture and Agent Structure

The main architecture of this agent-based application is presented in Figure 21. It stems from three main parts: a stationary agent on the client's machine, a mobile agent launched from the client's machine who will visit the e-shops, and stationary agents residing on the e-shop Websites.

The stationary agent that remains on the client's machine communicates with the user through graphic interfaces where clients enter their search criteria for a given product. It also manages the mobile agents sent over the network and receives the activity reports that will be used to make the appropriate decisions (end search, manage an unforeseen situation, etc.)

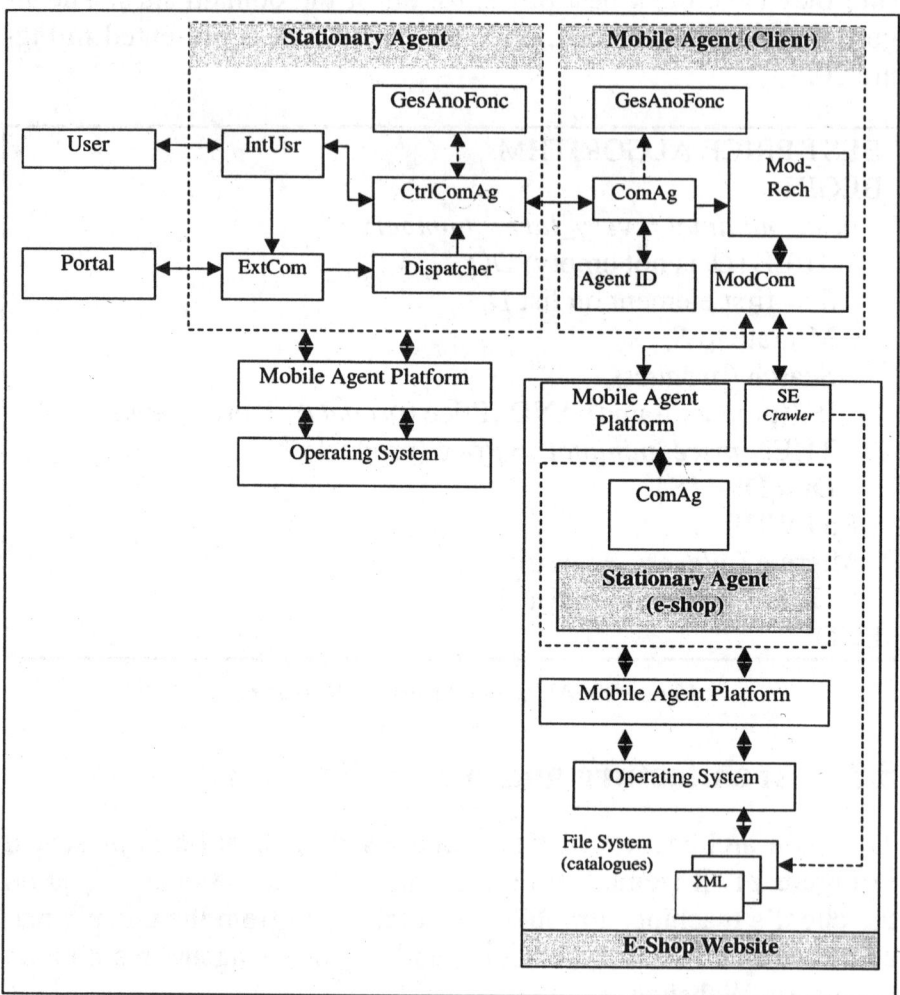

Figure 21. Functional architecture of stationary and mobile agents

The mobile agent(s) migrate over the network and travels through the e-shop Websites. Once on a Website, the mobile agent looks up the product catalogue and checks whether a given product corresponds to its search criteria. A stationary agent, assigned to each e-shop Website, communicates with the visiting mobile agent, suggesting services such as the search of specific products through catalogues.

The stationary agent is located on the user's machine. It includes the following modules:

- **IntUsr**: Graphic interface used to communicate with the user. It is used to input search parameters, and display results and possible errors.
- **ExtCom**: External communication module used to communicate with the portal. It obtains search parameters from *IntUsr* and sends them to the portal that returns a list of e-shop Websites on which the searched product could be found. The sites indicated on this list must be visited by the mobile agent.
- **Dispatcher**: Task distribution module which receives the site list from *ExtCom* and assigns, to each agent, a list of the sites it will visit.
- **CtrlComAg**: Module that manages and communicates with mobile agents. This module receives the search parameters from *IntUsr*, as well as the list assigned to each agent from the *Distpatcher*, and sends them to the agents.
- **GesAnoFonc:** Fault management module which makes decision in case of unforeseen situations, such as the lost of an agent, an unusually high response time, etc. For example, this module may decide to send an agent to a site where another agent was lost. *CtrlComAg* communicates with *GesAnoFonc* to reveal malfunctions.

Using a *proxy* limits the set of actions a client's agent may perform, restricting it to a limited set of available functions, thus preventing access to the e-shop resources and information. Hence, the proxy protects the e-shop Website from potential attacks.

Two-way communication between the stationary agent (*blackboard*) and mobile agents is also secured through a *proxy*. Communication among agents within the algorithm and inter-agent communication is also supported by a *group proxy* that allows sending *multicast* messages to all mobile agents who belong to the group.

5.4 Implementation and Performance Evaluation

In order to implement and gather measurement data, an Ethernet/Fast Ethernet 10/100 Mbps local area network (LAN) was used. The machines used to act as the e-shop Website are equipped with an 800 MHz Pentium III processor and 256 Mbytes memory. As the test network is not a dedicated network, tests were conducted at night while the network was not used in order to limit potential errors.

The software environment includes the following:

- Operating system: Windows 2000 Professional.
- Java environment and virtual machine: JDK 1.3.1 included with Borland Jbuilder 7.0.
- Development environment: Borland Jbuilder 7.0, including JVM (Java Virtual Machine), the graphic environment and development libraries.
- Mobile agent management platform: Grasshopper 2.2.4b from IKV++ Technologies AG.
- XML Xerces parser from Apache. It is a series of libraries based on API standards DOM (*Document Object Model*) that allows to analyze, edit and validate XML documents.
- EtherPeek 4.0.2 from *AG Group*. It is a network load measurement software. It allows to follow, in real time, the network load going through the node on which the software is installed. It also allows to filter the inducted load of a given protocol, or gone through a given node or port.

Using Agents for Information Retrieval and E-Commerce 437

To measure the network load, EtherPeek, a frame capture software, was used. In order to limit the measured load to that generated by this application, a filter was applied to the captured frames. In fact, the Grasshopper platforms communicate on certain well-known logical ports (7000, 7002, 7004, etc.). The frames from these ports were filtered to obtain more exact measurements from the load inducted by this application and avoid counting other packets moving through the network (TCP/IP signal packets, service packets, etc.).

To manipulate the agents, we used a Grasshopper 2.2.4b platform equipped with an API system (Application Program Interface) that allows to send, receive and execute agents. The agents come from pre-compiled Java class (with the extension *.class*) who must inherit from the mother class *de.ikv.grasshopper.agency.Agent* included in Grasshopper APIs. Two main methods manage an agent's life cycle and it is mandatory that they be implemented in the agent's body for the platform to execute the agent correctly:

- *Init (Object() Arguments)*: this method is executed upon the creation of the agent.
- *Live()*: this is the agent's main method. It is executed every time the agent arrives on a site, including the agent's creation site.

An agent is created, sent, and executed in a given *place*. A place is a virtual area within an *agency*. The agency is also a logical space that contains places. The concepts of places and agencies are provided by Grasshopper in order to logically organize the agents' execution spaces. An agent can only be created and executed in a place, which is located within an agency. The agency also acts as a server to greet and execute agents arriving from foreign sites. Grasshopper includes both textual and graphic interfaces to visualize and manage the agents.

Grasshopper also supports registration mechanisms for the agents and agencies of a *region*. A region is a public "yellow pages" service which allows agencies to register in order to be found by agents or foreign agencies. A foreign agent may seek information for a region, similar to the information provided in a phone book, to find registered agents and agencies and communicate with them. In Grasshopper, a region keeps track of all of the registered agents, as well as their position on the network, in order to facilitate communication among mobile agents.

In order to recognize certain events related to the platform components (such as the creation or deletion of an agent or an agency, the departure of an agent, etc.), Grasshopper uses listening mechanisms, called *listeners*, which capture the events generated by the platform. These events are the following:

- creation, deletion, migration and arrival of an agent;
- creation and deletion of a place;
- creation and deletion of an agency.

Actions which are triggered by each event may be implemented in the methods which correspond to specific events.

Figures 22, 23 and 24 indicate the total search time variations according to the number of agents, *individualistic* algorithms, *with blackboard* and *with inter-agent communication*, respectively. These figures show that an increased number of agents does not always generate a decreased searched time.

For example, in the algorithm with a blackboard, when the number of agents m increases, the global search time decreased up to $m = 5$. However, from this value, search time becomes consistent and it even shows a tendency to increase in certain cases, as displayed in the best and worst scenarios with the blackboard algorithm (Figure 23).

In fact, when the number of agents increases, the number of sites to visit for each agent decreases, and the agents visit the sites individually, which decreases the global search time. However, at a certain threshold (a certain number of m agents, $m = 5$ in this example), the time required to process, send and manage the source machine agents become increasingly considerable, negatively compensating for the time that was supposed to be saved by increasing the number of agents.

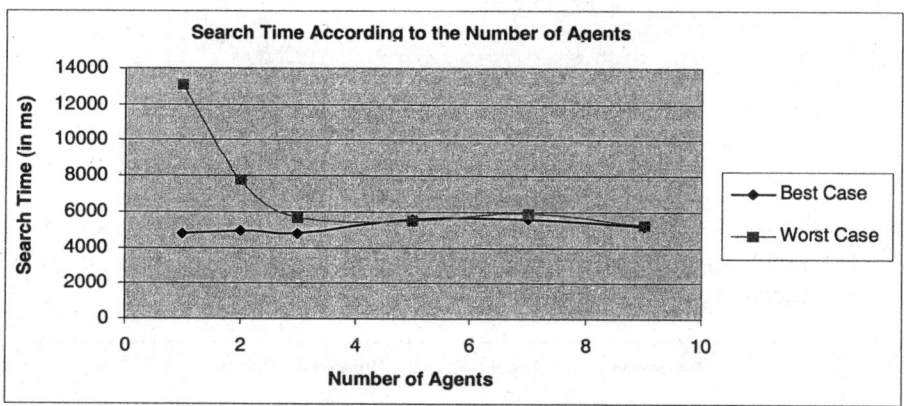

Figure 22. "Individualistic" algorithm: search time variation according to the number of agents

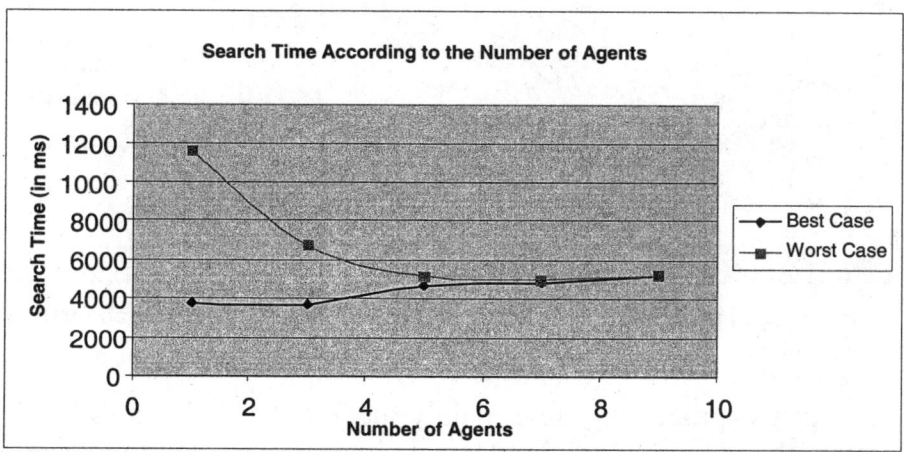

Figure 23. Algorithm with blackboard: time search variation according to number of agents

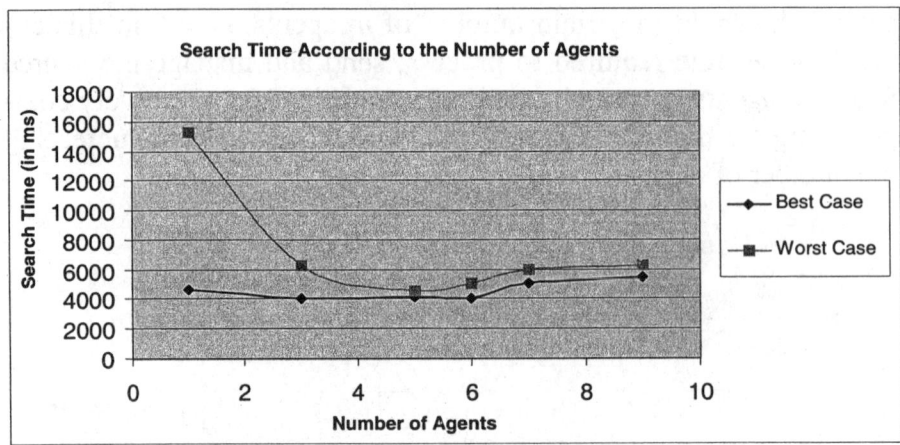

Figure 24. Algorithm with inter-agent communication: search time variation according to the number of agents

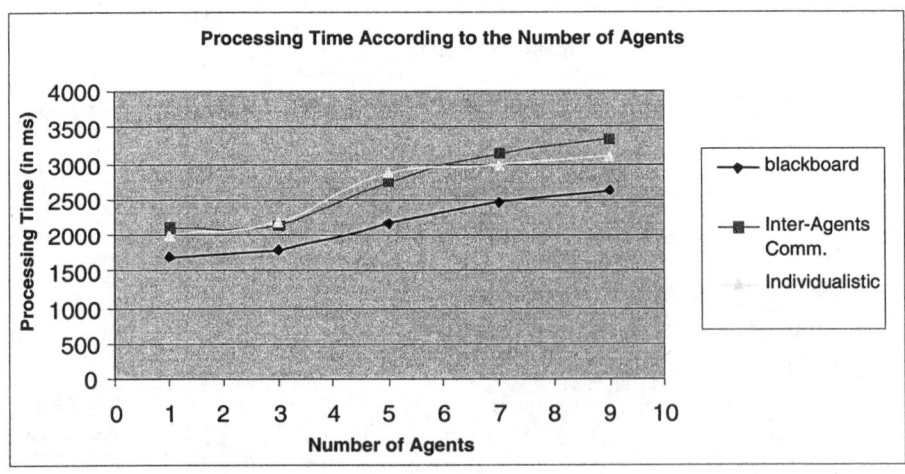

Figure 25. Processing time according to the number of agents for all three algorithms

The agents' processing time mainly include the time necessary to create the agents and implement inter-agents communication mechanisms before they can be launched. In fact, these mechanisms are analogous to two-way communication bridges (or proxies) between the mobile agents and the stationary agent blackboard in the case of the algorithm with blackboard, and a group proxy in the case of an algorithm with inter-agent communication. These times differ from one algorithm to another and they also vary according to the number of agents, as displayed in Figure 25.

Figure 25 shows that the agents' processing and sending time in the algorithm with blackboard and the one in the inter-agent communication are superior to those of the individualistic algorithm due to the implementation time of the communication mechanisms. After a certain number of agents (six in this example), the time required to process inter-agents communication surpass those of the blackboard solution, which could affect the global performances of both algorithms.

Also, note that when the number of agents approaches its upper bound (number of agents = number of sites), the "best" scenario leans towards the "worst" scenario. Moreover, in the extreme cases where the number of agents is identical to the number of sites ($m=N=9$), the "best" and the "worst" scenarios merge, as each agent visits a single site, and, if an admissible price is found, it can only be on this site.

Moreover, when comparing the mono-agent solution ($m=1$) and the multi-agent solution ($m >1$), we notice that after 4 agents, the "worst" case save 1/2.5 ms and 1/3 ms, depending on the algorithm. This represents a significant time saver for the multi-agent solution compared to the mono-agent solution. For this comparison, the "best" case is not very significant as it does not highlight this time economy. This can be explained by the simple fact that in the "best" case, the admissible prices were found on the first sites visited. Hence, whether there are one or many agents, it (they) will visit at the most k sites, that is to say 2 in this example. Hence, the difference between the mono-agent solution and the multi-agent solution is a single site more, visited by the mono-agent solution agent, which does not make much of a difference in the global search time.

The difference between the mono and the multi-agent solutions may have been more noticeable in the "best" case if k had been slightly higher than 2, as, in this case, a single agent would have visited k sites before finding k admissible price, while each one of the m agents of the multi-agent solution would have traveled $E(K/m)$ sites. This practical limitation is mainly due to the quantity of test machines available for this experiment.

In fact, if we wanted to increase k, we would have to increase the number of sites N not to obtain the same limitation as in the "worst" case. This being said, it would be reasonable to assume, by observing the "worst" case (significant case), that in general, a multi-agent solution would be highly likely to generate better search times than a mono-agent solution. Although results such as 1/2.5 ms or 1/3 ms may not be achieved, significant gross gains would certainly be obtained.

Hence, an optimal number of agents corresponds to a relative minimum (Figures 22, 23 and 24). At this point, the search time ceases to increase or decrease, according to the algorithm. Also, a multi-mobile-agent solution provides, for all algorithms, better research time than mono-agent solutions. This time economy certainly has some implications on the network load.

In all three algorithms, the network load increases with the number of agents, as displayed in Figures 26, 27 and 28. The network load results from the agents' movements and the messages exchanged. With a larger population of agents, the number of messages exchanged proliferates, causing the global load to increase.

The multi-agent architecture performance is defined according to two factors: the search time and the induced network load. If we consider the quality/price ratio analogy, we may refer, here, to the product of *time*network load* which would allow to compare the performance of many cases or many algorithms, as the time factors and the network load evolve disproportionally. Actually, the selection of an "optimal" number of agents depends on the users' project specifications and constraints. If they wish to obtain smaller search time without minding the load, they must select the optimal number of agents which minimizes search time. However, if there is a heavy network load constraint, such as is often the case in m-commerce, they may favor a mono-agent solution over short search times.

Generally, a compromise must be made by selecting the "right" product (*time * network load*). The number of agents will then take a value ranging from 1 to the optimal value for the search time. Obviously, the selected value for the number of agents must not be superior to the optimal value for the search time, as, beyond this value, the time and network load will both increase, thus hindering performance.

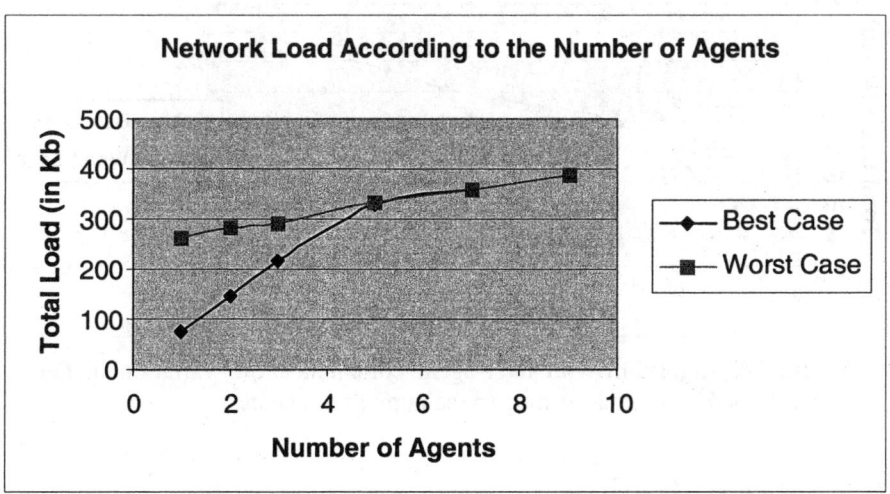

Figure 26. "Individualistic" algorithm: variation in the network load induced according to the number of agents

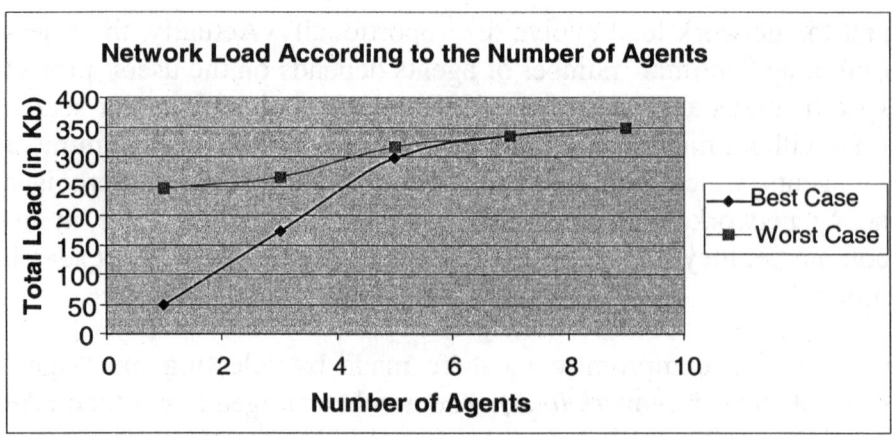

Figure 27. Algorithm with blackboard: variation in the network load induced according to the number of agents

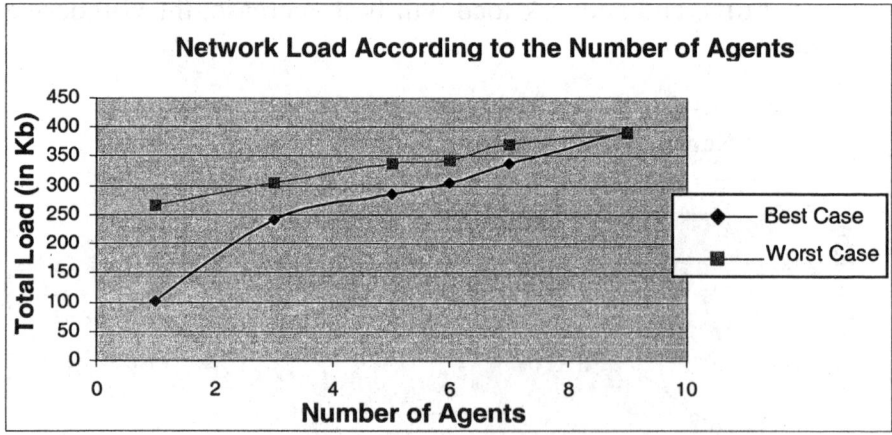

Figure 28. Algorithm with inter-agent communication: variation in the network load induced according to the number of agents

6 Conclusion

This chapter presented two mobile multi-agents architectures which support applications for IP telephony correspondent search, image searches on Internet, and e-commerce product search. The first architecture aimed to associate a mobile agent arriving on a machine with agents, already located on this machine, which can provide the services it requires. It links a service with a specific Java interface which corresponds to this service. Hence, the agents can then use this interface to establish a communication proxy.

In order for the agents' associations to remain as flexible as possible, a service search function was implemented in a specific agent of this architecture, the *Register*. This agent is somewhat of the extension of the mobile agent platform. The mobile agent system used, Grasshopper, already offers agent search functions through various criteria which remain useful, but which are insufficiently flexible, nor do they allow a single agent to offer many interfaces. Moreover, the register attends to finding services on nearby machines and all over the network, and to load them if they are not found locally.

This leads to the concept of a dynamic server for which the services and the interfaces presented by a server may change dynamically according to the requests of the running applications and the server's rules and limitation. The application requests are represented through the register loading services upon the agents' requests. The server's rules and limitations may be felt at the level of the register, of the mobile agent system, or both, or by another agent. The lower the interface levels offered by the server, the greater the modification and optimization possibilities according to each user service.

Performance measures demonstrated the validity of the architecture design. Due to the use of the Grasshopper cache and the Java machine, the agents' movements using either algorithm were comparable. As far as the number of necessary movements for the agent to find the right correspondent with each algorithm, all benefited from the feedback learning mechanisms to reduce their movements.

As for the second architecture designed for product search, the performance measures and the observation of the suggested algorithms yielded the following conclusions:

- Increasing the number of agents in a mobile multi-agents architecture does not necessarily improve search time, although this augmentation allows the reduction of individual search time for each agent, and thus, the global task search time. This is mainly due to the agents' processing time and communication mechanisms.
- In all cases, multi-agent solutions generate better search time than mono-agent solutions.
- Although individualistic, non-cooperative strategies have the advantage of lighting up the agent (not carrying communication mechanisms such as proxies) and decreasing communication time, cooperation always improves on the performances of a multi-agent architecture. Agents' cooperation prevents useless site visits, thus saving time and network load.

References

Bah, T., Glitho, R.H., and Pierre, S. (2002), "Schemes for updating mobile service agents in virtual home environment," *IEEE International Conference on Communications*, ICC, New York, pp. 2014-2018.

Baumannn, J., Kohl, F., Rothermel, K., and Strasser, M. (1998), "Mole – concepts of a mobile agent system," *Mobility Processes, Computer and Agents*, ACM Press, Addison Wesley, pp. 536-556.

Brewington, B., Gray, R., Moizumi, K., Kotz, D., Cybenko, G., and Rus, D. (1999), "Mobile agents in distributed information retrieval," in Klusch, M. (ed.), *Intelligent Information Agents*, chap. 15, Springer-Verlag, pp. 355-395.

Cabri, G., Leonardi, L., and Zambonelli, F. (2000), "Agents for information retrieval: issues of mobility and coordination," *Jour-*

nal of Systems Architecture, Elsevier Science, vol. 46, no. 15, December, pp. 1419-1433.

Chamam, A. (2002), *Architecture multi-agents mobiles pour le commerce électronique*, M.A.Sc. Thesis, Ecole Polytechnique de Montréal, December.

Chamam, A., Pierre, S., and Glitho, R. (2003), "A multi mobile agent-based architecture for price retrieval in electronic commerce," *IEEE Canadian Conference on Electrical and Computer Engineering*, CCECE'2003, 4-7 May, Montréal, Canada.

Dagupta, P., Narasimhan, N., Moser, L.E., and Melliar-Smith, P.M. (1999), "MAgNET: Mobile agents for network electronic trading," *IEEE Trans. Knowledge and Data Engineering*, vol. 11, no. 4, pp. 509-525.

Desrochers, S., Glitho, R., and Sylla, K. (2000), "Experimenting with PARLAY in a SIP environment: early results," *IPTS 2000*, Atlanta, GA, 11 September, pp. 3-8.

Emako Lenou, B., Glitho, R., and Pierre, S. (2003), "A mobile agent-based advanced service architecture for Internet telephony: implementation and evaluation," *IEEE Transactions on Computers*, vol. 52, no. 6, pp. 690-705.

Glitho, R.H. (2000), "Advanced services architectures for Internet telephony: a critical overview," *IEEE Network*, July/August, pp. 38-44.

Glitho, R.H., Olougouna, E., and Pierre, S. (2002), "Mobile agents and their use for information retrieval: a brief overview and an elaborate case study," *IEEE Network*, vol. 16, no. 1, pp. 34-41.

Glitho, R., Emako Lenou, B., and Pierre, S. (2000), "Handling subscription in a mobile agent-based service environment: an agent swapping approach", in Horlait, E. (ed.), *Proceedings of 2nd International Workshop on Mobile Agents for Telecommunication Applications, MATA'00*, Paris, September, *Lecture Notes in Computer Science,* vol. 1931, pp. 50-66.

Goutet, S., Pierre, S., and Glitho, R.H., (2001), "A multi-agent architecture for intelligent mobile agents," in Pierre, S. and Glitho, R.H. (eds.), *Proceedings of 3rd International Workshop on Mobile *Agents for Telecommunication Applications, MATA'01*, August 14-16, Montréal, *Lecture Notes in Computer Science*, vol. 2164, pp. 124-138.

Goutet, S. (2001), "Conception d'une architecture multi-agents supportant des agents mobiles intelligents," M.A.Sc. Thesis, Ecole Polytechnique de Montréal, April.

Hagen, L., Breugst, M., and Magedanz, T. (1998), "Impacts of mobile agent technology on mobile communications system evolution," *IEEE Personal Communications Magazine*, vol. 5, no. 4, pp. 56-69.

Karmouch, A. and Pham, V.A. (1998), "Mobile software agents: an overview," *IEEE Communications*, July, pp. 26-37.

Kotz, D. and Gray, R. (1999), "Mobile agents and the future of Internet," *ACM Operating Systems Review*, August, pp. 7-13.

Lander, S. and Lesser, V. (1997), "Sharing meta-information to guide cooperative search among heterogeneous reusable agents," *IEEE Transactions on Knowledge and Data Engineering*, vol. 9, no. 2, pp. 193-208.

Pelletier, S.-J., Pierre, S., and Hoang, H.H. (2000), "Une architecture multi-agents de recherche d'information," *INFOR*, vol. 38, no. 2, May, pp. 65-92.

Pelletier, S.-J., Pierre, S., and Hoang, H.H. (2003), "Modeling a multi-agent architecture for retrieving information from distributed heterogeneous sources," *Journal of Computing and Information Technology*, vol. 11, no. 1, pp. 15-39.

Pierre, S. (2003), *Réseaux et systèmes informatiques mobiles: Fondements, Architectures et Applications*, Presses Internationales Polytechnique, Montréal, 640 p.

Varshney, U. (2001), "Addressing location issues in mobile commerce local computer networks," *Proceedings LCN 2001, 26th Annual IEEE Conference*, pp. 184-192.

Varshney, U., Vetter, R.J., and KalaKota, R. (2000), "Mobile commerce: a new frontier," *Computer*, vol. 33, no. 10, pp. 32-38.

Printing: Strauss GmbH, Mörlenbach
Binding: Schäffer, Grünstadt